The Fundamental Physical Constants
and
the Frontier of Measurement

The Fundamental Physical Constants and the Frontier of Measurement

B W Petley

(National Physical Laboratory)

Adam Hilger Ltd, Bristol and Boston

British Library Cataloguing in Publication Data

Petley, Brian William
   The fundamental physical constants and the frontier of measurement
   1. Constants, Physical
   I. Title
ISBN 0-85274-427-7

Consultant Editor: **A E Bailey**

Published by Adam Hilger Ltd
Techno House, Redcliffe Way, Bristol BS1 6NX England.
PO Box 230, Accord, MA 02018, USA.

Photosetting by Thomson Press (India) Ltd, New Delhi.
Printed in Great Britain by J W Arrowsmith Ltd, Bristol.

Contents

Preface

'I believe that all electrons in the universe have the same charge and mass; that all photons travel with the same velocity in free space; that, given the frequency, the energy of a photon is uniquely defined; that all protons have the same charge and mass; that the gravitational attraction between two masses is characterised only by their mass and separation. I believe....'

How right are we in making statements such as the above? This book examines the present state of our knowledge, mostly from an experimental viewpoint, and in doing so takes the reader to the frontiers of what it is possible to measure meaningfully. Dimensioned quantities obviously cannot be measured more precisely than the SI units can be realised in practice, and so their accuracy and the properties of units must be examined. What are the consequences of changing to the use of an atomic constant to define the metre? Do the constants vary with time? How can we measure progress in science quantitatively? How square is the inverse square law of force? How neutral is matter? How well have the precepts of relativity been verified? These and many other questions are examined in this book.

Far from being a dull area of science, high-precision measurements are an increasingly exciting area in which to work, for they push theory and experiment to the very limits of which they are capable. This book examines the present state of the measurement arts and gives indications of likely future developments.

Most scientists, reasonably enough, prefer to work at the per cent level of precision, while others prefer the yes/no type of experiment: does electro-magnetic radiation have a measurable velocity (as Galileo sought to discover with his famous lantern experiment)? Others prefer the high-precision null experiment such as is required to set limits on the rest mass of the photon or neutrino, or to detect gravity waves. Such types of experiment are a valuable source of discovery. Beyond these lies the frontier of science where the metrologist operates. This is an area where theory and experiment are pushed and tested to the limit, where the fallibility of man is often all too apparent and where one's best is only just good enough.

Modern science has tended to become compartmentalised and it is not unknown to find that a scientist has spent the bulk of his or her career following the path determined by their PhD thesis topic. Precision measurement is, to a large extent, a specialised topic, and yet, as the content of this book demonstrates, it provides an opportunity to cover a wide range of physical phenomena. There are few other scientific disciplines where we are made more aware of the transient nature of our own contributions, or of the way in which we build on the work of others. If we recall that the rich oil deposits under the North Sea remained unknown until the advent of the very precise geophysical measurements, or that the spin of the electron was inferred from precise spectroscopic measurements, we see more clearly the class of discoveries which can come from precise measurement. At the very end, when all that is to be discovered is nearly discovered, most scientists will be operating at the nth decimal place. This may be a sobering thought to the man who works at the per cent level, but it shows that the metrologist is pioneering where all must finally follow.

In writing this book, I have become even more aware of the debt that I owe to the many dedicated metrologists: those that I have been privileged to meet and those whose work I have only read about—from the pedants to the Nobel prizewinners. They are too numerous to acknowledge individually and I extend to all of them my grateful thanks. To those whose contribution to the subject I may have distorted by my own ignorance I willingly apologise, and hope that they will continue to extend to me that patience and tolerance which I have found to be a characteristic of metrologists throughout the world. My wife, Janet, by her typing and constructive comments, has helped me throughout by providing impetus and encouragement when it was most needed. The mistakes yet to be discovered are, alas, my own and I can only hope that they will not lead anyone astray. In metrology so far there has always been someone else who can and does find a better way to do it. If, therefore, anyone is discouraged by what I have written I will be very sad, for it has provided myself and countless others with a very interesting and rewarding career.

Brian Petley
August 1983

1

The fundamental constants, their evolution and accuracy

1.1 Introduction

Measurements of dimensional physical quantities that are made in science and technology are normally expressed in SI units. This system of units is essentially a practical one in which each unit has a convenient magnitude for the proverbial man or woman in the street to use, but overall they do not have a natural basis. Consider how one would communicate an estimate of one's size and mass to someone on a remote planet in a distant galaxy. It is clear that metres and kilograms would not be very useful for this purpose. There are, however, physical quantities which come rather closer to being the natural units of science, and these are termed the fundamental physical constants. They include the velocity of electromagnetic radiation, the electronic mass and charge and a number of other constants (see table 1.1).

It is not unusual for a theoretical calculation to begin with the statement 'let $\hbar = e = c = 1$'. There is, of course, nothing wrong with this—the calculation is merely being performed with a system of units that differ from SI units. However, at some stage, the need will arise to compare theory with experiment, and it is at this point that it will be necessary to know the value of the fundamental constants in terms of the SI units. Different scientific disciplines, too, tend to use their own local units and the conversion from measurements which have been made in one discipline to those made in another, in order to intercompare or use the results, may also involve fundamental constants. Thus, atomic mass differences may be measured by means of nuclear reaction energies—in electron-volts, or by mass spectrometry—in atomic mass units, and the connection between the electron-volt and the atomic mass unit essentially involves the Faraday constant.

Since fundamental constants are so widely used, it is necessary that their values are known with greater precision than most scientists require. At first sight, this might appear unlikely, but it is important to remember that

Table 1.1 The fundamental constants of physics.

Fundamental quantity	Symbol
Velocity of electromagnetic radiation in free space	c
Elementary charge	e
Mass of the electron at rest	m_e
Mass of the proton at rest	m_p
Avogadro constant	N_A
Planck's constant	h
Universal gravitational constant	G
Boltzmann's constant	k

measurements which are being made today will in some cases still be referred to by scientists in the twenty-first century. In that time, the precision of measurement will have advanced by around an order of magnitude and scientists of that epoch will only use the earlier work if the precision remains adequate. It is therefore sensible to ensure that the units in which measurements are expressed remain relevant for some years. Thus, measurements of susceptibilities of materials which were made by Faraday in the nineteenth century were still being quoted in the 1959 edition of a well known reference book on physical constants, and the values were none the worse for their longevity.

Of course, fundamental constants should not be regarded as being merely units, and their utility extends far beyond this application. They play a key role in testing our understanding of physical phenomena. For example, a comparison of the Faraday determined from electrochemistry with that obtained from atomic measurements, via the evaluation of the charge to mass ratio of the proton (together with other indirect measurements), gives valuable information about the precision that is being achieved in reality in these separate disciplines. Measurements of the constants often have a classic simplicity about them which appeals to scientists if not to those students who find questions about them appearing on their examination papers! Certainly, as will be discussed, the measurement of fundamental constants has played an important part in testing the validity of both quantum electrodynamics and special relativity.

Modern physics is based on some intrinsic acts of faith, many of which are embodied in the fundamental constants. Thus we believe that all electrons in the universe have the same charge, mass and angular momentum. We believe that all like atoms have the same energy levels, whose separations are characterised by the fine structure and Rydberg constants, that the radiation emitted through transitions of electrons between these levels has the same frequency distribution for all atoms of the same kind, and that the photons all travel with the same velocity in free space. We believe that the energy fluctuations at a temperature T are characterised by the Boltzmann constant k, and that all atoms or large bodies are attracted by a force which is characterised by the same universal gravitational constant G.

Of course, we recognise that the above statements require some modification in the presence of other atoms or particles, or electromagnetic fields, etc. However, the above is broadly a statement of some of the underlying beliefs of modern physics. To state them, of course, is to invite them to be questioned and to question them is to be showing the first instincts which characterise the metrologist and the cosmologist. While the metrologist is concerned with the nth decimal place, the cosmologist may have to be content with identities of the form $1 \sim 10$, but both may be testing similar fundamental truths of the universe. Thus, discovering that something has changed by less than 10% since the universe began may be equivalent to measuring that something else has changed by less than 10^{-11} in a year: the former measures the average rate of change and the latter the instantaneous rate of change.

An important question to consider is the natural one of whether the constants are truly unchanging, or perhaps vary with time. This takes one into cosmology. Although such questions are answered by appealing to astrophysical or geological measurements, information may also be increasingly gleaned from laboratory experiments which have the necessary high precision. On the other hand, if the constants are constant then they may perhaps be calculated in terms of other constants. Such calculations have been performed since the role of the constants first became apparent. They vary in status from the highly respectable, such as calculations of the g values of the electron and muon provide, through to the more speculative, as represented for example by some of the calculations of the value of the fine structure constant. Of course, if a constant has dimensions associated with it, there is the added problem of expressing the constant in terms of an invariant unit, for otherwise it is meaningless to discuss whether or not it is constant with time. This is no easy task and we discuss this question and the experimental evidence further in chapter 7.

1.2 The number of fundamental constants

The situation with the fundamental constants of physics is in many respects analogous to that experienced with the SI units. Just as there are base and derived units, so there are some constants that are more fundamental than others. It is clear that the overall number of fundamental constants is not fixed. For example, the magnetic flux quantum $h/2e$, associated with superconductivity, was not a quoted constant before its major role became apparent through the Bardeen, Cooper and Schrieffer theory of superconductivity in 1957 and the tunnelling effects which were predicted by Josephson (1962).

The group of constants that are conventionally known as the fundamental physical constants include the elementary charge e, the electron mass m_e, Planck's constant h, the velocity of electromagnetic radiation c, the Avogadro constant N_A, and the proton mass m_p. The addition of two further constants, the Boltzmann constant k and the universal gravitational constant G, forms the larger group known as the fundamental constants of physics (table 1.1).

The use of the word 'fundamental' with respect to a constant is perhaps arbitrary and contentious and it might be better to use the phrase 'base constant' instead, much as we refer to the 'base SI units' (thereby avoiding stating that they are more basic than other physical quantities). The dimensionless fine structure constant, for example, might well have a strong claim for inclusion in the list and one might express the electronic charge in terms of α, h, c and μ_0 (or ε_0). This set of constants is characterised by their having been known for much of this century and also by their numerical values having in all cases been determined to better than 0.1%.

There are other constants which have not yet been measured with any precision and whose role in science is still under discussion. These may be expressible either in terms of one another, or of other constants. This group includes the coupling constant for the weak interaction G_F, the Cabibbo angle θ_C, which relates the strangeness-changing and non-strangeness-changing weak nuclear interactions; the strong interaction in quantum chromo-dynamics which is characterised by the constant Λ, and finally there is the Weinberg angle θ_W relating the electromagnetic and neutral weak currents. The value of θ_W is one of the predictions of the grand unified theories SU(5) and is in the process of being measured experimentally, for example by studying parity violation in spectroscopy and elsewhere. In turn, the Glashow–Weinberg–Salam electro-weak unification theory expresses G_F in terms of e, θ_W and a further constant m_W, the mass of the charged vector boson (see reference 16 in *Further reading*).

A further requirement of the grand unified theories is that some of the above 'constants' should be energy dependent. While this behaviour provides a theoretical basis for the discussion of the unification, it is of course a contradiction and serves to demonstrate that the subject is still being excitingly roughed out. It may therefore be some years before there is general agreement on how to treat these fundamental constants. It is salutary though to reflect that we may be measuring some of this group to parts per million in a hundred years time! For the present these constants remain outside the group of fundamental constants that are evaluated in the CODATA review by Cohen and Taylor and which feature in the book.

Returning to the group of constants in table 1.1, it is evident that it provides more fundamental standards in the sense of base units than we strictly require; for instance, we do not need two standards of mass. Similarly it might be possible, if we so wished, to eliminate G eventually, by expressing it in terms of some of the other constants. We see that physical science requires around the same number of constants as there are base SI units and, although this is interesting, it will not be discussed further here.

The proton mass, at present, must be regarded as having a questionable status as a fundamental constant, particularly since it is thought that quarks may be the fundamental nuclear particles. There have been suggestions (see, for example, Close 1981) that the proton might be unstable, albeit with a very long lifetime, $\sim 10^{32}$ yr and consequently the philosophical status of the proton as a truly fundamental particle may be questioned—we may soon decide to use quarks etc, instead. However, the proton is such a readily available particle that it is in any case likely to continue to have a useful metrological role as a reference standard for some time to come.

1.3 The evolution in the role of a fundamental constant

It is clear that some constants play a deeper role than others in science at present. This may be a consequence of the way in which the role of the constants in science evolves with time, so that some, particularly those constants of the twentieth century, have not yet reached their final stage of evolution. That constants have a changing role with the passage of time may be demonstrated by considering the classification suggested by Levy-Leblond (1979) in table 1.2, and discussing how the value of the velocity of electromagnetic radiation has evolved over the years.

Table 1.2 Classes of fundamental constants (after Levy-Leblond 1976).

Class of constant	Characteristics of class
Type A	*Properties of particular physical objects* e.g. masses of elementary particles, their magnetic moments etc.
Type B	*Characteristics of classes of physical phenomena* e.g. coupling constants of various fundamental interactions: nuclear, strong, weak, electromagnetic and gravitational
Type C	*Universal constants* e.g. constants entering into universal physical laws, for example Planck's constant

1.3.1 The changing role of the speed of light

When it was first introduced, c appeared as the velocity of light, at which stage it was a type A constant, being a property of light. We now know, of course, that it is the velocity not just of visible radiation (light), but of all electromagnetic radiation. This evolution into a class B constant began to come about the Kirchhoff noted that the combination $(\mu_0\varepsilon_0)^{-1/2}$ of the constants ε_0 and μ_0 of electrostatics and magnetism respectively had the dimensions of a velocity. Once this constant had been obtained from electrostatic and magnetic measurements by Weber and Kohlrausch (1856) as 3.1×10^8 m s^{-1}, which was equal to the velocity of light within the experimental errors, Kirchhoff was sufficiently impressed by the coincidence to remark upon it. Maxwell then went on to infer that 'light consists in transverse undulations of the same medium which is the cause of electric and magnetic phenomena' (figure 1.1). In this spectacular way c became a type B constant, which he so brilliantly applied to the whole of electromagnetism.

22   Prof. Maxwell *on the Theory of Molecular Vortices*

The velocity of light in air, as determined by M. Fizeau*, is 70,843 leagues per second (25 leagues to a degree) which gives

$$V = 314,858,000,000 \text{ millimetres}$$
$$= 195,647 \text{ miles per second.} \quad . \quad . \quad . \quad . \quad (137)$$

The velocity of transverse undulations in our hypothetical medium, calculated from the electro-magnetic experiments of MM. Kohlrausch and Weber, agrees so exactly with the velocity of light calculated from the optical experiments of M. Fizeau, that we can scarcely avoid the inference that *light consists in the transverse undulations of the same medium which is the cause of electric and magnetic phenomena.*

* *Comptes Rendus*, vol. xxix. (1849), p. 90. In Galbraith and Haughton's 'Manual of Astronomy,' M. Fizeau's result is stated at 169,944 geographical miles of 1000 fathoms, which gives 193,118 statute miles; the value deduced from aberration is 192,000 miles.

Figure 1.1 The moment in Maxwell's classic paper when the velocity of light was extended to include the velocity of electromagnetic radiation. There is much of interest in this extract: the variety of units, the missed proof correction and the appropriateness of the number assigned to the equation. (Courtesy CUP and NPL Library.)

This example also shows a further important role of many fundamental constants in unifying two previously separate disciplines: in this case light and electromagnetism. The stage of the evolution of the speed of light into a type C constant, where the role is a universal one, is illustrated by the part that c plays in relativity through the ideas of Einstein, in relating energy E and mass m, through his famous equation.

$$E = mc^2$$

and also through the earlier introduction of c in the Lorentz transformation which is used for transferring from one coordinate system to another.

There is a still later stage when the fundamental constant becomes so built into our physical system as to become a near invisible constant, and certainly one that is no longer measured experimentally. In some respects c has now entered this phase, with the re-definition of the metre and the implied fixed value for c.

In the final stage of the evolution it is evident that one has moved well away from c being merely associated with the velocity of light. Indeed, it is conceivable that the situation could arise whereby the constant which we denote by c might no longer be exactly the velocity of light; if the photon proved to have a non-zero rest mass, for example (we consider the evidence for this in chapter 7). In these final stages of the evolution we have arrived at a fundamental constant that is truly constant, for it is no longer measured experimentally, but instead has a fixed, defined, value. Even this is not the ultimate stage for the 'coup de grace' may then be delivered by so changing the unit system that the constant has the value of unity. It is too soon for this to have happened with the fundamental constants, but such indeed has been the fate of Joule's mechanical equivalent of heat J, with the introduction of SI units. (A constant such as J does not disappear totally with a change in definition, and indeed J is incorporated into the specific heat of water.) The type C constants group into the *modern* ones, such as J, which are still discussed, but which now appear as unit conversion factors, and the *classical* ones which have been so well assimilated as to be taken for granted and which have become invisible (as happened when the unit of volume for liquids and gases, the litre, became defined as $10^{-3}\,\mathrm{m}^3$ exactly).

It is apparent, therefore, that in a conceptual sense the constants evolve as our knowledge of science evolves and there is a certain degree of interchangeability between the total acceptance of concepts and the designation of a quantity as a fundamental constant. Thus the ratio of the circumference of a circle to its diameter was a fundamental constant which was measured experimentally by the ancient Greeks, but is now a precisely calculable constant. We have long since ceased to distinguish between a length which is the distance between two points measured as an arc or one which is a direct linear measure by expressing their measure in different units. It is interesting though that we do not consistently regard their quotient as dimensionless; for example, the radian and steradian have been assigned the status of secondary units in the SI which (as we agree to choose) can be regarded equally as dimensioned or dimensionless quantities. So too there is an element of arbitrariness in whether or not a quantity is regarded as a physical constant.

1.3.2 Classes of fundamental constants

The above discussion has implicitly assumed that the fundamental constants, as with any other physical quantity, are expressible in the form of the product of a numerical value and some combination of SI base units which are raised to

Table 1.3 Categories of fundamental constants and units.

| Type | Classification | |
	Dimensioned	Dimensionless
Not exact	Most fundamental constants	E.g. fine structure constant
Unity	Derived SI units. μ_0 in cgs system	Metrologically identical with unity e.g. 1 μV/10^{-6} V, 1 g/10^{-3} kg, 1 L/10^{-3} m^3 etc.
Or a simple integer	Units which are not coherent with SI e.g. bar, day, week (defined exactly or calculable)	Near unity: lepton magnetic moments in terms of Bohr magneton etc.

various powers. While most constants have both dimensions and a numerical value differing from unity, there are some which fall into the other categories shown in table 1.3.

Surprisingly perhaps at first, much of metrology is devoted to the realisation of constants which are both dimensionless and very close to unity. Thus it is necessary to ensure that 1 μV is 10^{-6}V, 1 g is 10^{-3} kg, etc, while the ratios of the maintained units to the values implied by their SI definitions must also be determined as accurately as possible. The metrological laboratories therefore have the task of ensuring that our measurement system properly incorporates physical concepts. The philosopher and theoretician may decide, for example, that conceptually the volume of a cube is simply the product of the lengths of the orthogonal edges, but the metrologist has to ensure that this is true in practice to the limits of the available technology.

1.4 A basic metrological problem

There is a basic metrological problem, which is equivalent to attempting to lift a bucket while standing inside it, which has been termed 'using Zanzibar units'. The mythical Zanzibar unit is illustrated by a story which Cohen et al (1957) attribute to Harrison of MIT (see figure 1.2). Apparently there was a retired sea captain who lived on a remote part of the island of Zanzibar and, as captains are wont to do, he had introduced a ceremonial raising and lowering of the flag each day, and also fired a cannon at noon. A visiting friend enquired how the captain knew the exact time, to which he replied that he reset his chronometer according to the time displayed by the clocks in the window of the jeweller in the local town. After he had left the captain, the friend was in the town and thought that he would pursue the matter further and enquired how the jeweller set his clocks. 'Oh, that is simple' he replied, 'there is a retired sea captain on the

Figure 1.2 The Zanzibar effect: an example of a circular argument in the use of units: (*a*) the captain fires a noon-day cannon and sets his watch by the time in the local jewellers; (*b*) the jeweller sets his clocks when he hears the noon-day cannon. (Based on a tale by Harrison, MIT)

other side of the island who always fires a cannon at noon and I use it to set my clocks.' Evidently, then, the island noon differed from the true noon by an unknown amount.

This simple tale, that recounts the obvious, underlines a very important aspect of metrology, for Zanzibar units are widely though inadvertently used. In fact, it is the problem of relating the local units to the base units that permeates the whole of measurement. The ways in which Zanzibar-type effects occur vary widely. A simple one arises when an instrument is used to obtain a measurement for a long period until for some reason the measurements are suspected of being in error. Then a second instrument gives a different reading under the same conditions. This may well be followed by a similar experience with a third instrument, etc, and for a while the local measurement system collapses completely. Sometimes these differences arise through the instruments being used incorrectly and sometimes because they need recalibrating, either locally or by a standards laboratory. National standards laboratories too experience the problem at a higher level. Thus the material British Imperial yard length standard was found, some years ago, to be shrinking, by about a part in a million every twenty three years, with respect to other length standards that were thought to be more stable. A similar experience occurred with the volt, for previously the units of potential were maintained in various countries by means of standard cells. These were intercompared, by each country sending a group of their standard cells to the BIPM (Bureau Internationale des Poids et mesures) and the EMFs of these were compared against the similarly maintained volt at the BIPM. This intercomparison, of course, was unable to reveal whether all of the maintained volts were drifting in an absolute reference frame and the absolute drift, $\sim -0.2\ \mu\mathrm{V\ yr^{-1}}$, could not be detected until the advent of the Josephson effects. Searches for changes with time of the values of the fundamental constants may similarly be masked if the units in which they are expressed are drifting in the same way (see chapter 7).

1.5 The uncertainty in the value of a fundamental constant

According to the 1973 list of the best values of the fundamental constants of physics the value for the speed of light in a vacuum is 299 792 458.0(12)m s^{-1}. Here we have a number together with the figure in parentheses where the latter is intended to act as a guarantee of the worth of the result, for 1.2 m s^{-1} represents a measure of the size of the region in which the correct result was thought to lie. The probability density within this region is frequently represented by the normal error curve or Gaussian distribution and the figure 1.2 m s^{-1} represents the standard deviation, or standard error, of the distribution. However, the assumption of the normal distribution is not mandatory. If, with this in mind, we look at the quoted value, we see that the final number is an 8, which presumably means that the value has been rounded and could be anywhere between 795 and 805. If we take the standard deviation assigned, which is 1.2 m s^{-1}, then we find that with this standard deviation the probability of the final digit of the result lying within 7.95 and 8.05 is only about 1 in 30 (assuming that the probability distribution for the uncertainty is a Gaussian one). Since this has such a small probability of being the number, one wonders why we publish it. Clearly then, in this sense, the constant which we have measured is very unlikely to remain numerically constant, but we must quote the number to this precision in order to define the centre of the 'bull's-eye' within which the number must lie. Of course the digits further along are much more likely to be correct. For example, if we look at the probability that the 5 should be a 6 (that is $c = 2.997 924 6 \times 10^8$ m s^{-1}) we find that it is 1 in 20. In view of the fact that we have now stabilised the value for the speed of light on the above number, the reader may wonder whether it would really have affected physics so very much if we had adopted the round 299 792 460.0 m s^{-1} value allowing us to drop one of the digits (anyone with an eight digit hand-calculator will do so anyway).

1.6 The method of arriving at the uncertainty

The uncertainty in the value is the combination of two major components. The first is the 'random' uncertainty which is obtained from the observed fluctuations in the measured value which occur during the experiment. The second results from an attempt to estimate the effect of all the factors which might have influenced the result, but which the experimenter was unable to investigate in a controlled manner, and is usually known as the 'systematic' uncertainty. Both quantities are expressed as standard deviations, and the final uncertainty is obtained by combining them as the root sum of their squares. In many experiments the conceptual classification of the uncertainties in this way is a considerable oversimplification and so it has been proposed to replace the concepts of 'random' and 'systematic' uncertainties by 'type A' and 'type B' respectively. We return to this subject in chapter 9, for it is an important aspect of the measurement of any experimental quantity. Throughout this book, the uncertainties are expressed as standard deviations. These are the numbers given in parentheses, and are to be interpreted as being the uncertainty in the last digit quoted. The standard deviation is a useful statistical measure which has application to many distributions, including the familiar Gaussian probability distribution.

1.7 The precision with which the fundamental constants are known

Just as the precision of the base units has improved with time, so too has the precision with which the fundamental constants are known. The improvement has been just as dramatic despite the relative youth of many of them. The measure of the uncertainty in the values of the fundamental constants has also changed since the first evaluation by R T Birge (1929). In the early days, the probable error was quoted. This has now been changed to the standard deviation, since this is a measure which is rather more meaningful in the statistical sense and applies to a wide variety of distributions, as well as the well known normal distribution. Cohen and Dumond (1963) published the uncertainties of the output values from their evaluation as three standard deviation uncertainties which, for a Gaussian distribution, would imply that there was only about a 1 in 300 chance that the 'correct' value was outside this uncertainty. Table 1.4 shows the estimated standard deviation uncertainties

Table 1.4 Precision in standard deviations with which the fundamental constants were thought to be known at the time of evaluation.

| Constant | Precision, parts per million (ppm) in: | | | | | |
	1929	1948	1963	1969	1973	1983
c	20	20	0.3	0.3	0.004	—
e	1660	156	15	4.4	2.9	1.8
h	1800	246	25	7.6	5.4	3.6
N_A	1000	100	15	6.6	5.1	3.6
m_e	1660	195	15	6.0	5.1	3.6
m_p	2744	100	16	6.6	5.1	3.6
k	279	116	43	43	32	25
G	1140	—	750	—	615	86
α^{-1}	175	73	5	1.5	0.8	0.2
m_p/m_e	450	163	8	6.2	0.4	0.06

for many of the fundamental constants for a number of evaluations since 1929. While most of the constants show a considerable change, it is apparent that there has been much less improvement in our knowledge of the universal gravitational constant. Overall, the uncertainties have decreased by about three orders of magnitude since 1929, corresponding to about an order of magnitude every twelve to fifteen years. For this reason, it is appropriate to review the values of the fundamental constants about every ten years. Interestingly, the improvement in the precision is quite regular with time and this allows us to extrapolate to give the probable accuracy by the year 2000.

1.8 How accurate are the output values from the evaluations of the fundamental constants?

It is possible to look back at the early values and compare them with the present ones in order to act as a guide to the ability of the metrologists and reviewers to estimate the experimental errors. (At this point it is emphasised that the estimated uncertainties consist of the statistical measures from the internal fluctuations in the value obtained for the result, together with the estimates of the unmeasured systematic uncertainties, which are all combined together as the root sum of their squares as discussed in § 1.6.) As a measure of the accuracy of the evaluations, therefore, one may take the difference between the earlier value and the present value, and normalise it by dividing by the originally estimated uncertainty. Table 1.5 shows the values which are obtained for three of the evaluations and for a number of the constants. (It

Table 1.5 Changes in the recommended 'best' values of the fundamental physical constants in terms of their associated standard deviations at the time of the recommendation, as expressed by (value (t_x) − value (1973))$/\sigma_{t_x}$.

| Constant | Time of evaluation | | |
	1929	1948	1963
c	0.6	− 2.7	0.4
e	− 4.6	− 0.8	0.6
h	− 6.6	− 1.7	− 3.5
N_A	7.0	2.4	5.2
k	− 7.0	− 2.1	− 2.0
m_e	− 5.0	− 2.3	− 3.2
α^{-1}	− 0.2	− 1.5	4.4
m_p/m_e	0.5	1.4	− 3.7
average |change|	3.9	1.9	2.9
average change	− 1.9	− 0.9	− 0.2

should be remembered here that the estimates of the values and the uncertainties were not all independent, and that the correlations differ from one evaluation to the next.)

The results are humbling for anyone who expects the uncertainties to obey the Gaussian law of errors, for one would expect that only half of the values would exceed 0.655 (the probable error), whereas half of the measurements were incorrect by more than 2.1 standard deviations. This rather suggests that the reviewers have tended to be optimistic in their estimate of the uncertainties by about a factor of two. In view of subsequent progress, which has revealed much about the unknown physics involved in many of the experiments, this performance is not as unsatisfactory as a strictly statistical approach would lead us to believe. Thus the Birge (1929) evaluation was led astray by the famous Millikan oil drop experiment which was later proved to be in error because an incorrect value was used for the viscosity of air (someone else's!). One may therefore take two points of view here: either to regard the constants as merely something to look up in a table, which would lead one to play safe by expanding the uncertainties, or, alternatively, one may regard their measurement as contributing to discoveries in physics as well and give as realistic an estimate of the uncertainty as one can.

It seems reasonable to expect that the output values from future least squares adjustments of the fundamental constants may well have an uncertainty assigned to the uncertainty, for example an output value e_0 may be written as $e_0\,(\sigma, \delta\sigma)$ instead of the present $e_0(\sigma)$. Alternatively, since this might well be too sophisticated for many users, the $\delta\sigma$ could be expressed in the form of a degree of freedom, much as is done for random uncertainties. Cohen (private communication) has gone some way towards this with the more sophisticated algorithms that he has been developing for use with the next adjustment. However, one should remember the computer adage: 'garbage in, garbage out', for no amount of statistical sophistication can improve inherently bad input data. On the whole, as long as only a few scientists are led astray by using values which are subsequently found to be slightly incorrect, the latter is much the more interesting situation—if only because it encourages the view that there are still things left to discover!

1.9 The units in terms of which the dimensioned fundamental constants are measured

The measurements of the dimensioned fundamental constants must all be expressed, ultimately, in terms of the base SI units. As will become apparent, many of the constants are measurable today with accuracies which approach, or even in principle can exceed, the precision with which the base units can be realised. One consequence of this is that the measurements must be made in close proximity to the base unit, so that there is no loss in precision, and a second is that the fundamental constants may have an important role in replacing the existing methods of defining or maintaining the SI units. As examples of the latter, the velocity of light has been incorporated in a new definition of the metre, the Josephson effects are used to maintain the SI volt with high precision, the gyromagnetic ratio of the proton is used to enable magnetic fields to be measured with high accuracy and the quantised Hall effect may well have an important part to play with respect to the unit of resistance.

There is an important direction in which modern metrology is moving which has been described by the term *quantum metrology* (Petley 1969, Cook 1974). This takes account of the fundamental role that time and frequency play in quantum physics, together with the fact that these quantities are the most accurately measurable of all of the physical quantities. If, therefore, a fundamental relationship can be found between the quantity which one desires to measure and frequency, the measurement of that quantity may subsequently be made with great accuracy. There is the added advantage that the user may in principle construct the SI unit system in terms of quantum physics by means of simple recipes without the need to refer back to a national standards laboratory through a hierarchical calibration chain. The fundamental constants are already playing a major role in this endeavour. It must be emphasised that this is coming about by a process of evolution and discovery and not simply by a systematic dedicated effort on the part of metrologists.

We must take care in using the fundamental constants in this way if the integrity of the SI is to be sustained. The major problem stems from the ampere definition, for it must be remembered that the definition is framed in such a way as to ensure that the mechanical units of force and energy are identical with the electrical units of force and energy, respectively. In addition, the desired simplicity of Maxwell's equations is ensured by simultaneously defining the permeability of free space μ_0 as $4\pi \times 10^{-7}\,\mathrm{A^2\,m^{-2}\,N^{-1}}$ which, after the henry is defined (to keep the energy units the same as above), allows us to use different units for μ_0, namely $\mathrm{H\,m^{-1}}$ and the impedance of free space is also simply $\mu_0 c\ \Omega$.

If, as some have suggested, we take a fixed value for the Josephson effect, $2e/h$, in order to define the volt, we find that the ampere definition must be reframed in order to define the kilogram. This method has the merit of allowing the arbitrary prototype kilogram to be replaced, but the disadvantage that some kind of ratio measurement must be used to derive mechanical forces from electrical ones. As we shall see in chapter 5, one gains nothing by the change, except perhaps that one may capitalise on the fact that, up to now, the electrical measurements have been more limited by the precision with which the units may be realised than have the mechanical ones. This is an active area of research and it is important to discuss possibilities, for these may lead to new experimental discoveries.

Time was the first unit to be defined by atomic phenomena and it is interesting to recall at this point that the direction of modern metrology was accurately foretold by Maxwell in his Presidential address to the British

Association more than a hundred years ago (Maxwell 1870):

> Yet, after all, the dimensions of our earth and its time of rotation, though, relatively to our present means of comparison, very permanent, are not so by any physical necessity. The earth might contract by cooling, or it might be enlarged by a layer of meteorites falling on it, or its rate of revolution might slowly slacken, and yet it would continue to be as much a planet as before.
>
> But a molecule, say of hydrogen, if either its mass or its time of vibration were to be altered in the least, would no longer be a molecule of hydrogen.
>
> If, then, we wish to obtain standards of length, time, and mass which shall be absolutely permanent, we must seek them not in the dimensions, or the motion, or the mass of our planet, but in the wave-length, the period of vibration, and the absolute mass of these imperishable and unalterable and perfectly similar molecules.

(It is clear that Maxwell used 'molecule' where we now use 'atom'.) Length, through the velocity of light, has just taken this step. As we have mentioned above, the electrical units already use the Josephson effect volt and we may use the quantised Hall effect ohm as well. It is still possible that the Avogadro constant may have a part to play in defining the kilogram, although this would involve the mole as well.

Temperature is involved in the gas constant and the Boltzmann constant and these may come to be used to define the unit of temperature in some way. It is certainly already necessary to know these constants with high accuracy in order to compare temperatures measured by acoustic, gas and radiation thermometry etc. At the present, temperature is defined in terms of a property of water and requires the use of a triple-point cell. At some future date, progress will be limited by the precision with which the unit of temperature may be realised by this method, and the solution adopted will depend on the state of science at that time—perhaps entropy may make a comeback.

The candela, as we will see in chapter 9, is the unit which is realised with the lowest precision and its role is an important one in measurements which are relevant to human vision. Lasers and optical communication have helped to lead to its definition in terms of the watt and it may be that in the future the candela will no longer be regarded as a base SI unit. This will depend on the view taken by scientists throughout the world at that date. The magnitudes of the units are unlikely to change very much, except within the present reproducibility.

1.10 The measurement of time, frequency and wavelength: the key to modern metrology

Since time, frequency and, to a lesser extent, length measurements are the apex of the measurement pyramid in modern metrology, and hence act as the ultimate frontier of precision, we end this chapter by a discussion of both the methods used to realise the definition of the second and to measure length.

1.10.1 Atomic clocks

The definition of the second as the duration of a specified number of periods of the radiation corresponding to the transition between the two hyperfine levels of the ground state of the caesium-133 atom (figure 1.3) is realised by means of an atomic beam standard. Caesium, in common with other alkali atoms, has a single valency electron and this may be aligned with its spin parallel or antiparallel ($J = \pm\frac{1}{2}$) to the direction of the nuclear spin ($I = \frac{7}{2}$). It is the transition between these two energy states $W(F, m_f)$ which must be observed experimentally: that is $W(F = 4, m_f = 0) - W(F = 3, m_f = 0) = hf_0$.

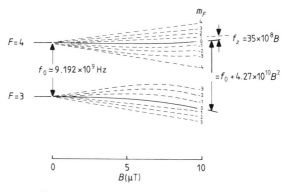

Figure 1.3 The Zeeman effect in the hyperfine transition of the level in the caesium-133 atom, which is used in atomic clocks.

A beam of caesium atoms (figure 1.4(a)) emerges from a slit in an oven which is heated to about 100 °C. The beam enters the gradient magnet (~ 1T) which is designed to produce a nearly uniform dB/dz gradient over the beam, and atoms are deflected upwards or downwards according to their spin direction. Atoms in one only of the two states can traverse the apparatus. The atoms then pass through a uniform field region where the flux density is in the region of $5 \, \mu$T and then pass through a second deflecting magnet before hitting the hot tungsten wire detector. In the uniform field, or C field, region the atoms pass through two microwave cavities whose microwave fields are almost exactly in phase. When the microwave frequency corresponds to the transition frequency, the spin direction of the atoms is reversed so that they are deflected in the reverse direction and there is a corresponding change in the detected signal. The use of a double cavity, or Ramsey technique (Ramsey 1956), leads to a narrow resonance (figure 1.4(b)) whose width depends on the separation of the cavities. The resonance shape also depends on the velocity distribution of the caesium atoms.

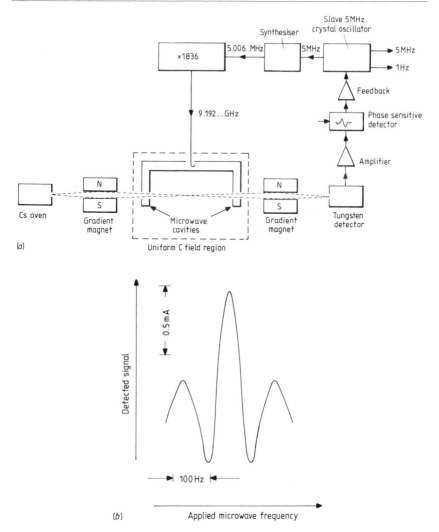

Figure 1.4 (*a*) Schematic of a caesium beam atomic frequency standard. The gradient magnets deflect the required spin states, before and after passing through the microwave fields and the uniform magnetic field region between them. In modern standards the oven source and the detector are interchangeable to eliminate possible direction-dependent systematic effects; (*b*) the central portion of the Ramsey resonance observed with an NPL 2.3 m long caesium-133 frequency standard.

In the *C* field, the two levels split into a number of levels (figure 1.3) and the frequency of the transition between the respective central hyperfine levels is almost field independent and the shift is given by

$$f = f_0 + (4.27\ldots) \times 10^{10} |B|^2$$

where $f_0 = 9\,192\,631\,770$ Hz and the flux density is in teslas, which means that in practical standards the correction is about 1.5 Hz. This shift is usually measured to a few millihertz by exciting some of the other hyperfine transitions.

The resonance width, with a cavity separation of about 2.3 m, is about 50 Hz (or about 5 parts in 10^9 of the caesium frequency) so that the centre of the resonance must be estimated very accurately if the desired precision of parts in 10^{13} is to be achieved. The problem is even more acute in the commercial atomic beam frequency standards, where the beam length may well be only a few tenths of a metre. It is important, therefore, to achieve a very good signal to noise ratio. For the highest accuracy, it is necessary to obtain a good estimate of the beam velocity v in order to evaluate the second-order Doppler frequency shift

$$\Delta f / f_0 = - v^2 / c^2$$

for this necessitates a correction of about 5×10^{-13}. Modern beam standards have provision for interchanging the source and detector which enables such effects as those from any small phase difference (milliradians) between the two microwave cavities to be measured and corrected for.

Aside from the pioneering work of Ramsey (1956), the first working caesium

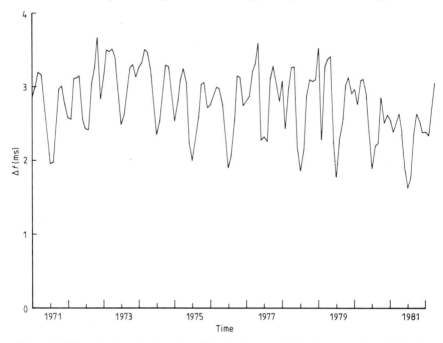

Figure 1.5 Showing how the duration of the day has varied during a decade and thereby illustrating that the period of the rotation of the Earth is no longer satisfactory for timekeeping purposes, although it remains of prime importance for many navigation purposes.

frequency standard was that of Essen (1958) at the NPL. Other laboratories rapidly followed suit and commercial instruments were made which had a very good performance. It had long been known that there were small variations in the rate of rotation of the Earth and it rapidly became apparent that the definition of the second in terms of the motion of the Earth was no longer adequate (figure 1.5). However, there were so many users of time who required it for navigation purposes, or for deriving the angular position of the Earth, that there was a delay before a suitable compromise was found which satisfied most of the users.

Although there is a unique definition of the second in terms of a caesium transition, there are still a number of different time scales in use. On the other hand, well known terms such as 'Greenwich mean time' have largely fallen out of use because they are no longer uniquely defined by the phrases used. These scales are summarised as follows:

A   Universal Time (UT)
In applications in which an imprecision of a few hundredths of a second cannot be tolerated, it is necessary to specify the form of UT which should be used:

UT0 is the mean solar time of the prime meridian obtained from direct astronomical observation;

UT1 is UT0 corrected for the effects of small movements of the Earth relative to the axis of rotation (polar variation);

UT2 is UT1 corrected for the effects of a small seasonal fluctuation in the rate of rotation of the Earth.

(UT1 corresponds directly to the angular position of the Earth around its axis of diurnal rotation. GMT may be regarded as the general equivalent of UT.)

B   International Atomic Time (TAI)
The international reference scale of atomic time (TAI), based on the second (SI), as realised at sea level, is formed by the Bureau International de l'Heure (BIH) on the basis of clock data supplied by cooperating establishments. It is in the form of a continuous scale, e.g. in days, hours, minutes and seconds from the origin 1 January 1958 (adopted by the Conférence Générale des Poids et Mesures (CGPM) in 1971.

C   Coordinated Universal Time (UTC)
UTC is the time scale maintained by the BIH which forms the basis of a coordinated dissemination of standard frequencies and time signals. It corresponds exactly in rate with TAI but differs from it by an integral number of seconds.

The UTC scale is adjusted by the insertion or deletion of seconds (positive or negative leap-seconds) to ensure approximate agreement with UT1.

D   Leap-Seconds

A positive leap-second begins at 23 h 59 min 60s and ends at 0 h 0 min 0 s of the first day of the following month. In the case of a negative leap-second, 23 h 59 min 58 s will be followed one second later by 0 h 0 min 0 s of the first day of the following month. First preference should be given to the end of December and June, and second preference to the end of March and September. The BIH decides upon and announces the introduction of a leap-second, and such an announcement is made at least eight weeks in advance.

Although the atomic time scale which is widely used, UTC, is kept within ± 0.5 s of UT1 (figure 1.6), corresponding to a navigation error of about 464 m at the equator, the broadcast time scales also contain marker pulses which enable the time to be obtained correct to 0.1 s (figure 1.7). Anyone requiring more accurate information can obtain it and corrections are published at regular intervals by the BIH.

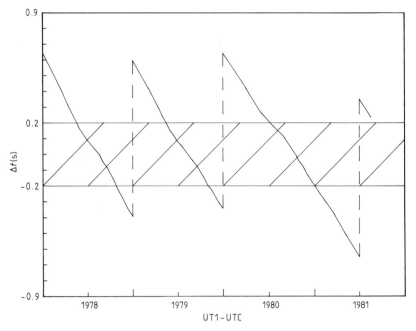

Figure 1.6 Showing how the disseminated atomic time scale is kept roughly synchronous with the time scale associated with the Earth's rotation by means of the introduction of leap seconds as required.

It was some years before the caesium clocks could be operated on a continuous basis for long periods without interruption and the system operated by Mungall *et al* (1980) at the National Research Council, Ottawa (NRC) has

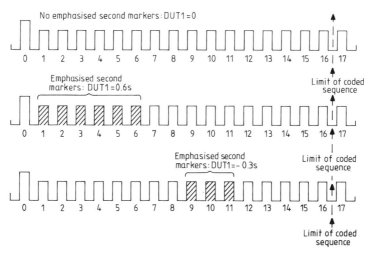

Figure 1.7 Examples of the coded marker pulses which are added to the transmission of time signals in order to achieve 0.1 s accuracy —more accurate information is available from the BIH for users who require it (from CCIR Report 460).

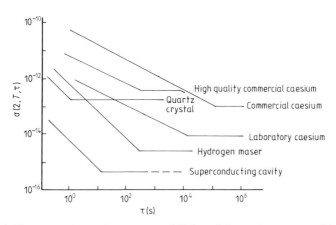

Figure 1.8 The approximate long-term stabilities of the various types of frequency standard —the potential superiority of the superconducting cavity for the shorter term is readily apparent. In the longer term, all of the standards are affected to varying degrees by various types of drift.

been probably the most successful to date. There has therefore been a need to have other frequency standards to act as a short-term 'flywheel' (figure 1.8) Quartz oscillators have been steadily refined for this purpose and rubidium frequency standards have also proved highly satisfactory as secondary standards. The hydrogen maser, which we return to in chapter 2, has better short-term stability than the caesium atomic clock, although it has suffered from the disadvantage of suffering a wall-shift correction which must be derived for each maser bulb. Superconducting microwave cavities, such as have been developed for the Stanford linear accelerator by Turneaure and others, promise well as short-term flywheels to clean up signals, which is an essential part of any frequency multiplication chain taking frequencies through to the visible region of the electromagnetic spectrum.

1.10.2 Length measurements

As we shall see in chapter 2, a definition of the metre which incorporates a fixed value for the velocity of light in some way has just been implemented. The need for this comes in part from the development of stabilised lasers (figure 1.9) for use

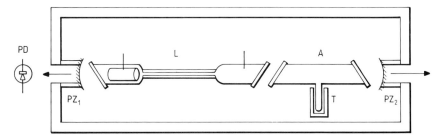

Figure 1.9 Example of a 633 nm helium–neon laser stabilised to an in-cavity transition in molecular iodine. The mirrors are mounted on piezo-electric ceramics (PZ_1, PZ_2), in order both to impose a frequency modulation and to adjust the length, thereby allowing stabilisation of the length of the cavity by a servo-system. The laser cavity contains the helium–neon discharge tube (L) and the iodine absorption cell (A), both having Brewster-angled windows to minimise losses. The thimble (T) serves to stabilise the temperature and hence pressure of the iodine cell.

as wavelength standards in the visible and infrared region of the spectrum. These lasers can be locked to a passive absorption line and, although there are still small shifts (due to intensity, cell pressure and temperature, cavity mode and so on) which mean that the operating conditions must still be carefully specified for high-precision work, the performance is orders of magnitude better than that given by a krypton-86 lamp (figure 1.10.) The developments in the visible region may well mirror those which happened earlier in the microwave region. Just as the microwave cavity wave-meter has now been replaced by direct frequency measurement for high precision work, so too may the Fabry–Perot inter-ferometer be replaced by some form of direct frequency instrument.

The advent of lasers has had a further advantage for accurate wavelength measurements for they have enabled ultra-high vacuum Fabry–Perot etalons to be used in a much more accurate way. Not only has the phase coherence of the radiation from lasers enabled much longer etalons to be used, but the

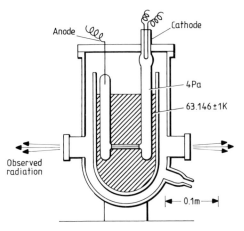

Figure 1.10 Example of a liquid nitrogen immersed, hot-cathode, krypton-86 standard lamp. The wavelength of the orange radiation from such lamps, under carefully specified conditions was, until recently, used to define the SI unit of length. The specified conditions included operation with 99% purity Kr-86 at a temperature within one degree of the triple point of nitrogen and a capillary internal diameter of between 2–4 mm. The lamp was observed end-on such that the light rays used travelled from the cathode end to the anode end and the current density was 0.1 ± 0.1 mm^{-2}. These gave reproducibilities within 10^{-8} of the unperturbed levels.

etalon length may be servoed using piezo-electric elements in the etalon spacers to be an integral number of vacuum half-wavelengths of one laser. The wavelength of a source which is close in frequency to the other radiation may then be servoed to provide an integral number of half-wavelengths for the etalon length. As long as the etalon is more than (say) a million fringes in length the beat frequency between this radiation and the unknown radiation is then conveniently well below 1 GHz—and hence easily measured with a frequency counter. Thus we have

$$n_1 \lambda_1 = n_3 \lambda_3 = 2l$$
$$\lambda_2/\lambda_3 = f_3/f_3$$

and

$$f_2 - f_3 \lesssim 1 \text{ GHz}$$

where λ_1 is one laser wavelength, λ_2 is the unknown laser wavelength, the etalon length is l (~ 30 cm) and f_3, λ_3 are the frequency and wavelength, respectively, of the laser which is close to λ_2. (It could equally be close to λ_1 with appropriate changes in the way that the measurement was performed.) A further small correction is required to allow for changes in the phase shift of the etalon mirrors as a function of wavelength, and this may be evaluated by using etalons of different lengths if necessary. It is also necessary to make diffraction corrections, to allow for wavefront curvature, and to ensure that the illumination of the mirrors is the same for both lasers so that the imperfections are averaged in the same way.

It does not follow that the adoption of a fixed value for the velocity of light as a method of defining the metre will necessarily be the end of the story. Consider, for example, the following situation. Suppose that the two-photon excitation of the 0.3 s lifetime of the Lyman-α 1S–2S transition in hydrogen could be used as a standard and that this 3 Hz linewidth could be split to better than 1%, thereby giving a wavelength (or frequency) standard which was good to one part in 10^{17}. Meanwhile, microwave atomic frequency standards might be realisable to a comparable precision. If this came to pass it might also be found that the gap between the microwave and vacuum ultraviolet could not be bridged without loss of precision. Thus we could envisage a situation with a microwave 'second' and a vacuum UV 'metre', with the velocity of light, *de facto*, once again being measured by experiment in the metrological laboratories. We consider the fundamental constants which require accurate time and length measurements in chapter 3.

2

The constants as natural units: are they constant?

We tend to take the units of science for granted, and so we consider here the properties of units, the number required and the two main types of units which are used. Many theoreticians, as we will see, use fundamental constants as units, and in this chapter we look at how such a system is developed. The fundamental constants would be totally inappropriate as units if they varied with time and so the chapter ends with a discussion of the present experimental evidence for their constancy. The way in which we define our base units affects the possible time variations and this is illustrated by looking at the consequences of the changes in the definitions of the metre. We begin by a consideration of what properties we require of base units.

2.1 The basic properties of a system of units

Any physical quantity is defined by a complete specification of the operations used to measure the ratio, a pure number, of two instances of the physical quantity. Each physical quantity so defined is given a name and is denoted by an algebraic symbol. If we make a number of such operations on the same physical quantity, then it may be convenient to treat it as being a unit. Thus all measurements of length are expressible in terms of the unit of length. Two entities E_1 and E_2 may be intercompared if they are of the same kind so that the question of whether E_1 is greater or less than E_2 is a meaningful one. Thus we cannot compare a length with a mass, for example the equation

$$4 \text{ atmospheres} + 0.3 \text{ litres} + 6 \text{ hectares}$$
$$= 405\,300 \text{ Pa} + 3 \times 10^{-4} \text{m}^3 + 6 \times 10^4 \text{m}^2$$

may be mathematically correct, but it is physically meaningless.

The units of the SI obey certain rules and one of the important rules, for the convenience of users, is that the derived units must be coherent with the base units. Thus one newton is the product of one kilogram and one metre divided by the second squared, while the unit of pressure, the pascal, is one newton per square metre. (On the other hand the cgs unit of pressure, known as the bar, which is 10^5 kg m^{-2} s^{-2}, is not a coherent unit (because a factor of 10^5 is involved) and hence may gradually be phased out from general usage—customers permitting!)

2.1.1 Units having extensive or intensive magnitude

The various quantities which are used in physics to express measurements fall into two distinct groups. The first group comprises those quantities which may be expressed in terms of the sum of smaller quantities of the same kind and hence they have an additive property. The second group comprises those quantities which have no such additive properties. The quantities of the first class are said to have *extensive* magnitude and those of the second, not having an additive nature, are termed as having *intensive* magnitude.

The distinction between the two groups is useful and is best illustrated by considering examples of the two types. Thus length, mass, volume, energy and entropy are all extensive physical quantities. For example, two volumes may be combined additively, and so can two masses. On the other hand, physical quantities such as temperature, density and magnetic permeability, cannot be regarded as being the sum of smaller quantities of the same kind and such quantities are said to have intensive magnitude.

Generally speaking, it is preferable that the quantities chosen to represent the base units are of the extensive class, for this permits simple addition of measurements. When the quantity under investigation is intensive, measurements are usually effected, almost automatically, by expressing them in terms of extensive units. Thus temperatures are intercompared by expressing them in terms of relative volumes of a fixed mass of gas or the lengths of a thread of mercury or the resistance of a platinum wire, and so on.

In the case of quantities with extensive magnitude, there are two groups: those in which there is a continuum, such as length, time or volume, and those in which the measurement of a quantity is composed of a finite number of distinct and identical parts, such as the number of molecules in a given volume. This latter class of measurement involves counting. Certainly measurements in this class may be effected with great rapidity; as has happened with the measurement of frequency with a digital frequency counter, or length with the aid of a laser interferometer. It may be, of course, that all measurement will eventually be shown to be comprised of this digital class. For example, there may be a natural quantum of length, time and so on. Indeed, the Heisenberg Uncertainty Principle implies this already, in a certain sense.

2.1.2 How many base units should there be?

There appears to be no natural law which sets a limit on the number of base units that should be used to describe the physical world and the number employed is largely a matter of convenience for the state of science and technology at any particular time—'units are made for man', not the other way around. It is probable that at least one magnitude must be required, but there is no upper limit. It is generally held, though, that the very useful 'method of dimensions' would not work very well if the number of base units were reduced below about five. The author has encountered scientists who have argued that only three are required—essentially a nineteenth century approach which led to the adoption of the cgs system of units. The lowest limit argued for has been one base unit. Interestingly enough this was mass, for at present the kilogram remains the only prototype artefact unit; others would argue strongly for time instead. However, it is clear that there can be confusion in such discussions between the definition of a base physical quantity and the operational definition of the magnitude of the unit.

There can be other problems too. Consider a definition such as: 'the metre is the distance travelled by electromagnetic radiation...'. In this definition, the concept of length appears to be derived from the prior concept of electromagnetic radiation travelling, i.e. length is derived from time and velocity. Does this allow velocity still to remain a derived unit? Yes, it does, for the above would be the operational definition of the unit of length, and statements about the physical quantities length and velocity can be made separately from these about the magnitude of the unit. There is, of course, nothing to stop us making velocity an apparently dimensionless unit. We could, for example, define the velocity of a body in terms of the relative Doppler shift $\delta\lambda/\lambda$, or $\delta v/v$ (such a measurement is indeed performed with radar velocity meters).

At the present, the SI comprises seven base units: mass, length, time, electric current, thermodynamic temperature, luminous intensity and amount of substance. The first five of the base units would be used by the majority of physicists, but the necessity for the last two would provoke more discussion. When the SI was first set up, the concept of amount of substance was not included. The use of luminous intensity too is very limited and it is conceivable that the candela might come to be a derived unit at some future date, especially when the present definition is considered. It is interesting, too, that there can be cases where the attitude to a quantity varies according to usage. Thus the units of angle and solid angle, the radian and steradian, are sometimes dimensioned and at other times dimensionless. This oscillation has been reflected by changes in their interpretation as supplementary units as experience with usage of the SI has built up.

The number of base units required depends on the particular scientific discipline. It is possible to illustrate this by making an hierachical build-up in the number of units which are required (table 2.1). Thus, in geometry there is only a requirement for length measurements. In kinematics both length and time are required, and extending this to mechanics involves the inclusion of mass. The science of electrodynamics involves mass, length, time and the electrical units, and extending this into thermodynamics brings in temperature.

Table 2.1 Illustrating how the number of units involved depends on the scientific discipline.

Branch of science	Physical quantities involved
Geometry	Length
Kinematics	Length, time
Mechanics	Length, mass, time
Thermodynamics	Length, mass, time, temperature
Quantum electrodynamics	Length, mass, time, electric current (temperature)
Electrochemistry	Length, mass, time, current, temperature, amount of substance
Photo-optics	Length, mass, time, current, temperature, amount of substance, luminous intensity

Electrochemistry adds the mole to the list and, in principle, laser spectroscopy of gases, or laser fusion, might invoke the candela as well, particularly now that it is firmly defined in terms of the watt.

It must be apparent from this that it is dangerous to argue from personal experience, for, although a particular SI base unit may not be required in some disciplines, it is quite possible that it would prove to be essential in another. In many respects, this is best illustrated by the mole, which is conceptually very necessary in chemistry as a base unit, but which many physicists find they use so rarely that they may argue that it is unnecessary. Certainly, the mole was not included among the SI base units which were adopted initially, and neither do standards laboratories find it necessary to derive and maintain the mole as they do other base units.

To a certain extent, the number of units which must be used depends both on our ability to perform the required metrology and also on how well particular concepts have been absorbed into our scientific culture. For example, the calorie was required for many years as a measure of heat energy and, because it was conceptually separate from the joule, it was necessary to measure the mechanical equivalent of heat experimentally. The introduction of the SI enabled the calorie to be phased out and experimental measurements of the mechanical equivalent of heat are no longer required, at least in the same form (it has reappeared in the SI as the specific heat of water). In navigation, for example, movement in the horizontal plane is much easier than in the vertical direction, and this may be emphasised by the use of a different unit, for example, in maritime use, by measuring depths in fathoms. We are familiar with thinking of volume as being derived from the product of three perpendicular lengths and so do not require a separate base unit for volume. Even so, through problems of metrology, the litre was only recently defined, as exactly 1 dm^3, in terms of the cubic metre.

Considerations such as those above lead one to speculate on whether the incorporation of a fixed value for the velocity of light into the definition of the metre might not be the beginning of the phasing out of the unit of length as a base unit separate from time—one's height might be expressed as 5.5 ns! Of course, the views of the layman will be as important in determining what happens as those of the scientist. The usage of units is therefore based in part on a well-defined set of rules and in part on satisfying a global customer demand.

2.2 Natural units

Most scientists feel, intuitively, that there must be some natural units against which all other measurements may be expressed. However, different natural units might well be appropriate to different disciplines, so that one might not end up with a universally agreed set. Certainly, scientists have a tendency to make their measurements with respect to some local unit and it is frequently the task of the metrologist to relate these local units to the SI units. Evidently, the electron-volt is a very convenient energy unit in nuclear physics or in x-ray spectroscopy, and the magnetic moment of the proton is a convenient magnetic unit in nuclear magnetic resonance. Over and above such practical considerations, there are the desires of some cosmologists for a more 'universal' set of base units than the SI provides at present.

The quest for natural units is not a new one; indeed the original cgs system was set up with the properties of the Earth in mind. Thus, the metre was defined in terms of a quadrant of the Earth's circumference passing through Paris, the gram as the mass of a cubic centimetre of water and the second in terms of the period of rotation of the Earth. In the position reached by science in the early nineteenth century, these were quite reasonable choices to take for natural units. It was recognised that these might not prove sufficiently constant and today we would think much more in terms of certain of the fundamental physical constants as envisaged by Rayleigh (1876) and it is quite usual to use certain of these as units in performing theoretical calculations. It is not at all unusual to find the statement 'let $\hbar = e = c = 1$' at the beginning of a theoretical paper, which sometimes leaves the experimentalist a little confused when trying to compute the magnitudes of expressions occurring later in the paper. However, this confusion is small in comparison with the uncertainty that frequently existed in the mind of the user in the days of electrostatic and electromagnetic units in deciding whether the electronic charge was intended to be in esu or emu and so on! The use of such units must be as old as the subject of fundamental constants. Perhaps the earliest example is that of Johnstone-Stoney (1881) who considered the units in the system

$$e = c = G = 1.$$

According to Millikan (1916), this was the first example of the postulated existence of a unit of elementary charge[†].

[†] Note that there is a class of problems for which we cannot use $c = 1$: for example, when one is discussing tests of the Einstein Equivalence Principle (EEP), when the value of c would depend on the method used to measure it (Will 1981).

In some cosmological theories, there is a natural time in the evolution of the early universe when quantum mechanical effects must become important in gravitation. This is characterised by the Planck time,

$$t_P = (G\hbar/c^5)^{1/2} \sim 10^{-43} \text{ s.}$$

There is also a corresponding Planck length

$$l_P = (G\hbar/c^3)^{1/2} \sim 10^{-35} \text{ m}$$

and a Planck mass

$$m_P = (\hbar c/G)^{1/2} \sim 10^{-5} \text{ g.}$$

(These may be our natural quanta of time and length.) Recent proposals for the unification of the electromagnetic, weak and strong interactions have suggested that these apparently distinct interactions arose from one single interaction by spontaneous symmetry breakdown at this early stage in the evolution of the universe. These units have a considerable usefulness in the particular area of cosmology where they are relevant, but it is apparent that as units they differ by an extremely inconvenient amount from the present base SI units.

It is likely that any departure from the present magnitudes of the SI base units would be very strongly resisted, and the author is *not* advocating change. It is a fundamental requirement of the SI for any change in the operational definition of any natural unit to be made in such a manner that it is imperceptible to the bulk of those making accurate measurements; that is, there must be continuity of the disseminated unit. The time at which one might have moved to a definition of the metre in which the velocity of light was $3 \times 10^8 \text{ m s}^{-1}$ exactly, without having a considerable impact on technology and the proverbial man-in-the-street, has long since passed (there are even some unrepentant Imperial types who point out that light travels about one foot in a nanosecond!).

When Dirac (1937) first generated his dimensionless 'big' numbers he considered a number of possible natural atomic units. Thus, for time, one may take

$$\left(\frac{\mu_0 c^2}{4\pi}\right)\frac{e^2}{m_e c^3} \quad :\left(\frac{\mu_0 c^2}{4\pi}\right)\left(\frac{e^2}{m_p c^3}\right) \quad :\frac{h}{m_e c^2} \quad :\left(\frac{h}{m_p c^2}\right) \quad :\frac{h}{m_e c^2} \quad :\frac{h}{m_p c^2}$$

$$1 \qquad\qquad :5 \times 10^{-4} \qquad :8.5 \times 10^2 \quad :4.6 \times 10^{-1} \quad :137^{-1} \quad :4 \times 10^{-6}$$

as possible time units; the one which he chose (the first given) lay around midway in the relative magnitudes of the possible contenders. This unit is often termed the tempon or chronon. (We should also mention Sir Arthur Eddington's endeavours in this direction, although his work is not cited today as often as that of Dirac.)

If we consider length units, we can readily see several which prove useful on an atomic scale. Thus, we could choose the classical radius of the electron r_e, the

Table 2.2 The Bohr ruler: natural units of length as powers of the fine structure constant.

Combination	Atomic property	Characteristic length (m)
$\alpha^3/4\pi R_\infty$	The classical electron radius, r_e	$2.817\,938\,0(70) \times 10^{-15}$
$\alpha^2/2R_\infty$	The Compton wavelength of the electron, λ_c	$2.426\,308\,9(40) \times 10^{-12}$
$\alpha/4\pi R_\infty$	The Bohr radius of the hydrogen atom, a_0	$5.291\,770\,6(44) \times 10^{-11}$
$1/R_\infty$	The reduced wavelength of hydrogen radiation, $1/R_\infty$	$9.112\,670\,34(83) \times 10^{-8}$

Compton wavelength of the electron, the Bohr radius a_0 of the hydrogen atom or the reduced wavelength of hydrogen radiation $1/R_\infty$ as our natural unit. Which of these would be the most natural? Evidently, it would depend on what we were using it for, and we see from table 2.2 that our range of units would cover four orders of magnitude. This ruler, known as a Bohr ruler, is scaled in terms of the fine structure constant. This involvement of α with our natural units is at first sight rather unexpected. It is interesting that if we follow Carr and Rees (1979) and introduce another dimensionless scaling constant, the gravitational fine structure constant α_G

$$\alpha_G = Gm_p^2/\hbar c \sim 5 \times 10^{-39}$$

and also the weak fine structure constant α_w

$$\alpha_w = g_w m_e^2/\hbar^3$$

where g_w is the weak coupling constant and

$$\alpha_G \sim \alpha_w^5,$$

then we may scale up from atomic dimensions to the whole universe by using combinations of these constants with m_p/m_e, as illustrated in figure 2.1. These play an important part in the anthropic principle (see § 2.5) which may lead to estimates of the magnitudes of many of the fundamental constants.

It is already apparent that there is a considerable choice in possible magnitudes for natural units, and it is not proposed to discuss these further. Whatever the ultimate choice for the natural units of science may be, it is apparent that we are steadily moving over to a system of units which bears a constant relationship to these units, despite the fact that at present we may not be able to derive explicit expressions in all cases (thus the atomic definition of the second is not at present directly derivable in terms of fundamental constants).

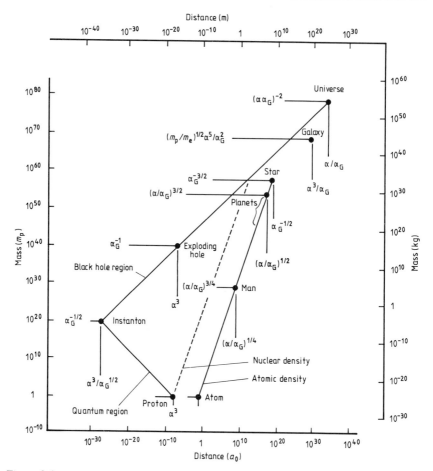

Figure 2.1 Illustrating how the mass and length scales, varying from atomic dimensions to those of the universe, may be expressed in terms of the mass of the proton and the size of atoms together with the dimensionless fine structure constant $\alpha(= \mu_0 c^2/4\pi(e^2/hc) \sim 137^{-1})$ and $\alpha_G(= Gm_p^2/hc \sim 5 \times 10^{-39})$ respectively. Evidently, the metre and kilogram occupy a reasonably central position as far as symmetry in positive and negative powers of ten is concerned, emphasising that the SI units are natural anthropic units as well. (Based on Carr and Rees 1979.)

2.2.1 A system of natural units

As was remarked earlier, the statements

$$\text{'let } \hbar = m_e = c = 1\text{'} \quad \text{or} \quad \text{'let } \hbar = m_e = e = 1\text{'}$$

are perhaps the most frequently encountered use of atomic units, and in this section we follow the treatment of Hartree (1926) and McWeeny (1973) to show

how the system of mechanical, electrical and magnetic units can be developed from these. The conventional approach is to take $h/2\pi$ as the unit of action, rather than the Planck constant, and also to take the constant k_0 as the unit of permittivity where $k_0 = 4\pi\varepsilon_0$, and ε_0 is the usual constant of permittivity used in the SI system. The values are given explicitly in the tables so that it will be readily possible to make the conversion to h and ε_0 as units, rather than $\hbar$ and $k_0(4\pi\varepsilon_0)$, if these are preferred.

Our base units in this system are given in table 2.3(a), and from these the units of length, time and energy may be derived (table 2.3(b)). The unit of length conveniently has the magnitude appropriate to a natural unit of length, for it is the radius of the first Bohr orbit in the quantum mechanical treatment of the hydrogen atom in the fixed nucleus approximation. The unit of energy too is the

Table 2.3(a) The base units in the system $h = m_e = e = 1$.

Quantity	Dimensions	Natural unit	SI equivalent (1981 values)
Mass	M	electron mass m	9.1091×10^{-31} kg
Charge	Q	electronic charge	1.60210×10^{-19} C
Action	MLT^{-1}	Planck constant $\hbar$	1.05450×10^{-34} J s
Permittivity	$Q^2W^{-1}L^{-1}$	$k_0 = 4\pi\varepsilon_0$	$4\pi \times 8.8542 \times 10^{-12}$ Fm^{-1}

Table 2.3(b) The derived units in the system $\hbar = m_e = e = 1$ (after McWeeny 1973).

Quantity	Dimensions	Natural unit (name)	SI equivalent
Length	L	$k_0\hbar^2/m_e^2$ (Bohr, a_0)	5.29167×10^{-11} m
Time	T	$k_0^2\hbar^3/m_e e^4$	2.41889×10^{-17} s
Velocity	LT^{-1}	$e^2/k_0\hbar$	2.18764×10^6 m s^{-1}
Force	MLT^{-2}	$m_e^2 e^6/k_0^3\hbar^4$	8.23831×10^{-8} N
Energy	ML^2T^{-2}	$m_e e^4/k_0^2\hbar^2$ (Hartree, E_0)	4.35944×10^{18} J
Power	ML^2T^{-3}	$m_e^2 e^8/k_0^4\hbar^5$ ($E_0^2/\hbar$)	1.80225×10^{-1} W
Electric current	QT^{-1}	$m_e e^5/k_0^2\hbar^3$	6.62329×10^{-3} A
Electric potential	WQ^{-1}	$m_e e^3/k_0^2\hbar^2$ (atomic volt)	2.72108×10^1 V
Capacitance	Q^2W^{-1}	$k_0^2\hbar^2/me^2$	5.88774×10^{-21} F
Magnetic permeability	MLQ^{-2}	$k_0\hbar^2/e^4$	1.87797×10^{-3} N A^{-2}
Magnetic flux	WTQ^{-1}	$\hbar/e$	6.58199×10^{-16} Wb
Magnetic flux density	$MT^{-1}Q^{-1}$	$m_e^2 e^3/k_0^2\hbar^3$ (atomic tesla)	2.35055×10^5 T

Table 2.3(c) The natural units of mass, length, time and electric current in the system $\hbar = m_e = e = 1$.

Quantity	Natural unit	Approximate value in equivalent SI units
Length	$4\pi\varepsilon_0 h^2/m_e e^2$	5.292×10^{-11} m
Mass	m_e	9.109×10^{-31} kg
Time	$(4\pi\varepsilon_0)^2 h^3/m_e e^4$	2.418×10^{-17} s
Electric current	$m_e e^5/(4\pi\varepsilon_0)^2 h^3$	6.623×10^{-3} A

magnitude of the electronic energy in the hydrogen ground state (tables 2.2, 2.3). The derived electrical and magnetic units follow once the base units are fixed. Of the magnetic units, the unit of magnetic moment is equal to twice the Bohr magneton, and the unit of magnetic flux is a factor of π smaller than the magnetic flux quantum $h/2e$. The latter serves to illustrate that at least some of the derived 'natural' units have less ideal magnitudes than we might desire! The unit of flux density at a value equivalent to 2.35×10^5 T is inconveniently large for many purposes, since flux densities even in atoms are commonly ~ 20 T, or 10^{-4} times smaller than the 'atomic' tesla.

Although this system is logically consistent, and may be made a coherent system (by choosing $\hbar = e = m_e = k_0 = 1$), it is far from being coherent with, or rationally related to, the SI units, and indeed it is not intended to be. The approximate magnitudes of these units have been given in terms of their equivalent SI units. It should be remembered that the values change slightly with each adjustment of the fundamental constants and that where the desired quantity is not quoted with the other output values (chapter 9) the uncertainties must be evaluated using the full error matrix. Incidentally, in this system the velocity of light becomes equal to the reciprocal of the fine structure constant. It brings out the relation between the time unit in this system and the unit in the alternative system

$$m_e = e = c = 1$$

where it is a factor of α smaller.

There is little purpose in using the fundamental constants as our base units if they vary with time in some way. We therefore devote the rest of this chapter to a consideration of the theories which require time variations of the constants and to the experimental evidence concerning their constancy.

2.3 The constancy of the constants

It is a fundamental and frequently asked question as to whether or not physics might be different in different locations or different epochs. While it is difficult to conceive a smooth variation in the mathematical form of a physical law from one event to another in space–time, e.g. from linear to square law, it is much easier to imagine a continuous change in the value of a physical constant. The assumption of constancy is of key importance in astronomy, where, by interpreting the red shift as an indicator of time in studying the emissions from distant galaxies, one is able, in effect, to move back through time. There are too, considerable assumptions involved in the extrapolation of physical laws which have been verified in the laboratory to the conditions which pertain on an astrophysical scale. Clearly, one risks having a distorted view of the universe if the possibility of variations in the constants is ignored totally, and if the constants really vary in space–time, then, once the nature of these changes had been established, appropriate corrections would have to be applied. It behoves scientists to investigate the possibilities for such changes continually, as the frontier of measurement advances. In the meantime, it is safest to begin by assuming that the constants are constant and to look for evidence to the contrary.

2.3.1 Possibilities to be considered

In searching for changes in the fundamental constants, there are a number of possibilities which must be considered. From the properties which are ascribed to units we know that any physical quantity x may be expressed as the product of a dimensionless number k_1 and the product of some combination of the base units raised to various powers:

$$x = k_1 \prod_1 (\text{m, kg, s, A}, \dots).$$

In the first instance, the metrologist might suppose that the units would be the base SI units, but if we are considering the variation of physical constants it is probable that the variation would have to be with respect to some natural base units, in which case we would express the above physical quantity X as

$$x = k_2 \prod_2 (h, e, m_e, c, \dots)$$

taking only sufficient constants as base units to form a consistent set of units. The establishment of the constancy of the constants is therefore strongly related to the problem of finding some natural units against which the variation may be revealed. Thus

$$dx = \frac{\partial k_1}{\partial t} dt + \frac{\partial \Pi_1}{\partial t} dt$$

or

$$dx = \frac{\partial k_2}{\partial t} dt + \frac{\partial \Pi_2}{\partial t} dt$$

We can only measure $\partial k_1/\partial t$ or $\partial k_2/\partial t$ and, in order to make inferences concerning dx/dt, it is necessary to be sure that our chosen unit system does not change, i.e. either

$$\frac{\partial \Pi_1}{\partial t} = 0$$

or

$$\frac{\partial \Pi_2}{\partial t} = 0.$$

Evidently, if we cannot find some basic units for our system of measurement whose constancy in space–time is assured, it may be a meaningless question to ask whether or not the constants may vary with time. The only solution would be to postulate some omnipotent set of unchanging base units of measurement which were in some sense outside our universe and completely unaffected by events inside it. In the absence of such units we avoid the proliferation of the possibilities which would arise from considering whether the masses of the electron or proton are constant in some absolute sense. Instead, we are restricted to considering, for example, whether the electron mass varies with respect to the prototype kilogram at the BIPM at Sévres, or relative to the mass of the proton. The first possibility is of considerable interest to the metrologist (who thereby could derive information about the stability of the prototype kilogram), and the second is more the concern of the cosmologist.

While we are free to define units arbitrarily at different times, as we will consider, and this might, for example, introduce apparent space–time changes in the value of the electron mass, such changes would be unlikely to have any real physical significance, except to reveal the imperfections of the chosen base unit (i.e. the metre in the UK in 1984 might differ from the metre in the USA in the year 2000, etc). On the other hand, no formal definition of the units would affect our examination of the invariance of the ratio of the electron mass to that of the proton, for our chosen unit in this case would be the mass of the proton. Even here, further consideration might show this to be a very unsatisfactory unit to choose if we were searching for significant space–time variations in the mass of the electron, for the proton may have a half-life $\sim 10^{32}$ yr and its mass might consequently vary in time as well.

However, as pragmatists, we can only seek to make the best use of the units which are available to us and be careful in any interpretation which we may place on our discovery of the constancy, or inconstancy, of the electron mass with respect to that of the proton. This point is illustrated in the following section by a discussion of the effects of the different definitions of the metre on the possible time variations of certain of the physical constants.

2.3.2 Constraints imposed by the definitions of the metre, kilogram and second

There have been three different definitions of the metre, and in this section we follow through the implications of the changes in the definitions on the permissible variations of certain of the dimensioned constants. It must be emphasised that the conclusions depend on the assumptions which are made as to which constants might vary, and consequently they should be regarded as illustrating the fundamental difficulty rather than necessarily proving a particular set of hypotheses. In addition to the definitions of the metre, the definitions of the other SI units also enter into the discussion.

The three definitions of the metre which will be considered are (i) the proto-type platinum–iridium metre bar, (ii) the definition embodying the specified wavelength of the orange radiation from krypton-86 atoms, and (iii) the incorporation of a constant value for the velocity of light into a definition of the metre.

Following Atkinson (1963), the length of the prototype metre bar depends on the interatomic spacings of the platinum–iridium used in its construction. These spacings in turn might reasonably be expected to depend to first order on the Bohr radius of the atom,

$$a_0 = (\mu_0 c^2/4\pi)^{-1}(\hbar^2/m_e c^2).$$

Thus, in defining the metre via the constancy of the length of the metre bar, we are ensuring that the quantity a_0 must be invariant.

In the case of the krypton-86 metre definition, it is of course not yet possible to predict the wavelengths of the emissions from heavy atoms with high precision, but it would be reasonable to assume that, to first order, the wavelengths would depend on the Rydberg constant and also on the reduced mass of the krypton-86 atom (in an analogous way to the expressions found for hydrogen, deuterium, singly ionised helium and heavier atoms when stripped of all but a single electron). In any case, the Rydberg constant may now be monitored with very high precision, $\sim 10^{-10}$, with respect to this definition. In consequence, the krypton-86 definition of the metre fixes the quantity $m_e c\alpha^2/2\hbar$ as constant.

The third possibility is that the incorporation of an implied value for the velocity of light into a definition of the metre will ensure that the quantity c must remain constant. The dependences set by the three definitions are summarised in table 2.4.

Table 2.4 Constancy of combinations of the fundamental physical constants assumed by the definitions of the SI units

Definition	Combination required to be constant
Metre	
(i) metre bar	$(\mu_0 c^2/4\pi)\hbar/m_e^2 c^2$
(ii) krypton-86	$m_e c\alpha^2/2\hbar$
(iii) speed of light	c
Kilogram	m_e/β
Atomic second	$m_e c^2\alpha^4\beta/\hbar$

As has been remarked earlier, it only makes sense to invoke possible time variations of dimensionless constants, and consequently we assume that there might be time variations of the fine structure constant α and the ratio of the electron mass to that of the proton β and see how the dimensioned constants would be affected by changes in the values of these constants. Aside from the definitions of the metre, the definitions of the prototype kilogram and the mole have the effect of making the number of atoms of hydrogen in a kilogram an invariant, thereby fixing m_e/β as constant (we have of course assumed that higher-order effects, such as changes in the binding energies, may be neglected. Any continuous creation of matter might complicate the situation further and the evidence for this is considered elsewhere in this book). The final unit to be considered, apart from the assumed constancy of μ_0, is the implication of the definition of the atomic second.

As with the krypton-86 definition of the metre, the adoption of the caesium-133 hyperfine frequency as a method of defining the atomic second has implications for the invariance of a combination of the fundamental constants. Unfortunately, we cannot calculate the transition frequency explicitly in terms of fundamental constants, but we may reasonably suppose that, to first-order, as with hydrogen, it will depend on the product of the Bohr and nuclear magnetons divided by the cube of the Bohr radius of the electron. (In any case, we could monitor the ratio of the caesium and hydrogen maser frequencies to ensure that the required degree of constancy was maintained). Therefore, the caesium-133 frequency definition of the atomic second fixes the combination $m_e c^2 \alpha^4 \beta/h$ as invariant.

2.3.3 Consequences of the constraints imposed by the definitions

The above assumptions enable the dependences of m_e, h, c and e on α and β to be derived explicitly, and these are shown in table 2.5. It is clear from this table that the possible variations of the dimensioned constants depend very markedly on the definitions which we use for our units of length. It follows too that changing our definitions of the other SI base units would also have a marked effect on the dimensioned constants. Perhaps the point of greatest

Table 2.5 Dependence of m_e, h, c and e on α and β for different definitions of the metre.

Definition	Dependence of constant on α and β			
	m_e	h	c	e
Metre bar	β	α^{-2}	$\alpha^3 \beta^{-1}$	$\alpha^2 \beta$
Krypton-86	β	constant	$\alpha^{-2} \beta^{-1}$	$\alpha^3 \beta$
Speed of light	β	$\alpha^4 \beta^2$	constant	$\alpha^5 \beta$

interest is that as long as the krypton-86 definition of the metre was used, it was meaningless to look for changes of Planck's constant. However, the fact that we have already, with this definition, in effect set

$$h = \text{constant}$$

may serve to reassure those who are uneasy about imposing the constraint that the velocity of light be an invariant quantity through its being incorporated into a definition of the metre. One accepts, of course, that fixing the velocity of light, or the group velocity of a photon, might have a deeper significance, through its key role in relativity, than fixing Planck's constant, which is a measure of the energy carried by a photon (Breitenberger 1971, Levy-Leblond 1979).

2.3.4 Possible time variations of the constant

In order to get some idea of what variations to look for experimentally, one needs to have some theory of how the constants may vary with time, and this takes one into the realms of cosmology. The subject begins for most people with Dirac (1937), although others, notably Eddington, were working on it at about the same time. The first of his classic papers was reportedly written while the Diracs were on their honeymoon (Alpher, 1969). (Perhaps this is a good time to get the sort of grand vision that the subject demands: it is interesting that Joule took a thermometer on his honeymoon to measure the temperature rise of waterfalls.)

In his 1937 paper, Dirac wrote that he was disturbed by the coincidence of general dimensionless ratios. The first difficulty stemmed from considerations of a time origin and a time axis, both of which seemed to be at variance with the requirements of general relativity. Further, his numerical ratios appeared to require a time dependence for the gravitational constant, again at variance with relativity.

From a study of the dimensionless ratios, Dirac proposed his cosmological principle, namely that any two of the dimensionless numbers occurring in nature are connected by a simple relationship in which the coefficients are of order unity. Moreover, if one of these relations varied with time then so must they all. These relations involve numbers of the order 10^{40} or $(10^{40})^2$ (table 2.6). In some theories (Teller 1948), the fine structure constant is also included as involving the logarithm of these numbers. The dimensionless quantities considered by Dirac involved what were thought to be natural ratios involving force, length, mass, energy and time.

(1)   Force
For force, Dirac took the ratio of the Coulomb electrostatic force to the gravitational force between electrons and protons; that is

$$\left[\left(\frac{\mu_0 c^2}{4\pi}\right)^2 \left(\frac{e^2}{r^2}\right)\right]\left(\frac{G m_p m_e}{r^2}\right)^{-1} = \left(\frac{\mu_0 c^2}{4\pi}\right)\left(\frac{e^2}{G m_p m_e}\right) \sim 0.2 \times 10^{40}.$$

Table 2.6 The Dirac 'big numbers'.

Property	Expression	Approximate Value
Force: coulomb, $p \leftrightarrow e$: $(\mu_0 c^2/4\pi)e^2 r^2$ / gravitation, $p \leftrightarrow c$: $Gm_p m_e/r^2$	$\dfrac{(\mu_0 c^2/4\pi)e^2}{Gm_p m_e}$	$\sim 0.2 \times 10^{40}$
Length: characteristic scale in the universe / classical radius of electron (or range of nuclear forces)	$\dfrac{R_0(\text{or } ct_{age})}{(\mu_0 c^2/4\pi)e^2/m_e c^2}$	$\sim 4 \times 10^{40}\,(t_{age} = 13 \times 10^9 \text{ yr})$
Mass: $\dfrac{\text{mass of visible universe}}{\text{mass per particle}}$ (or number of particles)	$\dfrac{\rho_0 c^3 t_{age}^3}{m_p}$	$\sim 10^{78}(\rho_0 = 7 \times 10^{-28} \text{ kg m}^{-3})$ $\sim 10^{79}(\rho_0 = 2 \times 10^{-26} \text{ kg m}^{-3})$ or $(10^{40})^2$ $\left(\begin{array}{l}\text{Note: The Eddington Number is} \\ \frac{3}{2} \times 2^{16^2} \times 136 = 10^{79}\end{array}\right)$ $\sim 0.2(\rho_0 = 2 \times 10^{-26} \text{ kg m}^{-3})$
Energy: gravitational potential energy of rest of the universe in the field of a nucleon / rest mass energy of a nucleon	$\dfrac{Gm_p \rho_0 c^3 t_{age}^3/ct_{age}}{m_p c^2}$	$\sim$ unity $= (10^{40})^0$
Time: age of universe, t_{age} / characteristic atomic time, tempon	$\dfrac{t_{age}}{(\mu_0/4\pi)(e^2/m_e)}$	$\sim 4 \times 10^{40}$
reciprocal of the fine structure constant, α^{-1}	$(\mu_0 c^2/4\pi)e^2/hc$	~ 137 $\sim 1.5\ln(10^{40})$

(2) Length

For length, he took the ratio of the characteristic scale of the universe to the classical radius of the electron. These were essentially of the order of the largest and smallest distances that one was likely to encounter (we now have the Planck length of course, which is even smaller). The former is the radius of the universe, on the assumption that it has been expanding with the velocity of light. The classical radius of the electron is also roughly the range of nuclear forces. Taking the universe to be 13 eons (13×10^9 years) old, then the ratio is

$$Gm_p\rho_0 c^3 t_{age}^3/m_p c^2 = G\rho_0 ct_{age}^3/m_p \sim 0.2 \times 10^{40}$$

for

$$\rho_0 = 2 \times 10^{-26} \,\text{kg m}^{-3}.$$

(3) Mass

For mass, Dirac considered the ratio of the mass of the universe to the mass per particle—which is about the number of particles in the universe. This requires an assumption about the density ρ_0 of matter in the universe, which is uncertain to a factor of a hundred or so. However, one obtains as the number of particles, N,

$$N = \rho_0 t_{age}^3 c^3/m_p \sim 10^{78} \quad \text{(if } \rho_0 = 7 \times 10^{-28}\,\text{kg m}^{-3})$$
$$\sim 10^{79} \quad \text{(if } \rho_0 = 2 \times 10^{-26}\,\text{kg m}^{-3})$$
$$\text{i.e.} \sim (10^{40})^2.$$

There is a similar number, known as the Eddington number (which survives from Eddington's theory), which is

$$N_E = \tfrac{3}{2} \times 2^{16^2} \times 136 \sim 10^{79}$$

(this number retains his original 136 for α^{-1}).

(4) Time

The question of time formed the crux of Dirac's argument. The largest time that we encounter is the age of the universe, $\sim 10^{17}$ s, and the smallest is about the time that it takes light to travel a distance equal to the classical electron radius, the tempon, or chronon, and which for many scientists represents a natural unit of time, $\sim 0.49 \times 10^{-23}$ s. Taking the ratio of these again yields about the same number $\sim 10^{40}$.

(5) The Dirac theory

The essential point made by Dirac was that the dimensionless ratios all came to this value, not as the result of an accidental coincidence, but because they depended on the age of the universe in chronons. This therefore led to his theory for the change of these ratios with time. Thus the force ratio led to a prediction that the gravitational constant would vary with time.

There have, of course, been a number of theories proposing time variations of certain combinations of the constants since that of Dirac. These constants are as follows:

(1)   the fine structure constant α,
(2)   the constant β' where

$$\beta' \sim g_\beta m_\pi^2 c/h^3 \sim 1.7 \times 10^{-7}$$

(The constant g_β appears in Fermi's theory of beta decay and is a measure of the intensity or strength of the interaction between the nucleonic and electron–neutrino fields; its value is very small $\sim 1.4 \times 10^{-62}\,\mathrm{Jm^3}$),

(3)   the constant γ, which we have encountered, is the ratio of the coulomb force to the gravitational force: $\gamma \sim 2 \times 10^{39}$ and
(4)   the quantities δ and ε

$$\delta = Hh/m_e c^2 \sim 10^{42}$$

$$\varepsilon = G\rho_0/H^2 \sim 2 \times 10^{-3}$$

for

$$\rho_0 = 7 \times 10^{-28}\,\mathrm{kg\,m^{-3}}.$$

Both δ and ε involve the Hubble constant H, which characterises the variation of the red shift of stellar light with distance and is roughly the reciprocal of the age of the universe. The different cosmologies predict different variations of the parameters with time, and the fact that none of them have been entirely successful is an indication that the subject is still an open one.

There are other dimensionless numbers involving 10^{36} which rely on time measured from a different origin. The origin is the point at which the density of the radiation from the 'Big Bang' was equal to the density of matter. These numbers are obtained by considering the range of nuclear forces and the interactions between nucleons. These reduce Dirac's numbers by a factor of 1836, and the large numbers $\sim 10^{40}$ are thereby reduced to $\sim 10^{36}$.

2.4 Experimental limits on possible time variations

In the following sections, we discuss briefly the experimental evidence which is used to make deductions about possible time variations of the constants. This is a rapidly developing and highly complex subject, and it is suggested that any reader who needs to incorporate the results into a theory examines the associated theoretical arguments very thoroughly—the other experts may have missed something, or a more recent development may have rendered their arguments invalid.

2.4.1 Fine structure constant

The evidence for the constancy of the fine structure constant comes from both terrestrial and astronomical sources. Bahcall et al (1967) studied the absorption spectrum of the quasi-stellar object (QSO) 3C-191 having a red shift $z = \delta\lambda/\lambda$ of 1.45. They studied the fine structure of three doublets of silicon-iron and obtained

$$\alpha(z = 1.45)/\alpha_{lab} = 0.98 \pm 0.05$$

whereas from a dependence on t_{age} a value of $\frac{1}{3}$ would be expected. However, one could question the use of the red shift as a measure of the distance of QSOs. Later, Bahcall and Schmidt measured the fine structure doublets in the emission line of oxygen in five radio galaxies with $z = 0.2$ and found that

$$\alpha(0.2)/\alpha_{lab} = 1.001 \pm 0.002$$

whereas one would expect 0.8 on the $\alpha^2 \sim t$ hypothesis. Thus, if one again believes the distance–time speed of recession scale the limit to time variations of α is at the one part in 10^{12} level per year.

There have been suggestions that α might vary close to a large mass. Dicke (1964), for example, suggested that it would vary with the radial distance r from the Sun as

$$\alpha \sim GM_{\odot}/rc^2$$

where $M_{\odot}$ is the mass of the Sun. However, this type of variation was ruled out by the results of the Eötvös experiments which we discuss elsewhere.

2.4.2 Superconducting cavity test of the constancy of α

The development of superconducting microwave cavities for use in the Standford linear electron accelerator led, somewhat unexpectedly, to a test of the constancy of the fine structure constant. The cavities, being superconducting, can be made to have a very high quality factor or Q. The very carefully fabricated cavities of Turneaure and Stein (1974) had a Q of about 4×10^{10} and a resonant frequency of about 8.6 GHz. They built three microwave oscillators having the cavities as the frequency-controlling element and phase-locked Gunn-diode oscillators. The power incident on the cavities was less than a microwatt but, even so, a correction was required for the frequency pulling due to the coupling of the incident microwave power. The frequencies of the cavity-controlled oscillators were compared over a period of 12 days with that of a caesium beam atomic clock and the relative drift rate was $(-0.4 \pm 3.4) \times 10^{-14} \, d^{-1}$.

Their result has significance for the fundamental constants as follows: the dimensions of the cavity depend on the Bohr radius of the atom, while the atomic clock frequency, being a hyperfine frequency transition, depends on the same constants as those in the hydrogen maser. In consequence, the ratio of the caesium frequency v_{Cs} to cavity frequency v_{scco} (where scco stands for superconducting cavity oscillator) depends on the g value of the proton, the ratio m_e/M_p and α, thus:

$$v_{Cs}/v_{scco} = (\text{constant}) \times g_p(m_e/M_p)\alpha^3.$$

The result of Turneaure and Stein (1974) therefore set a limit on the instantaneous variation of the product of these constants of less than $4.1 \times 10^{-12} \, \text{yr}^{-1}$, and this for a measurement time of only twelve days! Unfortunately, it is difficult to extend the measurement time much beyond this since it becomes necessary to refill the cryostat and, however carefully done, this may perturb the measurements.

2.4.3 Changes in e^2

Following the tight limits on possible changes in G (which will be discussed later), Gamov (1967) suggested $e^2 \sim t_{\text{age}}$ as a possibility. This would lead to a uniform red shift in spectral lines of distant astronomical objects and mean that isotopes which are unstable now would have been stable in the distant past. It also has implications for the luminosity of the Sun, which varies as t^{-3}, instead of t^{-7}, which would follow from G being dependent on t^{-1}. Gamov's theory therefore caused a careful study of the available evidence, for example by Dyson and Perles (1967).

The limits in the relative changes in the values of $1/e^2(\text{d}(e^2)/\text{d}t)$ come from the relative abundance of isotopes. Thus a change of e^2 with time would lead to a different change with time of α decay and β decay rates. This would effect the chain relationship of the three radioactive families known as ThC, RaC and AcC. Wilkinson (1958) concluded from his study of AcC that the change was less than 2 parts in $10^{12} \, \text{yr}^{-1}$.

Dyson (1967) studied the terrestrial abundances of the nuclei rhenium-187 and osmium-187. The first isotope decays by β decay to the second with a half-life of 4×10^{10} yr. The argument was based on the assumption that the decay rate would have been sensitively affected by a change in e^2 to such an extent that the survival of ^{187}Re would be unlikely if its decay rate had been as small as 2×10^8 yr during the early history of the Earth. Dyson proceeded from this to consider the implications for the difference of the coulomb energies and so was able to infer a limit for variations of e^2 of $\lesssim 3 \times 10^{-13} \, \text{yr}^{-1}$. He also obtained higher limits from considerations of the relative abundances of other isotopes (notably ^{40}V, ^{87}Rb, ^{123}Te, ^{138}La and ^{176}Lu.

Peres (1967) has analysed the possibility of a change in e^2 from geological evidence. Thus, the ratios of the neutron number to proton number of the stable heavy isotopes would be much closer to unity than those met at present (essentially because the deviation of N/Z from unity is due to the coulomb repulsion between protons). In particular, 3×10^9 years ago ^{208}Pb would have existed as radon, ^{208}Rn, which is a gas, and so as a consequence lead ores would be uniformly distributed throughout the world, which they are not. More generally, the stable isotopes at the present time would have had their origin in different sources. A consequence of this would be that there would be much wider fluctuations in the isotopic composition of the elements than are observed today.

A further method has been used by Chitre and Pall (1968). They compared two methods of geological dating, the uranium–lead method and the potassium–argon method. The ^{238}U decays by α decay with a half-life of 1.5×10^{-10} yr^{-1}, while the ^{40}K decays by capturing a k-shell electron to ^{40}Ar with a decay constant 0.6×10^{-10} yr^{-1}. The two decay rates differ significantly in their dependence of e^2 and hence the two methods can be used to set a limit on the variation of e^2. Thus an estimation of the age of meteorite samples by the uranium–lead method, taking into account a maximum age of the universe of 10 eons leads to $\lesssim 5 \times 10^{-12}$ yr^{-1}. If the age by the potassium–argon method is used, instead of the 'guestimate' of the age of the universe, then a further lowering of the limit to $\lesssim 5 \times 10^{-13}$ yr^{-1} is obtained.

Gold (1967) has estimated the upper limit on the variation of e during the last 2×10^9 years by looking at the measurements of the decay constant for spontaneous fission of ^{238}U by two methods. In the first, the decay rate was measured for traces of ^{238}U in old rocks whose age could be estimated by the decay of ^{40}K and ^{87}Rb. This result was then compared with the value deduced from direct measurements in the laboratory, using solid state detectors, of ^{238}U on pre-etched mica. By comparing the results he concluded that the change had been less than 4.7×10^{-4} during the last 2 eons, or less than 2.3×10^{-13} yr^{-1}.

Thus, all of these methods yield upper limits at about the same level and really they are all compatible with there being no change at all. It is, of course, not always easy to distinguish between changes of e^2 and changes in α^{-1} and the evidence for the one may, to a large extent, be exchanged with evidence for the other.

2.4.3 Planck's constant

Bahcall and Salpeter (1965) suggested looking for a possible cosmic time variation of Planck's constant. Their proposed method was to look at the light from a QSO, which was thought to be a large distance away in view of its large red shift. They argued that, since a prism deflects light according to its energy and a diffraction grating according to its wavelength, then by comparing the wavelengths of a particular line, any difference could be attributed to a change in the value of Planck's constant. Needless to say, their work gave a null result in terms of the experimental error. A positive result would certainly have been difficult to interpret.

Noerdlinger (1973) used an alternative method which essentially measures the product of Planck's constant and the velocity of light. Essentially, Noerdlinger's method was to look at the background black body radiation at 2.3 K, which is supposedly left over from the 'Big Bang' or some similar event. He argued that the intensity of the Rayleigh–Jeans portion of the long wavelength tail of the spectrum would essentially determine the product of the Boltzmann constant and the absolute temperature. On the other hand, a measurement of the turnover point would yield $h\nu/kT$. Thus a comparison of the two methods enables kT to be eliminated. Since their measurements were made in terms of wavelength rather than frequency, the measurements yield information about the product hc rather than h. The results were consistent with hc not having changed by more than 30% between an age of 10^{-2} eons and the present, which implied that hc changes by less than 0.3 in 10^{10} yr^{-1}.

Further evidence came from the work of Baum and Nielson (1976) and Solheim *et al* (1976). Both experiments made use of the fact that the energy of the electrons in a photomultiplier depends on the energy hv of the incident photon. Usually, photomultipliers are made so that this does not affect their performance, but Solheim *et al* (1976) used a photomultiplier which had an additional grid so that it was possible to discriminate between electrons of different energies. This enabled them to test the constancy of the product $E\lambda = hc$.

The principle of the methods used in both experiments was to pass the light received from near and distant stars through a narrow-band interference filter, thereby defining a narrow band of wavelengths. Light from nearby galaxies was compared with light from distant galaxies ($z = 0.19$) and Solheim *et al* (1976) concluded that the variation in h was less than 5×10^{-13} yr^{-1}. Incidently, they also concluded, from the fact that light from near and distant stars and galaxies shows the same aberration, that c varies by less than a part in 10^{12} yr^{-1}.

Further evidence for the constancy of h was claimed to come from the absorption of light in the atmosphere. Solheim *et al* (1976) found that two oxygen absorption bands, the strong one at 760 nm and a weaker one at 687 nm, both occur at the same point in the spectrum for light from either galaxies at $z = 0.03$ or for sunlight, to within 0.2 nm. This set a limit of 5×10^{-13} yr^{-1}. QSO spectra show a similar concordance, which, if one believes the red shifts as indicating a distance scale, sets 2×10^{-14} yr^{-1} for a QSO of $z = 2.69$, corresponding to a look-back time of 16 eons. The work with photomultipliers by Solheim *et al* (1976) on quasars up to $z = 1.6$ set a limit of 10^{-12} yr^{-1}. It is interesting that Pegg (1977) considered that the apparatus used in the above types of measurement had the ability in effect, to see into the future.

The assertion that it is possible to infer something about the possible time variation of a dimensioned constant appears at first to run counter to the statements made earlier in this chapter and by Dicke (1962) or Cook (1957), and by others, that it is not meaningful to consider the time variations of such quantities. However, Bekenstein (1980) has criticised the meaning of these experiments on rather fundamental grounds. Thus both measurements of the product $E\lambda$ were interpreted as meaning that $hc = E\lambda$ had the same value for all of the observed objects. Bekenstein argued that the implicit assumption in these experiments was that the four-wave vector and four-wave momentum of the photons were propagated parallel to themselves, and that this could only be achieved if their proportionality factor, hc, was constant along the path of the photon. Consequently, the measurements were certain to yield a null result since the constancy of hc was assumed from the beginning! We have also seen that while the ^{86}Kr definition of the metre was being used we might well be in the system where h must be an invariant quantity through the way that the units are defined.

It is clear, therefore, that the interpretation of the results of such measurements is likely to be open to question. This should not necessarily deter experimenters from undertaking them, for at the very least they serve to clarify thinking and lead to a deeper understanding of the basic assumptions that are made in interpreting the meaning of observed physical phenomena.

2.4.5 The ratio of m_e/m_p

The dimensionless ratio m_e/m_p figures in many of the cosmological theories, and consequently any limits of possible time variations are of great usefulness. The nuclear stability and radioactive lifetime arguments impose very tight limits on the variability of the strong, weak and electrostatic interactions and analyses have been made by Wilkinson (1958), Dyson (1972) and Yahil (1975). Slyakhter (1976) has imposed even tighter limits, in particular that the fine structure constant is constant to one part in 10^7 over 10^{10} yr. These, taken together with the limits on $\alpha^2 g_p m_e/m_p$, tell us about the limits on the variation of $g_p m_e/m_p$.

An attempt has been made to look at the possible variations in m_e/m_p. Yahil (1975) placed a limit of 120% per 10^{10} yr, which did not exclude a direct proportionality with time. The method was to compare the K–Ar and Rb–Sr methods of geochemical dating, based on the fact that one method relies on k-electron capture and the other on β-decay. Pagel (1977) used an astronomical method relying on evidence from quasars. The basis of this method is that the wavelengths of the hydrogen lines depend on the Rydberg constant and $(1 + m_e/m_p)^{-1}$. The second term is essentially unity for heavier atoms (C, N, O, Ne, Si etc). Consequently, if β varies, the red shifts of the hydrogen lines differ from those of heavier atoms. Pagel, by taking the age of the universe since a red shift of 2 to be 3 eons, derived a three sigma limit on the variation of m_e/m_p as less than $5 \times 10^{-11}\,\mathrm{yr}^{-1}$, and hence $g_p \lesssim 5 \times 10^{-11}\ \mathrm{yr}^{-1}$.

2.4.6 Time variations of the universal gravitational constant

Any variation of G with time would, of course, have a profound effect as far as astronomers and cosmologists are concerned and considerable attention has been focused on setting limits on the possible variation of G. The possibilities of any time variation of G of the order of that predicted by Dirac in his original theory was eliminated by arguments presented by Teller (1948). He argued, from palaeontological evidence about the temperature of the Earth during the pre-Cambrian era and the fact that the radius of the Earth's orbit and temperature of the Sun both depend on the value of G, that 0.4 eons ago the value of G must have been within 10% of the present value, or a variation of less than 2.5 parts in $10^{10}\ \mathrm{yr}^{-1}$. There are doubts too as to whether the Sun could have survived until now if G had changed very much. Further evidence for the constancy of G has come from the slopes of petrified sand-dunes and the degree of compaction of the London clay.

Continental drift could be explained in terms of an expanding Earth of around 0.6 mm per year over the last 4.5 eons. Wesson (1973) has argued that the required overall expansion of 3600 km could only be accounted for in part, that is by about 180 km, by the Teller limit on changes in G. Shapiro (1971) set a limit of less than 4 parts in $10^{10}\,\mathrm{yr}^{-1}$ from radar observations of the orbit of the planet Mercury. He has obtained rangings of Venus as well, but unfortunately, the precision of the method is limited by the large and unknown area of the portion of the planet that is reflecting the radar. A change in G of 1 part in $10^{10}\,\mathrm{yr}^{-1}$ corresponds to a change in distance of about 10 m in the range.

The story can rapidly get quite complicated: Creer (1965) suggested that a change in ε_0 was required to explain the apparent expansion of the Earth. However, McElhinny et al (1978) made new estimates of the palaeoradius of the Earth and set limits on the expansion of the Moon, Mars and Mercury, and concluded that $G \lesssim 8 \times 10^{-12}\,\mathrm{yr}^{-1}$ in constant mass cosmologies and $\sim 2.5 \times 10^{-11}\,\mathrm{yr}^{-1}$ in Dirac's multiplicative creation cosmology. Canuto (1981) has argued that some cosmologies would result in the radius of the Earth being larger in the past if G was larger at that time. A constant G is required by the Einstein theory of gravitation, but Dirac (1979) has shown that the two theories could be reconciled if the Einstein space metric differed from the atomic metric.

The Apollo Moon program led to the placing of a cube-corner array on the surface of the Moon and the distances of the Moon can be measured by measuring the time of flight of a laser pulse which is launched and detected by a telescope. The Earth–Moon distance can be monitored at present to about 15 cm. This program has provided a more accurate ephemeris timescale than could be obtained by transit telescope measurements. It is hoped to use the method for the direct measurement of continental drift, and also to detect any lifting of the Earth's surface locally before an earthquake. In addition, the lunar laser ranging program has already allowed a search to be made for any elongation of the Moon's orbit towards the Sun which was of a non-geodesic origin. This indirectly enabled Williams et al (1976) and Shapiro et al (1976) to set a limit of less than 1 part in $10^{12}\,\mathrm{yr}^{-1}$ on the variation of G. Meanwhile, Newton (1975) had looked at changes in the length of the day, from atomic clock measurements, which set a limit of 1.5 parts in $10^9\,\mathrm{yr}^{-1}$. More recently, Dearborn and Schramm (1974) argued from a study of the clustering of stellar galaxies that, because globular clusters have a finite limit to their dimensions, one can set a lower limit of 4 parts in $10^{11}\,\mathrm{yr}^{-1}$. Van Flandern (1975) looked at the timings of lunar occultations of stars, which can be quite abrupt as a star passes behind a lunar mountain, and concluded that the variation, although consistent with the decrease required by Dirac's theory, was less than $10^{-10}\,\mathrm{yr}^{-1}$. Muller (1975) studied ancient eclipses from 1374 BC to 1715 AD and concluded $2.6(15) \times 10^{-11}\,\mathrm{yr}^{-1}$. Mansfield (1976) studied the slowing down of pulsars and obtained $5.8(10) \times 10^{-11}\,\mathrm{yr}^{-1}$. Eichendorf and Reinhardt (1977) re-examined the palaeontological evidence considered by Teller in the light of changes in the assigned values for the age of the universe, and recent palaeonotological discoveries in South Africa set a limit of less than 1.7 to 2.9 parts in $10^{11}\,\mathrm{yr}^{-1}$.

Blake (1977) examined palaeontological evidence from corals and bivalves for changes in the length of the solar day and synodic month and obtained $0.5(20) \times 10^{-11}\,\mathrm{yr}^{-1}$.

Since 1955, Van Flandern (1981) has studied the motion of the Moon about the Earth by timing the occultation of stars by the Moon. The practical difficulty is that there is a contribution from tidal friction which must be allowed for before looking for a statistically significant remainder which is attributable to a variation of G. It is necessary, too, to be certain that the proper motions of the occulted stars have been properly taken into account. If the angular velocity of the Moon is ω the total lunar tidal acceleration results suggest

$$\dot{\omega}_{\mathrm{tidal}} = -28.8(15)''\,\mathrm{century}^{-2}$$

whereas the total variation obtained from lunar ranging and occultation results gives

$$\dot{\omega}_{\mathrm{total}} = -23.2(12)''\,\mathrm{century}^{-2}.$$

The difference between these leaves a residual of $+3.2(11)$ parts in $10^{11}\,\mathrm{yr}^{-1}$, which, if of cosmological origin, enables limits to be set on $\dot{G}/G$. The interpretation depends on the theory, thus

$$\dot{G}/G = f\dot{\omega}/\omega$$

where on the Dirac theory, with multiplicative matter creation, $f = +1$, with additive matter creation, $f = -1$ and in the simplest case, with only G varying, $f = \frac{1}{2}$. By interpreting the results in terms of the covariant cosmology of Canuto and Hsieh (1978), in which $f = -2$, Van Flandern concluded that

$$\dot{G}/G \sim -6.4(22) \times 10^{-11}\,\mathrm{yr}^{-1}.$$

Reasenberg and Shapiro (1978) have obtained results of a similar sign from radar planetary ranging of Mercury, Mars and Venus, but the uncertainties are such that the results are compatible with a constant G. Hellings et al (1983) have analysed the range data from the Viking landings on Mars, which enabled a value of $0.2(4) \times 10^{-11}\,\mathrm{yr}^{-1}$ to be set for the time variations for G. Most of the uncertainty in the measurement arises from the difficulty in estimating the masses of the asteroids and their effects on planetary orbits. A reanalysis of primordial nuclear synthesis data by Rothman and Matzer (1982), if correct, sets one of the lowest limits to date. More work is required before the G variations can be regarded as firmly established, but the present results appear to be incompatible with the time variation of G set by the Dirac theory.

It seems clear, therefore, that providing the cosmological time scales can be relied on and that our understanding of gravitation is sufficient, the variations of G are less than around 1 part in $10^{10}\,\mathrm{yr}^{-1}$. This conclusion is supported by a range of different evidence, some of it from quite short-term experiments. If the change postulated by Dirac is ruled out, one is left with changes in G to some small power of time, or perhaps to postulate a cyclic variation with little variation at the present epoch. Evidently, future cosmological theories must be more sophisticated than a simple linear variation with time.

2.4.7 Conclusion

Overall, the experimental evidence for the time invariance of the fundamental constants is as summarised in table 2.7. It is evident that any claimed changes in the terrestrially measured values of the fundamental constants are unlikely to be a genuine result of time-dependent effects, for these limits are well below the accuracy of most of the present measurements. The inference is clear: if there do

Table 2.7(a) A selection of some of the limits set on time variations of the fundamental constants.

Authors	Method	Combination	Fractional change per year	per Hubble time[†]
Dyson (1972)	^{187}Re decay over geological time scale	α	$< 8 \times 10^{-14}$	4×10^{-4}
Dyson (1972)	^{187}Re and ^{40}K decay rates: weak interaction constant	$\beta = g_{\mathrm{f}} m_{\mathrm{p}}^2 c / h^3$	$< 4 \times 10^{-10}$	2
Reasenberg and Shapiro (1975)	Planetary ranging	Gm^2/h_{c}	$< 1.5 \times 10^{-10}$	1
Slyakhter (1976)	^{149}Sm: ^{147}Sm	α	$< 5 \times 10^{-18}$	3×10^{-8}
Slyakhter (1976)	^{149}Sm: ^{147}Sm	$Gm^2 c/h^3$	$\sim 10^{-12}$	10^{-2}
Turneaure and Stein (1976)	Superconducting cavity	$g_{\mathrm{p}}(m_{\mathrm{e}}/m_{\mathrm{p}})\alpha^3$	$< 3 \times 10^{-11}$	10^{-1}
Wolf et al (1976)	Red shift in absorption of radio and optical waves	$\alpha^2 g_{\mathrm{p}} m_{\mathrm{e}}/m_{\mathrm{p}}$	$< 2 \times 10^{-14}$	10^{-4}
Pagel (1977)	Mass shift in quasar spectral lines, $z = 2$	$m_{\mathrm{p}}/m_{\mathrm{e}}$	$< 2 \times 10^{-10}$	1
van Flandern (1981)	Lunar motion	Gm^2/h	$-6.4(22) \times 10^{-11}$	0.3

[†] $\sim 2 \times 10^{10}$ yr ($H_0 = 55\,\mathrm{km s}^{-1}\,\mathrm{Mpc}^{-1}$)

Table 2.7(b) Limits on the variation of $(1/G)dG/dt$.

Authors	Limit on the decrease of $-(1/G)dG/dt$ per year	Method
Teller (1948)	$< 2.5 \times 10^{-10}$	Palaeontological evidence for temperature of the Earth
Shapiro et al (1976)	$< 4 \times 10^{-10}$	Radar ranging of orbit of Mercury
Newton (1973)	$< 1.5 \times 10^{-9}$	Change in the length of the day relative to atomic time
Dearborn and Schram (1974)	$< 4 \times 10^{-11}$	Clustering of galaxies and globular clusters
van Flandern (1975)	$8 \pm 5 \times 10^{-11}$	Lunar occulations
van Flandern (1981)	$-6.4(22) \times 10^{-11}$	Lunar motion
Muller (1975)	$2.6 \pm 1.5 \times 10^{-11}$	Ancient observations of solar eclipses, 1374 BC to 1715 AD
Mansfield (1976)	$5.8 \pm 1 \times 10^{-11}$	Slowing down of pulsars
Williams et al (1976) and Shapiro et al (1976)	$< 10^{-12}$	Lunar ranging: analysis of Moon's orbit for elongation towards the Sun of non-geodesic origin
Hellings et al (1982)	$0.2(4) \times 10^{-11}$	Viking Lander on Mars
Blake (1977)	$0.5 \pm 2 \times 10^{-11}$	Examined palaeontological evidence of corals and bivalves for variation of length of solar day and synodic month
Eichendorf and Reinhardt (1977)	$< (1.7 \text{ to } 2.9) \times 10^{-11}$	Re-examined Teller's method in the light of recent palaeontological discoveries and revised age of the universe
McElhinny et al (1978)	$< 8 \times 10^{-12}$	Limits to expansion of Earth, Moon, Mars and Mercury.

appear to be changes with time, then either, as experimenters, we have somehow failed to be good enough at estimating our uncertainties or, if they are dimensioned, the units have changed. This does not include, of course, experiments (as of Turneaure and Stein 1974) which are particularly designed to show up time variations.

2.5 The anthropic principle

The large-number coincidences, $\sim 10^{40}$, remarked upon by Dirac (1937) still exist and, although the constants appear to be more constant than Dirac's explanation required, they still deserve some explanation. Consideration of these numbers led Dicke (1961), Carter (1971), Carr and Rees (1979), Wheeler (1966) and others to propose and discuss theories which have been included in the term 'anthropic principle'.

These theories make use of the fact that 'man exists' to extrapolate backwards and make predictions about the early universe. For example, for man to exist, stars must be old enough for supernovae to occur so that the heavy elements could evolve, and yet not so old that our own Sun would have stopped shining. These considerations set lower and upper limits on the present age of the universe or the Hubble constant. Quite a narrow range of temperatures is required to support life on Earth as we know it. This depends on the radius of the Earth's orbit about the Sun and on the luminosity of the Sun, both of which depend critically on the value of G. In this way, one can make arguments about the size of the various constants. As Carr and Rees (1979) discuss, the evolution of life as we know it depends on the values of relatively few basic constants and depends quite sensitively on their values.

Such theories are essentially of a *post hoc* nature, and the anthropic principle would undoubtedly be more satisfactory if it could be given a more physical foundation and used to predict forward rather than backward in time. Not all cosmologists are happy with the anthropic principle (Roxborough 1981) and it may well be that its long term role will be seen to have been to enable the relevant facts requiring explanation to have been put in a convenient package to pave the way for a more satisfying and predictive theory.

The anthropic principle takes two forms. The Weak Anthropic Principle (WAP) asserts that the universe must be *consistent* with the existence of intelligent life (Dicke 1961) and the second, the Strong Anthropic Principle (SAP) asserts that intelligent life *must* evolve somewhere in any physically realistic universe (Carter 1971 and Wheeler 1966).

2.6 The role of α in dimensional metrology

Although one might begin by thinking that there should be a unique system of natural units, we have seen already, from the consideration of the Bohr ruler in § 2.2.1, that our natural length units are related by powers of the fine structure constant. This is a problem which has long been experienced by theoretical physicists, but is one which the pragmatic and practical metrologist might well dismiss as being academic. However, when one examines the changes in definitions of the base units which have occurred in the last thirty years, it is apparent that the fine structure constant, although dimensionless, has been playing an important underlying role in our practical metrology (table 2.8). Thus, in changing from the krypton-86 definition of the metre to a fixed value for the velocity of light (depending on the atomic second) we are changing by a factor of α^2. Similarly, the change from the prototype metre to the krypton-86

Table 2.8 The underlying role of α in metrology.

Involvement of α	Length definition	Conductance	Current, voltage
α^0	Krypton-86	Free space, $(\mu_0 c)^{-1}$	ECE of silver, $N_A e$
α^1	Prototype metre bar	Quantised Hall effect, $1/R_H$	Josephson effect, $h/2e$
α^2	Speed of light	—	—

definition involved a change by a factor of α. Equally, we can see that the quantised Hall resistance is related to the impedance of free space by a factor of α. Interestingly, too, the earliest use of the Faraday to maintain the unit of current constant is related to the Josephson effects by a factor of α. We could invoke powers of α as well in our changed definitions and methods of realising the unit of time.

It seems reasonable, therefore, to infer that powers of α will continue to manifest themselves in our practical metrology, and that as we progress we may switch from one 'best method' to another —just as the theoreticians do. It might be a mistake therefore to regard the adoption of a particular atomic unit as a final step and just as we abandoned the prototype metre in 1960 or the krypton-86 definition in 1983 so too the fixed speed of light definition may not endure. Indeed, it is salutary to recall that although the dream of Michelson and others of using the wavelength of light to realise the metre took some 80 years to bring to fruition it only endured for some 23 years.

2.7 Conclusion

The above, and the general content of this chapter, have served to emphasise that precision metrology has been going through an exciting period of rapid progress and that changes will probably be made in the future as our understanding of the universe is refined further.

We have seen in this chapter that the SI units are intimately bound up with the fundamental physical constants. In considering how constant the latter are, we have invoked evidence from a wide range of science, particularly of an astronomical or cosmological nature. Much of this evidence was of low inherent precision and the very tight limits were set by the very long timescales over which the evidence accumulated.

In the following chapters, we move on to consider metrological measurements which are made in the laboratory over much shorter timescales and to much higher accuracy, and yet form part of the same picture. We may do so, secure in the knowledge that the unit system is just about adequate for our immediate purposes, is capable of further refinement, and that the constants are established as being sufficiently constant for our present needs.

3

The fundamental constants and the units of length, mass, the mole and temperature

3.1 Introduction

There is an increasingly close relationship between the fundamental constants and the SI units. We illustrate this by a discussion of the constants which are involved with the units of length, mass, the mole and temperature. These constants are the velocity of light, the Rydberg constant and the lattice spacing of silicon, which are all associated with length, and the Avogadro constant and the gas constant which are involved with mass, the mole and temperature. The fundamental constants which relate to the electrical units are considered in the next chapter.

3.2 The velocity of light

The velocity of light is the oldest of the physical constants and the changing role of this constant has already been discussed in chapter 1. The earliest known attempt to make a measurement is the famous lantern experiment of Galileo in 1638. Römer in 1676 succeeded in deducing a finite value after his study of the systematic variation (having an amplitude of about 20 s) of the apparent period of Jupiter's first moon around its parent planet (about 42 hours). This thirteen month variation was just the time for the Earth to move in its orbit from one closest approach to Jupiter to the next. His result was $214\,200\ \mathrm{km\,s^{-1}}$ (it contains rather more digits than the precision of his measurements would justify) and the major error was due to an inadequate knowledge of the radius of the orbit of the Earth. Römer's conjecture that light had a finite velocity was confirmed by Bradley in 1727. He observed the apparent change in the position of the stars as a result of the motion of the Earth about the Sun. These gave a uni-directional measurement of the velocity of light whereas the terrestrial measurements use round-trip propagation or standing waves.

In 1849, Fizeau made his famous toothed wheel (720 teeth) measurement between two hills near Paris which were about 8.6 km apart. These provided the first terrestrial confirmation of the finitude of the velocity of light. His measurements were improved by Cornu in 1874, Young and Forbes in 1881, and by Perrotin in 1900 and 1902. The method was limited in precision by the difficulty in deciding on which angular velocity of the wheel corresponded to complete extinction of the returned light. The rotating mirror method was developed to overcome some of the limitations of the toothed wheel method. This method was derived from a suggestion of Wheatstone in 1834 which was developed by Arago in 1839 and 1850, and by Foucault in 1850 and 1853.

Newcomb in 1885 also used the method, and it reached its ultimate perfection in the work of Michelson (figure 3.1) in his series of measurements between 1879

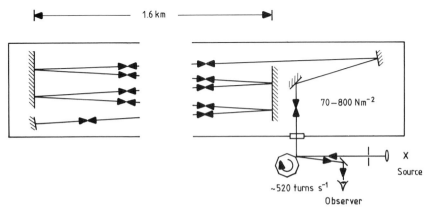

Figure 3.1 Schematic diagram of the rotating mirror method used by Michelson to determine the velocity of light.

and 1927 over a 70 km light path, together with his later collaboration with Pease and Pearson. This incorporated a mile long evacuated pipe in order to remove the effects of refraction etc of the air and a 32-sided mirror which was rotated at an angular velocity such that light reflected from one face was reflected from an adjacent face on its return. This measurement was completed after Michelson's death in 1931 and the results were published in 1935. Unfortunately, this classic series of measurements appears to have had an unsuspected systematic error. It is not entirely clear what this was, and it may have been due to refraction effects in the parts of the pipe which were not evacuated, or some instability in the measured baseline.

It is strange that the above measurement, together with those of Anderson (1937), Huttel (1940) and Anderson (1941), all yielded values which were some 10 km s^{-1} lower than the present-day values. Certainly, they appear at first sight to be an example of the phase-locking effect discussed in § 9.2.4. The last two mentioned measurements both used the Kerr cell method to modulate the light beam. This method had been steadily improving in precision since it was used by Gutton in 1912 to compare the velocity of visible radiation in air with that of electromagnetic waves on a transmission line.

The values obtained for the velocity of light have been frequently evaluated over the years and it is not the intention to perform yet another critical review here. It is of interest to compare the average of the experimental results for the periods shown in table 3.1 with present-day results, since these give an indication of the performance of the many metrologists and their ability to estimate their uncertainties.

Table 3.1(a) The principle determinations of the velocity of light prior to the twentieth century.

Date	Authors	Method	Result ($km\,s^{-1}$)	Standard deviation ($km\,s^{-1}$)
1676	Römer	Jupiter's satellites	214×10^3	
1727	Bradley	Stellar aberration	308×10^3	
1849	Fizeau	Toothed wheel	314×10^3	
1860	Foucault	Toothed wheel	298×10^3	
1875	Cornu Helmert	Toothed wheel	299 990	300
1879	Michelson	Rotating mirror	299 910	75
1883	Newcomb	Rotating mirror	299 860	45
1883	Michelson	Rotating mirror	299 853	90

Table 3.1(b) Values obtained between 1900 and 1941 for the velocity of light.

Date	Authors	Method	Result ($km\,s^{-1}$)	Standard deviation ($km\,s^{-1}$)
1907	Rosa and Dorsey	Ratio of esu to emu	299 784	15
1923	Mercier	Lecher wires	299 782	15
1924	Michelson	Rotating mirror	299 802	30
1926	Michelson	Rotating mirror	299 798	22
1928	Karolus and Mittelstaedt	Kerr cell	299 786	15
1935	Michelson, Pease and Pearson	Rotating mirror	299 774	6
1937	Anderson	Kerr cell	299 771	15
1940	Huttel	Kerr cell	299 771	15
1941	Anderson	Kerr cell	299 776	14

Table 3.1(c) The measurements between 1947 and 1967 of the velocity of light.

Date	Authors	Method	Result (km s^{-1})	Standard deviation (km s^{-1})
1947	Essen and Gordon-Smith	Cavity resonator	299 792	4
1947	Smith *et al*	Radar (in air)	299 695	50
1947	Jones	Radar (in air)	299 687	25
1949	Aslakson	Radar	299 792.4	2.4
1949	Bergstrand	Geodimeter	299 796	2
1949	Jones and Cornford	Radar (in air)	299 701	25
1950	Essen	Cavity resonator	299 792.5	1
1950	Bol	Cavity resonator	299 794	1
1950	McKinley	Quartz modulator	299 780	70
1950	Houston	Quartz modulator	299 775	9
1950	Hansen and Bol	Cavity resonator	299 789.3	0.8
1951	Bergstrand	Geodimeter	299 793.1	0.26
1951	Aslakson	Shoran radar	299 794.2	1.4
1951	Froome	Microwave interferometer	299 792.6	0.7
1952	Rank *et al*	Spectral lines	299 776	6
1954	Froome	Microwave interferometer	299 792.75	0.3
1954	Rank *et al*	Spectral lines	299 789.8	3
1954	Florman	Radio interferometer	299 795.1	3.1
1955	Scholdstrom	Geodimeter	299 792.4	0.4
1955	Plyer *et al*	Spectral lines	299 792	6
1956	Wadley	Tellurometer	299 792.7	2
1956	Rank *et al*	Spectral lines	299 791.9	2
1956	Edge	Geodimeter	299 792.4	0.5
1957	Wadley	Tellurometer	299 792.6	1.2
1957	Rank *et al*	Band spectra	299 793.7	0.7
1958	Froome	Microwave interferometer	299 792.5	0.1
1964	Rank *et al*	Band spectra	299 792.8	0.4
1966–7	Karolus and Helmberger	Ultrasonic modulator	299 792.44	0.2
1967	Karolus and Helmberger	Ultrasonic modulator	299 792.5	0.15
1967	Simkin *et al*	Microwave interferometer	299 792.56	0.11

3.2.2 Determinations between 1940 *and* 1970

The impetus given to the development of pulsed microwave measurements following the development and use of radar in the 1939–45 war led to the application of these techniques to measure the velocity of light. The first published result was that of Essen and Gordon-Smith (1948) at the NPL who measured the wavelength of microwave radiation of known frequency by using an evacuated cavity resonator. In a sense, this was an extension of the Lecher wire measurements of Mercier at 47 MHz in 1924. The wavelength of electromagnetic radiation λ_g in an evacuated cavity resonator is related to the free-space wavelength by

$$1/\lambda^2 = 1/\lambda_g^2 + 1/\lambda_c^2.$$

Here λ_c is a parameter which depends on the dimensions of the cavity and the resonant mode being excited. It was eliminated by changing the length of the cavity, the length separation between successive modes being $\lambda_g/2$. These measurements, together with knowledge of the particular mode excited and the precisely measured dimensions of the cavity, enabled c to be obtained from

$$c = f \, \lambda$$

where f was the precisely measured microwave frequency.

The publication of the Essen and Gordon-Smith (1948) result led to the publication of some earlier measurements by Aslakson and by Hart. They had both been using the Michelson and Anderson result for distance surveying, the former by the American Shoran method and the latter using a development of the British Oboe blind-bombing methods. In both cases the distance computed from the time of passage of a pulse of radio waves was compared with the surveyed distance. However, such was the stature of the accepted value for c that they had not felt sufficiently sure of their results to question it publicly. Determinations made by Bergstrand (1952), using the Kerr cell method, and Aslakson in 1951, using the Shoran method, provided confirmatory results at higher precision. At about this time Froome (1952) developed his 1.25 cm wavelength Michelson interferometer. This technique was also used by Florman (1955) at the NBS at a much lower frequency of 172.8 MHz.

Froome subsequently increased the frequency of the microwaves used in his interferometer and made measurements at a frequency of 24 GHz and 72 GHz (wavelength ~ 4 mm). A diagram of his four-horn interferometer, which operated in the Fraunhofer diffraction region (waves which are intermediate between the spherical and plane wave diffraction conditions), is shown in figure 3.2.

The dominant uncertainty was that involved in the measurement of the length of the end standards which were used to measure the displacement of the receiving carriage. The refractive index correction was measured with a cavity resonator refractometer operating at 72 GHz and which was mounted in close

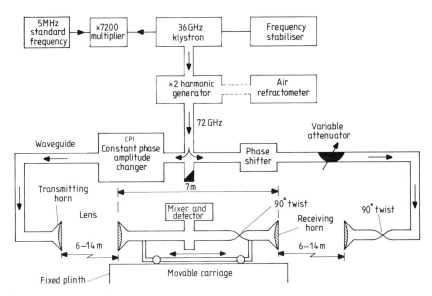

Figure 3.2 Schematic diagram of the microwave interferometer method used by Froome (1952) to measure the velocity of electromagnetic radiation.

proximity to the interferometer. This instrument was described by Froome (1958) and was based on the 24 GHz design used earlier by Essen and Froome (1951). The length measurement was the dominant uncertainty and this was followed by the uncertainty in the refractive index correction. The overall uncertainty of Froome's determination was 100 m s^{-1}.

His remained the most precise measurement until the advent of the new methods of measuring the velocity of light during the 1970s. There were a number of measurements in the interim, with a slightly lower precision, and these confirmed Froome's result. These included the Kerr cell determinations of Bergstrand (1957), Scholdstrom (1955) and Mackenzie (1954) with 8 MHz modulation, the acoustic modulator determination by Karolus and Helmberger (1964) and Baird (1967) using modulation at 10 MHz and a 46 m baseline, and Simkin *et al* (1967) (who improved on the moving reflector method used earlier by Froome) measured the wavelength of 37 GHz microwaves and displaced the reflector over a distance of 4.47 m (this may be compared with Froome's use of 24 GHz microwaves and 1.62 m reflector displacement).

In the Kerr cell measurements the intensity of the light beam is modulated and the received signal intensity I varies as

$$I = I_0[1 + a \sin 2\pi v(t - x/v_p)]$$

where $a \lesssim 1$, v is the modulating frequency and v_p is the phase velocity. The phase term

$$\phi = 2\pi v x/v_p$$

represents the relative phase of the modulation envelope between the Kerr cell and the point of observation.

Bay and Luther (1972) reported a measurement which marked the end of the old ways of measuring the velocity of light. They used some of the techniques which have been incorporated into the major improvement of the technology which is discussed in the next section. Their technique was to frequency modulate a 633 nm He–Ne laser at 5 GHz, thereby producing sidebands which were separated by 10 GHz. These, together with the piezo-electrically adjusted length of a Fabry–Perot etalon, were tuned until both of the sidebands were at a transmission peak of the etalon. A measurement of the length of the etalon in

Table 3.2(a) Post 1970 values of the speed of light in vacuum.

Date	Author (laboratory)	Laser system	Value of c ($m s^{-1}$)	Standard deviation ($m s^{-1}$)
1972	Bay et al (NBS)	Lamb-dip stabilised and microwave-modulated He–Ne laser (0.63 μm)	299 792 462	18
1972	Evenson et al (NBS)	3.39 μm He–Ne laser stabilised by saturated absorption of methane	299 792 457.4	1.1
1974	Blaney et al (NPL)	9.3 μm CO_2 laser stabilised by saturated fluorescence of CO_2	299 792 459.0	0.6
1976	Woods et al (NPL)	Remeasurement of 9.3 μm wavelength	299 792 458.8	0.2
1980	Baird et al (NRC)	3.39 μm He–Ne laser stabilised by saturated absorption of methane	299 792 458.1	1.9

Table 3.2(b) Internationally agreed values for the velocity of light in vacuum.

Date		Value of c ($m s^{-1}$)	Standard deviation ($m s^{-1}$)
1957	Conventional value adopted (uncertainty revised later) by International Union of Radio Science (URSI) and International Union of Geodesy and Geophysics (IUGG)	299 792 500	100
1973	Consultative Committee for the Definition of the Metre (see Terrien 1973)	299 792 458	1.2

terms of a standard wavelength, together with the sideband frequency, then gave a measurement of c. Their measurement of

$$c = 299\,792\,462(18)\,\mathrm{m\,s}^{-1}$$

had an uncertainty which was about five times smaller than Froome's measurement. However, even as this measurement was published, results were being obtained which were a further ten times more accurate (see table 3.2).

3.2.3 The internationally agreed value

As a result of a consideration of all of the available results, the International Scientific Radio Union (URSI) and the International Union for Geodesy and Geophysics (IUGG) recommended in 1958 the international use of the value of

$$c_0 = 299\,792.5(4)\,\mathrm{km\,s}^{-1}$$

for the velocity of light *in vacuo*. This lasted until the adoption of the CCDM value of

$$c_0 = 299\,792\,458.0\,(1.2)\,\mathrm{m\,s}^{-1}$$

in 1972. The methods by which the latter value was obtained will be discussed in the next section, but the reason for the adoption of the latter value was that both terrestrial and certain astronomical distances have increasingly been measured in terms of the propagation time of electromagnetic waves. These methods extend beyond the use of radar techniques. Thus the Kerr cell methods of modulating the light beam were incorporated into commercially manufactured instruments, such as the Geodimeter, or the Tellurometer of Wadley, or the Mekometer of Froome and Bradsell, which were used for rapid geophysical surveying purposes over terrain where the use of conventional surveying tapes would have been impossible. Such surveys have been important for the study of ground movements in volcanic and earthquake-prone regions of the Earth such as Iceland. Shapiro (1967) and others made use of radar pulse methods for lunar and planetary distance measurements. The Earth–Moon distance is regularly monitored with a precision of about 10 cm by measuring the transit time of pulses of laser radiation from a laser mounted on a telescope on the Earth and cube-corner arrays on the Moon. These arrays were placed on the Moon as part of the space programs of the USA, USSR, France and other countries. All of these measurements are part of long-term experiments and are made in terms of time rather than distance. They therefore require a value for the velocity of light in order to convert them into length measurements. The increasing use of such techniques meant that there was a steady increase in the scientific and commercial demand for more accurate values of the velocity of light.

There was a need in electrical metrology, too, with the advent of the calculable capacitor method for the realisation of the ohm. (Thus, although the advent of the SI has meant that the cgs magnetic and electrostatic units, which were related by the square of the velocity of light, have increasingly fallen into disuse, this constant still appears in the appropriate formulae.) This demand provided the stimulus for the metrological laboratories to devote the required effort towards improving the precision by some two orders of magnitude on the earlier measurements of Froome and others.

3.2.4 Post-1972 determinations of the velocity of light

In order to improve on the precision of the earlier measurements, it was necessary to reduce the diffraction correction. This required working at the highest possible frequency which could be compared accurately with that of a caesium beam standard. A major problem in the earlier measurements was the difficulty of effecting the length measurements; indeed, it is remarkable how many of the results were subsequently corrected after a recalibration of the length standards. It was clear that an interferometer method would have to be used for the length measurement. The experience with Michelson's work, where the results were at one time erroneously reduced to *vacuo* by using the phase refractive index of air instead of the group refractive index, served to emphasise the desirability of making the measurements of length *in vacuo* rather than in the laboratory environment. Meanwhile, too, the use of the prototype metre bar to define the metre had been discontinued in favour of the krypton-86 definition, and this meant that distances could be meaningfully measured to about 2 parts in 10^9. The need to use higher frequencies was apparent to everyone, but the method of measuring the higher frequencies was not at all certain at the commencement of the experiments. One of the difficulties was that the efficiency of harmonic mixers decreases very rapidly at higher frequencies and the intensity of the higher harmonics usually falls well short of the $1/n^2$ (where n is the harmonic order number) response that one might hope for.

The point-contact type of Josephson junction played a key part in the frequency measurement. This superconducting point-contact shows an excellent non-linear response to ~ 1 THz which makes it ideal for very high-order harmonic mixing. With the aid of this, it was possible to measure the frequency of the HCN laser at 0.8 THz and also that of the water vapour laser at 10.7 THz. These lasers operate in the far infrared region, and so it is difficult to compare their wavelengths with that of the orange light from a krypton-86 standard lamp because the diffraction corrections are much larger. Fortunately, a number of lasers were developed in the period around 1970 and some of these were capable of being stabilised very accurately, so that they could be operated as secondary wavelength and frequency standards. The problem remained of measuring their frequency absolutely. It was discovered by Javan at MIT and Evenson at NBS, Boulder, that careful attention to the focusing of the various radiations onto the point-contact enabled harmonic mixing of laser and microwave radiation to be extended to much higher frequencies than had initially been thought possible. The fragile contact between a tungsten point and a tungsten post played an invaluable role in this process. Gradually, therefore, the upper limit to the frequency measurements was extended through the CO and CO_2 laser lines towards the near infrared. Evenson and his colleagues at the NBS, Javan at MIT, and others working at MIT, the NPL, NRC and elsewhere made important contributions to this work.

The above development left the wavelength measurements as the remaining problem. It was clear, of course, that ultimately the wavelength would have to be measured in terms of the wavelength of the krypton-86 orange line, for this wavelength was incorporated in the definition of the metre. However, for the wavelength comparison, use was made of the iodine or methane stabilised helium–neon lasers as secondary standards. The wavelengths of these transitions had already been determined with a precision which was limited in any case by the imprecision of the realisation of the krypton-86 lamp in various laboratories. These were combined to obtain a set of values which were recommended for general adoption by the CCDM in 1973. There were, therefore, essentially two possible laser wavelengths against which the CO_2 laser wavelengths could be compared: either the infrared methane-stabilised helium–neon laser, at 3.39 μm or the 633 nm iodine-stabilised helium–neon laser. The former wavelength was favoured by the NBS and the latter by the NPL. The NBS method had the advantage that two infrared wavelengths were being compared, but the NPL method was to mix the CO_2 laser radiation with that from a stabilised helium–neon laser thereby imposing two visible sidebands on it. This enabled a conventional 1 m long, plane-plate, ultra-high vacuum Fabry–Perot interferometer to be used for the wavelength comparison. The method of imposing sidebands was used earlier in the measurements of Bay et al (1972) who imposed 9.2 GHz sidebands on a visible helium–neon laser and achieved a factor of ∼ 5 improvement in the precision of the Froome determination. In the NPL work, the radiations from a 9.3 μm (1–10 W) CO_2 laser and a 633 nm, 10 mW, He–Ne laser (800 mm length discharge tube) were mixed in a liquid nitrogen cooled non-linear crystal of proustite. Since the harmonic generation occurred over a region, rather than at a point, and the refractive index varied with wavelength, it was important to ensure that both the mixing frequencies and the generated harmonic radiation travelled with the same velocity. This was achieved by arranging that the laser radiations propagated as e-rays and the mixed radiation, namely the dark red light, 100 nW at 679 nm, was propagated as the o-ray. The wavelength of this radiation λ_d was related to that of the CO_2 laser λ_c, and the helium–neon laser λ_s, by the equation

$$\lambda_c/\lambda_s = 1/(1 - \lambda_s/\lambda_d).$$

Thus the ratio of the infrared wavelength to the visible wavelength could be effected by comparing the wavelengths of the two visible radiations λ_s and λ_d. The radiation λ_d was in the visible region (679 nm) and the wavelength was close to that of the 633 nm helium–neon laser. This method had the advantage of reducing the diffraction effects over those in the infrared region, and also enabled fused silica to be used for the optical components of the interferometer. (Baird *et al* (1973) similarly mixed 3.39 μm and 0.633 μm laser radiations in a proustite crystal and measured the wavelength 0.533 μm of the sum-frequency radiation.)

The CO_2 laser used for the upconversion technique was frequency stabilised by reference to the centre of a saturated absorption feature (R(12) transition) observed in the fluorescence from an external CO_2 cell (Woods and Jolliffe 1976). The wavelength ratio was determined by using an evacuated, 1 m long plane Fabry–Perot interferometer, the 50 mm diameter plates of which were silvered to a reflectivity of about 90% at 633 nm. After paying particular attention to the alignment of the optics and of the two light beams, thereby ensuring that both radiations were incident on the optics in the same way, the wavelength ratio was measured as

$$\lambda_d/\lambda_s = 1.072\ 889\ 518\ 61(3)$$

i.e. the uncertainty was 2.9 parts in 10^{11}. This was increased by a factor of about 14 in the estimation of the CO_2 laser wavelength since the upconversion method effected some loss of accuracy.

The greatly improved accuracy has, as we will discuss in § 3.1.7, important consequences for our standard of length, and so the techniques required for these measurements will now be discussed in more detail. These conveniently divide into a discussion of the methods of frequency multiplication and measurement and the method of measuring the wavelength.

3.2.5 Method of frequency multiplication

The first requirement in the frequency multiplication chain (figure 3.3) is for a source of high spectral purity, because the sidebands are mixed in as well at each multiplication stage. Klystron oscillators have generally proved suitable for this purpose when suitably operated and are normally phase-locked to a harmonic of a 5 MHz standard quartz crystal oscillator. (Such crystal oscillators are used as the basic element in the frequency synthesis chain for a caesium-beam atomic clock.) An alternative method might be to use a superconducting niobium microwave cavity as the frequency determining element in a stabilised oscillator. This technique has led to oscillators having the best frequency stabilities ever reported for averaging times between 10 and 100 s. Thus Turneaure and Stein (1978) achieved stabilities of 3 parts in 10^{16} in this time interval. The superconducting cavities are designed to have very high quality factors Q, and values which are of the order of 10^{10} to 10^{11} have been obtained. These cavities are made from niobium which is annealed at a high temperature, under high vacuum conditions, and essentially becomes a large single crystal of niobium.

These microwave frequencies are a factor of about 10^4 lower than the frequency of oscillation of the CO_2 laser and unfortunately, even with these very high stability microwave oscillators, direct frequency multiplication to the

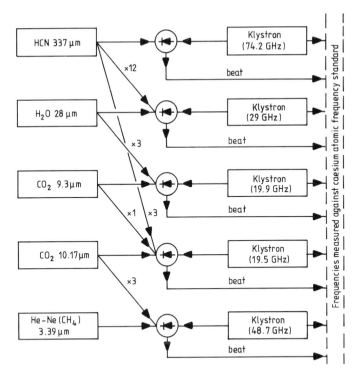

Figure 3.3 Example of a frequency multiplication scheme used to measure the frequency of a methane stabilised helium–neon laser in terms of a microwave atomic frequency standard via the frequency of an HCN laser (NPL/NBS).

visible region is not yet feasible. One reason for this is that the noise sidebands increase as n^2, where n is the harmonic order number. Consequently, it is necessary to use intermediate frequency oscillators in order to clean up the signal effectively and maintain an adequate signal to noise ratio and it is necessary to measure their frequency as an intermediate step. The general principle in laser frequency measurement is to obtain the unknown frequency from an equation of the form

$$f_2 = lf_1 + mf_2 + nf_3 \pm f_b$$

where f_1 and f_2 are the laser frequencies, f_3 is a stabilised microwave frequency and f_b is the measured beat frequency. The beat frequency is usually arranged to be around 30 MHz or 60 MHz, for low-noise amplifiers are readily available at these frequencies owing to their extensive use in radar. The integers l, m and n may be positive or negative, and are arranged to be as small as possible.

The HCN and water vapour lasers are less suitable as transfer oscillators than either a methane-stabilised helium–neon laser, or a CO_2 laser which is stabilised to an absorption feature in CO_2, for the lack of a suitable absorption feature means that they are less stable. The above equation assumed that there was a suitable non-linear device in which the harmonic mixing could take place.

The most generally used harmonic generating and mixing device is the point-contact in one form or other, although some planar devices are under development. Different arrangements have different upper frequency limits (or cut-off frequencies).

The point-contact may be either a normal conductor or a superconductor, and of these the point-contact Josephson junction has given the highest order of harmonic mixing. Thus, mixing of the 825th harmonic of the lower frequency with the higher frequency has been obtained (MacDonald *et al* 1971). The cut-off frequency is in the region of 1 THz, although effects have been seen at rather higher frequencies than this. Both the point and the post are made from niobium and are operated at a temperature of around 4 K.

Although providing a much lower order of harmonic mixing, the metal to insulator to metal type of point-contact has been the most successful in the far infrared region, operating through to at least 200 THz. This was first used by Hocker *et al* (1968) at MIT. It generally comprises a whisker of tungsten wire, which is from 10 to 25 μm in diameter, and the tip is electrochemically sharpened to a tip radius of around 50–100 nm. This tungsten wire is in contact with a polished nickel post. The surface of the nickel is covered with a natural oxide some 1–2 nm thick. The straight portion of the whisker from the tip to a right-angle bend acts as an aerial which helps to concentrate the focused laser radiation down to the tip (Mataresse and Evenson 1970). It is interesting that the polarity of the rectified voltage reverses at wavelengths shorter than about 1.5 μm and continues in this condition through to the visible region. The polarity at longer wavelengths is normally negative (with the nickel post grounded) although the reversal may occur at lower frequencies as well, depending on the method of diode construction. It is hoped that more complete understanding of the causes of this polarity reversal may lead to the use of this type of point-contact through the visible region of the spectrum, probably with either an oxide or vacuum as the insulator.

Other successful devices for mixing in the far infrared have been the GaAs to metal planar Schottky barrier diode, which shares excellent low noise performance with the Josephson junction when used in heterodyne receivers, and other semiconductor devices. These include silicon to metal, and germanium to metal point-contacts, and the indium arsenide to metal point-contact hot-carrier diode. At present, the final elements in the frequency multiplication chain to the visible tend to be bulk materials, such as non-linear optical crystals made from lithium niobate or proustite. Similarly, bulk crystals have already been used for multiplying visible laser radiation towards the vacuum ultraviolet. In the next few years, therefore, one can anticipate an increasing extension of direct frequency measurements throughout the portion of the electromagnetic spectrum lying between the microwave and ultraviolet regions.

3.2.6 The servo-lock technique for the interferometer

The considerable advance in the precision of wavelength comparisons achieved in the NBS, NPL and NRC results stemmed from the use of the servo-lock technique. In earlier measurements, it was necessary to scan through one or several fringes of the interferometer by changing the optical length in some way. This was achieved, for example, by inserting piezo-electric elements between the etalon plates and the spacer or, more traditionally, by changing the pressure of the gas in the interferometer.

The advent of lasers, with their very high spectral purity, has allowed the use of much longer interferometers than was possible hitherto without significant loss in the visibility or contrast of the fringes. The servo-lock method of laser wavelength comparison (illustrated by the NPL work) involved stabilising the etalon length to be an integral number n_1 of half wavelengths of a powerful helium–neon laser. This laser in turn was locked, so that the upconverted light was simultaneously at a transmission peak for the interferometer corresponding to n_2 half wavelengths. The difference frequency between the two helium–neon lasers was then obtained by beating them together. In this way, the Fabry–Perot interferometer was used in the completely passive mode.

Thus, for an etalon of length l the determining equation was

$$2l = n_1 \lambda_1 + \phi_1$$
$$= n_2 \lambda_2 + \phi_2.$$

The standard laser used in the NPL work was a helium–neon laser stabilised to component d of the hyperfine structure of the R (127) line of the (11-5) band of the $B^3II(O_u^+) \rightarrow X^1(O_g^+)$ electronic transition of molecular $^{127}I_2$. The phase changes at the mirrors were investigated very carefully by changing the length of the Fabry–Perot etalon by using different spacers.

The use of this technique made a major contribution to the two orders of magnitude advance in the precision which was achieved in the interferometry side of the velocity of light measurements. The precision of earlier work (Terrien 1973) was typically 2 parts in 10^9. The interferometer used in the NBS determination also incorporated the servo-lock technique. In their measurements the precision was limited to 2 parts in 10^{10} by the imperfections in the shape of the curved surface of the mirrors.

3.2.7 Incorporation of the velocity of light into a new definition of the metre

Astronomers have long used the light year as a convenient measure of interstellar distances. Similarly, lunar, planetary and terrestrial ranging measurements have been in terms of the time of flight of laser or radar pulses. As a result of these developments there have been an increasing number of scientists, technologists and surveyors who have had to assume a value for the velocity of light in order to convert their measurements of distance into metres. The light second is a distance approximately equal to about three quarters of the distance to the Moon. The use of a fixed value for c, when combined with a 3 parts in 10^{14} realisation of the ^{133}Cs clock, would mean that this distance could be specified with an accuracy of about ± 0.01 mm. There is no immediate application requiring a knowledge of such distances with such high accuracy. However, the present accuracy of the Earth–Moon distance measurements is already ~ 10 cm, or only 10^4 times greater than this, and gravity wave detectors in space over comparable distances requiring a better resolution (though not accuracy) than this are already under discussion (Bender and Faller 1982). It is clear that the krypton-86 definition of the metre has to be abandoned, but there is a choice: either to abandon the krypton-86 definition of the metre in favour of some laser-defined wavelength, or to change to a fixed value for the velocity of light. The first solution would present very real problems, for each time that a new improved type of stabilised laser was discovered it might well be necessary to change the formal wording of the metre definition. The second would allow such lasers to be phased into or out of use without affecting the continuity of the unit of length or the definition. The adoption of a fixed value for c is, therefore, conceptually highly attractive; however, it implies that some form of direct frequency measurement is available for the wavelength standards which are used in practice and operate in the visible or near infrared regions of the spectrum. This requirement will extend into the ultraviolet region of the spectrum as well, as the technology advances further. Fortunately, rapid progress is being made in the direct measurement of laser frequencies, and it is reasonable to assume that measurements of laser frequencies in terms of a microwave frequency standard will be routinely possible in the near future—provided that there is sufficient effort on an international basis. Along with this, careful studies are being made of the performance of the lasers in order to improve their reproducibility and thereby increase the time between checks of their absolute frequency.

The adoption of a fixed value for the velocity of light requires the assumption of certain properties for light, i.e. of photons. If these properties are not correct it will be necessary to consider the implications for the definition. Thus (i) the velocity of light might change with time, or (ii) have a directional dependence in space, or (iii) be affected by the motion of the Earth about the Sun, or motion within our galaxy or in some other reference frame or (iv) it may depend on frequency. Bay (1972) has argued that the first three are irrelevant for a choice between wavelength standards or the unified system, although they might be relevant to the consistency of measurements which were made at different parts of the spectrum. The properties of photons are discussed further in chapter 8.

3.3 The Rydberg constant

Hydrogen is the simplest of atoms, consisting of a single electron and proton, and consequently it has the simplest spectrum. The spectrum is not that easy to record, and the most prominent line was detected by Ångstrom in 1853.

Although work continued in research laboratories, the first extended series of hydrogen lines was identified in stars by Sir William Higgins in 1881. The difficulty of observing atomic hydrogen lines in the laboratory stems from the need to dissociate the normally diatomic molecules of H_2 into atomic hydrogen, the spectrum of molecular hydrogen being much more complicated than atomic hydrogen. This dissociation energy is rather higher than the energies encountered in thermal excitation, such as in flames, and a more energetic method, such as excitation in a gas discharge, is required before the atomic hydrogen lines are observed.

In 1885, Balmer found from a study of the astronomical measurements that he could account for all of the known lines by applying a simple empirical formula. This set of lines in the visible region of the spectrum has since come to be known as the Balmer series ($m = 2$). There is another group of lines which lies in the ultraviolet, known as the Lyman series ($m = 1$). There are further series, too, in the infrared region. The brightest of the Balmer lines is a red line, known as the Balmer-α line, at 656 nm. It is frequently termed the hydrogen-α or H-α line.

In 1889, Rydberg found, from a study of alkali and other spectra, that many series of lines could be represented by a simple formula, a later version being of the form:

$$\frac{1}{\lambda} = R\left(\frac{1}{(m+b)^2} - \frac{1}{(n+c)^2}\right)$$

where n and m were integers and b and c were constants which depended on which line was being measured. A particular series corresponded to one value of m, but R was essentially the same for all of the elements. Rydberg was delighted to discover that Balmer's formula for atomic hydrogen was a simple variant of his general formula with $m = 2$ and having both b and c essentially zero (at the precision appropriate to that time). When the appropriate values for n and m are substituted in this formula it yields all of the lines in the hydrogen spectrum. Thus $m = 3$, 4 or 5 yield the Ritz-Paschen, Brackett and Pfund series respectively (figure 3.4). Some time later, Bohr, in 1912, while working with Rutherford in Manchester, came to hear almost accidentally of this formula and soon showed that it had a fundamental and simple basis, for the expression $(1/m^2 - 1/n^2)$ is proportional to the energy difference between two energy states of an atom and the integers serve to label the states themselves. In addition to explaining the empirical equation, and showing that spectral lines had a simple and fundamental basis, Bohr derived an expression for the constant R in terms of fundamental constants and showed that for a nucleus of infinite mass.

$$R = R_\infty = (m_e e^4/4\pi\hbar^3 c)(\mu_0 c^2/4\pi)^2$$

(where the final term converts Rydberg's expression to SI units). The discrete energy states were a consequence of the quantisation of angular momentum, with the quantum being closely similar to Planck's constant, namely $h/2\pi$.

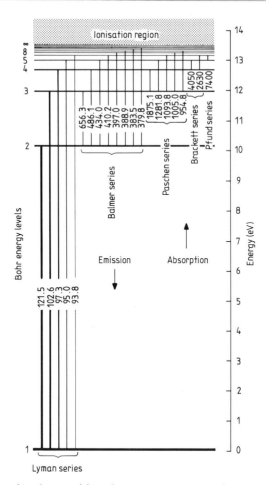

Figure 3.4 Illustrating the transitions between energy states in atomic hydrogen and the corresponding wavelengths in nanometres of the lines of the various named series.

Even with atomic hydrogen, the beautiful simplicity of the Bohr theory soon began to be lost, for Michelson in 1892 had already studied the Balmer-α line and showed that it had two partially resolved components which were only 0.014 nm apart. Bohr suggested that the electron orbit might be slightly elliptical and that the consequential relativistic corrections might split each stationary state into a group of states. Just as Kepler and others rescued planetary motion from the complications imposed by epicycles, so too quantum theory built on the simple ideas of Bohr, notably through the work of Schrödinger, and Heisenberg and Born in the 1920s, leading to the introduction of the fine structure constant α. A further sophistication was the inclusion of the electron and nuclear spins in the theories by Dirac in 1928 and of the Lamb shift in 1947.

3.3.1 Earlier measurements

The simplicity of the formula led spectroscopists to test it and to measure the Rydberg constant with high precision. Rydberg obtained the value 109 721.6 cm^{-1}, where his wavelengths were measured in air. Bohr showed that the wavelengths of the lines in the spectrum of a single electron atom, with a nuclear charge Z and mass M, are given by

$$\frac{1}{\lambda} = \frac{R_\infty Z^2}{(1 + m_e/M)}\left(\frac{1}{m^2} - \frac{1}{n^2}\right).$$ (3.2.1)

Houston (1927) measured the wavelengths of lines in singly ionised helium and hydrogen, Chu (1939) in ionised helium and Drinkwater *et al* (1940) in hydrogen and deuterium. These measurements were made during the period when the Dirac theory was thought to give an adequate description of the energy levels of these atoms and the results were later corrected by Cohen (1953) in the light of the discovery of the Lamb shift. There was some disagreement among the results, which was traced to the use of Merrill's (1917) measurement of the helium $2^1S_0 - 3^1P_1$ line as a standard by both Houston and Chu. The wavelength of the line was re-measured by Field and Series (1959) who confirmed that this was the cause of the discrepancy between the results.

The experimental measurements tended to concentrate on the Balmer-α and Balmer-β lines at 656 nm and 488 nm respectively from atomic hydrogen and deuterium, and of the corresponding lines in helium. They now require the use of a much more sophisticated theory than that of merely applying equation (3.2.1) in order to obtain a value of the Rydberg constant, notably using the work of Garcia and Mack (1965) and Erickson (1974). It should be pointed out that even today, as far as absolute wavelengths are concerned, the quantum theory can only be applied with really high precision to the spectrum of single electron atoms, although advances are being made, particularly in the study of high Rydberg states—the outermost levels of complex atoms—and studies of fine structure.

In principle, it should have been possible, given adequate spectroscopic resolution, to derive a very precise measurement of the Rydberg constant from the experimental observations, but the situation was complicated by the unresolved fine structure, which was first found by Michelson. The causes of the structure were, of course, soon well understood, but the difficulty was that in order to produce the Balmer-α radiation, the atoms were excited in a discharge and it was fairly certain that the relative intensities of the fine structure components differed from that expected theoretically (see Condon and Shortley 1935). That is, the H-α line was not a doublet, but a blend consisting of about seven components.

The resolution of these components was limited, not by the lifetimes of the states, but by the Doppler broadening due to the motion of the emitting atoms, and so attempts were made over a number of years to reduce the Doppler width of the Balmer-α line. The solution was to concentrate on the spectrum from the heavier deuterium, tritium and helium atoms, which provided better resolution of the two major components, and also to try to cool the atoms as far as possible while sustaining the high electron temperature which is necessary in order to excite the atoms. The most successful attempts in this direction were those of Kibble *et al* (1973) and Kessler (1973) on singly ionised helium, but other comparable work included the measurements of Csillag (1965) on deuterium using a ^{198}Hg standard and of Masui (1971) on hydrogen. The results, with the Cohen and Taylor (1973) revisions, are given in table 3.3(*a*). These measurements marked the end of an era during which the uncertainty in the value of the Rydberg constant had hardly changed by much more than an order of magnitude. The reason was simply that the period represented one of careful refinement of experimental measurements and a new technique was required. This advance was provided by the advent of the tunable dye laser.

Table 3.3(*a*) Values obtained for the Rydberg constant to 1973.

Date	Workers	Lines studied	Final corrected result for Rydberg constant (m^{-1})	
1927	Houston[†]	H-α, H-β, He$^+$ (468)	10 973 733.6	(16)
1939	Chu[†]	He$^+$ (468)	10 973 731.4	(20)
1940	Drinkwater *et al*[‡]	H-α, D-α	10 973 731.4	(9)
1968	Csillag[‡]	D-α,..., D-η	10 973 730.60	(60)
1971	Masui[‡]	H-α	10 973 731.88	(45)
1977	Kessler[‡]	He$^+$ (468, 656, 1012)	10 973 732.08	(85)
1973	Kibble *et al*[‡]	H-α, D-α	10 973 732.53	(77)

[†] Taylor *et al* (1969) and earlier revisions.
[‡] Cohen and Taylor (1973) with revisions, where made.
The figures in parentheses after the He$^+$ measurements refer to the wavelengths in nm of the lines which were studied.

Table 3.3(*b*) Values obtained for the Rydberg constant by laser spectroscopy.

Date	Group	Transition	Rydberg constant R_∞ (m^{-1})	
1973	Taylor and Cohen	Review value	10 973 731.2	(11)
1976	Hänsch *et al*	$\begin{cases} 2^2P_{3/2}-3^2D_{5/2}, \text{H-}\alpha \\ 2^2P_{3/2}-3^2D_{5/2}, \text{D-}\alpha \end{cases}$	10 973 731.43	(10)
1978	Goldsmith *et al*	$2^2S_{1/2}-3^2P_{1/2}$, H-α	10 973 731.476	(32)
1979	Petley and Morris	$\begin{cases} 2^2P_{3/2}-3^2D_{5/2}, \text{H-}\alpha \\ 2^2S_{1/2}-3^2P_{3/2}, \text{H-}\alpha \\ 2^2P_{1/2}-3^2D_{3/2}, \text{H-}\alpha \end{cases}$	10 973 731.513	(85)
1981	Amin *et al*	$\begin{cases} 2^2S_{1/2}-3^2P_{3/2}, \text{H-}\alpha, \text{D-}\alpha \\ 2^2S_{1/2}-3^2P_{1/2}, \text{H-}\alpha, \text{D-}\alpha \end{cases}$	10 973 731.521	(11)

3.3.2 Dye laser measurements

The width of a spectral line comes from two main sources: one is the width due to the motion of the atoms (the so called Doppler width), and the other is the width set by the natural lifetimes of the two states between which the transition occurs, since this represents an uncertainty in the value of the energy. This natural linewidth sets a limit on the spectroscopic resolution which may be achieved, but until recently it had rarely been possible to achieve such resolution with lighter atoms.

One method of reducing the linewidth would be to measure the spectrum of the radiation emitted perpendicularly from a collimated beam of atoms, which would show only a small Doppler shift and hence a correspondingly narrower linewidth. However, as the collimation of the beam is improved, the intensity of the radiation decreases and there is the attendant difficulty that it is difficult to prepare the atoms in the beam into the excited state. Experiments for measuring the fine structure constant have made use of this type of technique by using the longer lived S-states, but the method was not successfully applied to the measurement of the absolute wavelengths as was required to measure the Rydberg constant.

It was suggested by Series (1970) that the advent of the tunable dye laser might provide better precision in the measurement of the Rydberg constant and this prediction was soon to be proved correct. Whereas most lasers have a limited tuning range, which is set by the Doppler width and gain profile of the transition concerned, the tunable dye laser has a very much wider tuning range (although it may require some re-adjustment to achieve fine tuning over a wide range). The very broad tuning range of dye lasers meant that initially it was very difficult to adjust them to emit a single frequency for an extended period. However, it was soon found that by introducing highly wavelength-selective elements into the laser cavity the losses could be increased for all but a single mode, notably through the use of Fabry–Perot etalons and birefringent filters. An attendant problem was that of finding a suitable dye to allow operation at 656 nm, this being a particular problem for cw dye lasers which are operated much closer to their lasing threshold (this is evidenced by the large number of published studies on the sodium-D line which lies within the tuning range of the very efficient Rhodamine 6G dye).

As far as the scope of this book is concerned, the tunable dye lasers have been applied very successfully to overcome the problems of the Doppler width, with the results given in table 3.3(b). Essentially, three techniques have been applied, the first two to measure the Rydberg constant and the third to measure the hydrogen ground-state Lamb shift. This latter technique is also described since it could be applied to yield a measurement of the Rydberg constant as well.

3.3.2 (a) Laser saturation absorption spectroscopy.

With the first method, which is known as laser saturation absorption spectroscopy, the technique (Hänsch et al 1972), was to divide the laser radiation into two counterpropagating beams which crossed at a small angle, $\sim$ mrad, in an atomic hydrogen discharge (figure 3.5(a)). Of these, one beam, the probe beam, was about ten times weaker than the other—the saturating beam. When the dye laser frequency was scanned through the Balmer-α line, both beams were absorbed by those hydrogen atoms which were in the $n = 2$ state and hence showed the normal Doppler absorption profile. However, for atoms which were moving in a direction perpendicular to the two counterpropagating beams, the two laser radiations appeared to be at the same frequency. Most of the atoms were pumped up to the $n = 3$ state by the saturating beam and consequently these atoms absorbed the probe beam to a reduced extent. Hence, by detecting a second probe beam, which had the same intensity but was offset spatially from the other beam, and then subtracting the photodetected signal, the remaining signal was that corresponding to the reduced absorption due to those atoms having zero Doppler effect. In order to use phase sensitive detection techniques the saturating beam was chopped (by a rotating opaque disc containing rectangular shaped holes around a circle).

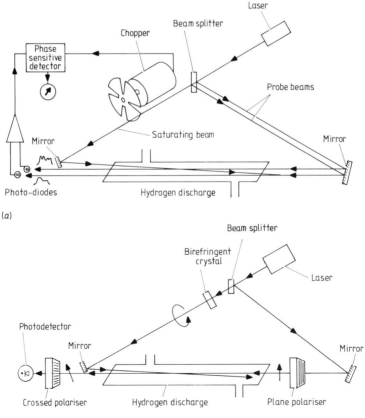

(a)

(b)

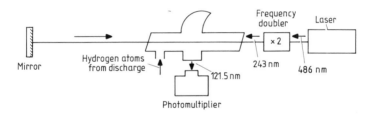

(c)

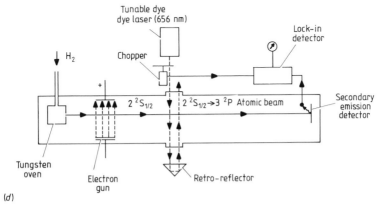

(d)

Figure 3.5 In (a), (b) and (c) are shown examples of the various techniques of laser spectroscopy which were used at Stanford University, notably by Schawlow and Hänsch: (a) laser saturated absorption spectroscopy, (b) laser polarisation spectroscopy (c) laser two-photon spectroscopy. (d) shows the atomic hydrogen beam method used by Amin *et al* (1981). Methods (a), (b) and (d) were used by the Stanford group to determine the Rydberg constant and (c) to deduce the ground state Lamb-shift of atomic hydrogen.

The method was applied by Hänsch *et al* (1974) to measure the Rydberg constant for atomic hydrogen and deuterium using a pulsed nitrogen pumped dye laser, and by Petley and Morris (1979) and Petley *et al* (1980) (figure 3.6) for atomic hydrogen using a cw argon ion, laser-pumped dye laser. The technique is also usable inside the laser cavity where the method is known as the 'Lamp dip'. The power inside the dye laser cavity is rather too high for the technique to be used to measure the Rydberg constant.

3.3.2(b) Laser polarisation spectroscopy. A second technique, which is known as laser polarisation spectroscopy, results from the induction of a small amount of birefringence in the absorbing gas by the saturating beam. In this method, counterpropagating beams are also used, but this time the probe beam passes between crossed plane-polarisers at either end of the discharge and the saturating beam is circularly polarised (figure 3.5(b)). Since light is only received at the photo-detector when the plane of polarisation of the probe beam is rotated

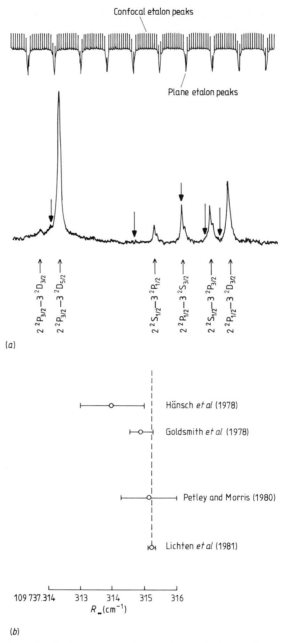

Figure 3.6 (*a*) Example of a broad scan of the hydrogen Balmer-α line by the saturated absorption technique (Petley and Morris, NPL). The arrows indicate positions of possible crossover resonances. (*b*) The measurements of the Rydberg constant by laser spectroscopy.

slightly, this method yields a much better signal to noise ratio than the saturation method described above (where fluctuations in the dye laser power, flicker effect noise in the two photo-detectors and drifts and differences in the gains of the photo-amplifiers all contributed to a greater noise level). The plane-polarised beam may be resolved into two counter rotating circularly polarised waves. When this probe beam passes through the absorbing gas, one of the components is absorbed more strongly because the saturating beam has already induced a prevailing orientation of the non-Doppler population of atoms. This effectively rotates the plane of polarisation and so some light is able to reach the photo-detector. The plane of polarisation is rotated only for those atoms which form part of the same absorbing population for both beams, i.e. those atoms moving across the beams having essentially zero Doppler shift in the direction of the laser beams. The circularly polarised saturating beam is selectively absorbed by those atoms which have a particular orientation. It is, of course, very important to avoid birefringence in any optical components which are located between the two polarisers, and care is required in avoiding strain in the windows at the end of the discharge tube. The method was used by Goldsmith et al (1978) to measure the Rydberg constant in hydrogen. Helium was used as a buffer gas in order to increase the proportion of atomic hydrogen, and the buffer gas produced pressure-dependent shifts which were carefully measured.

3.3.2(c) Two-photon spectroscopy. The third method has so far only been applied, as far as the fundamental constants are concerned, to measure the ground-state Lamb shift in atomic hydrogen by Hänsch et al (1978). In this method, the Doppler motion is eliminated in a different way and all of the atoms are able to contribute to the observed signal. The counterpropagating beams were at about one half of the difference frequency between the energy levels and the transition was induced by the simultaneous absorption of two photons, one from each of the counterpropagating beams (figure 3.5(c)). Since an atom sees one beam decreased in frequency by the Doppler effect and the other increased by an equal and opposite amount, the sum of the two frequencies is a constant for all of the atoms and so they can all contribute to the absorption. The absorption of two photons is of course much less probable than the absorption of a single photon. In some cases, but not in the experiment under discussion, the cross section may be enhanced by the presence of a virtual transition at about half the frequency of the desired transition. The absorption of two photons has the advantage that it is possible to excite transitions which would be forbidden for single photon absorption. Thus Hänsch et al (1978) were able to excite the normally forbidden $1^2S_{1/2}$–$2^2S_{1/2}$ transition in atomic hydrogen and, by comparison of the laser frequency with that required to induce the Balmer-β transition at 488 nm, they were able to measure the ground-state Lamb shift. The width of these lines under ideal conditions should approach 1 Hz in $\sim 10^{14}$ Hz. It is apparent that the method has potential applications for use as a frequency or wavelength standard as long as recoil and second-order Doppler shifts can be evaluated with the required precision.

3.3.2(d) Atomic beam techniques. Barger *et al* (1976) suggested measuring the Rydberg constant by using an atomic beam (see also Roberts and Fortson (1973), Lee *et al* 1975 and Weber and Goldsmith (1979) and Baklankow and Chebotaev (1974)). This technique is highly desirable since an atomic hydrogen discharge is far from being an ideal environment in which to perform high-precision spectroscopy. Thus Weber and Goldsmith (1979) studied the pressure shifts induced through the use of helium as a buffer gas and found that the shift was far from varying linearly with pressure. There are Stark shifts to consider as well (these were calculated by Blackman and Series (1974)). Lensing effects in the laser beam, which ideally has a gaussian profile, can produce some asymmetries in the observed absorption curve (Flack *et al* 1981), which are particularly important if some portions of the light in the probe beam fail to reach the photo-detector, or the sensitivity of the photo-detector varies across the receiving area. In a discharge too, the relative intensities of the various components differ from those expected from the simple theory and, in some cases, there can be crossover effects as well between unresolved hyperfine components which share common upper or lower levels. The latter was avoided in the Goldsmith *et al* (1978) polarisation spectroscopy work.

Although one could envisage interacting the laser light with an atomic hydrogen gas at a very low temperature in a container (atomic hydrogen is expected to remain gaseous right down to absolute zero temperature), the simplest solution at room temperature is to use a beam of atomic hydrogen. Such a measurement of the Rydberg constant was made by Lichten and his colleagues at Yale in 1981 (Amin *et al* 1981; figure 3.5(*d*)). The experiment was the optical analogue of the classic Lamb–Retherford experiment (Lamb and Retherford 1950). Their atomic hydrogen beam was produced by heating a tungsten oven to 2850 K and the atomic hydrogen beam emerged through a small hole. The beam passed through a transverse electron beam and about 10^{-6} of the atoms were excited into the metastable 2S state. Provided there were no electromagnetic fields to mix the 2S and 2P levels the metastable atoms traversed the apparatus and were detected by causing secondary electron emission from the detector (typically $\sim 3 \times 10^{-14}$ A with a 300:1 signal to noise ratio). A cw dye laser beam, having an intensity of about $2 \, \mathrm{mW \, cm^{-2}}$, crossed the beam and quenched the metastable atoms with Balmer-α light (2^2S–3^2P at 656 nm). The laser beam was chopped so that lock-in detection techniques could be used. The Rydberg constant was measured for both the $2^2S_{1/2} - 3^2P_{3/2}$ and $2^2S_{1/2} - 3^2P_{1/2}$ transitions in hydrogen and deuterium.

The second-order Doppler shifts were avoided by using a retroflector to reflect the light back across the atomic beam. The linewidths were close to theoretical (~ 50 MHz) and this measurement achieved an overall experimental uncertainty of 1 part in 10^9. The wavelength measurements were made using 0.314 m and 0.1225 m Fabry–Perot etalons against the wavelength of an $^{129}I_2(B)$ stabilised helium–neon laser. The high precision already achieved promises well for future experiments.

The next stage is likely to be the combination of an atomic beam with the excitation of a hitherto 'forbidden' transition by a two-photon absorption technique mentioned above, and 2S–3S and 1S–2S experiments are already in progress. The high precision will make the provision of a sufficiently stable wavelength standard increasingly difficult. Helium–neon lasers stabilised to internal iodine absorption cells show intensity-dependent shifts which depend on the mirror curvature etc (Rowley 1983) and shifts of ~ 5 kHz in 5×10^{14} Hz are difficult to avoid. This will mean that hydrogen–deuterium isotopic shifts will be studied initially, for they do not require high-precision absolute wavelength measurements.

3.3.3 Usefulness of determinations of the Rydberg constant

The values obtained for the Rydberg constant are given in table 3.3(b). It is seen that the agreement with the earlier 1973 value of Cohen and Taylor and also among the results themselves is quite good so that, unless there are some systematic errors common to all of the measurements, the Rydberg constant is almost certainly already known to much better than 1 part in 10^8. The Rydberg constant has in the past served as an auxiliary constant for the evaluation of the fundamental physical constants and in this connection an uncertainty of 1 part in 10^8 should suffice for the needs of the present decade. It has also been used in the past, by Cohen (1952) using the H-α and D-α measurements of Robinson (1939), Shore and Williams (1935) and Williams (1938), to evaluate the ratio m_e/m_p. In order to obtain this constant to 1 part in 10^6 measurements at the 1 part in 10^9 level are needed. It is evident that even to be useful today a further factor of ten in precision is required, and rather greater accuracies will be required in the near future. The method will undoubtedly be used to perform checks on such quantum electrodynamic calculations as the Lamb shifts and second-order Doppler effects, etc. The two-photon method should shortly be extended to positronium (Mills 1982).

The adoption of the new definition of the metre will be accompanied, and usefully supplemented, by measurements of the Rydberg constant. It is, after all, itself a 'natural' wavelength, or frequency unit which is widely used in spectroscopy.

It is not yet clear how the techniques made available through the new types of spectroscopy which have been made possible by the advent of the dye laser will be applied to the measurement of other spectroscopic constants such as the fine structure constant, but the use of forbidden transitions between metastable states by two-photon spectroscopy to overcome lifetime-broadening effects is a likely area of development. As the dye laser becomes more reliable, and the linewidth is made narrower, it will become an even more useful metrological tool.

At present our unit of atomic time is based on a hyperfine energy transition and efforts will undoubtedly be made to explore the possibilities of using the main energy transitions in some possibly exotic heavy atom. However, Crampton *et al*(1979) and Hardy *et al*(1978) have found that atomic hydrogen remains a stable gas at liquid helium temperatures and this will surely have some metrological applications. Looking back, one can see that in the past the spectrum of atomic hydrogen has proved to be a fruitful source of discovery each time that it was examined more closely. However, as Hänsch *et al*(1979) remark, the greatest surprise may yet be the absence of anything new from these exciting techniques—a prediction which everyone would like to see disproved—very soon.

3.4 The Siegbahn constant

It is quite usual in physics for a local system of units to evolve, and the metrologist has the task of ensuring that the conversion factor between these units and the SI units is established with adequate precision. It was usual to measure the wavelengths of x-rays by measuring the diffraction angle θ from a crystal of known lattice spacing d by making use of the Bragg equation

$$2d \sin \theta = n\lambda. \tag{3.4.1}$$

This was because it becomes difficult to fabricate diffraction gratings for use at grazing incidence at such short wavelengths. Assuming that the angle could be measured with adequate precision, it is clearly necessary to know the lattice spacing in metres in order to know the value of λ absolutely. The measurement of the lattice spacing had always been difficult and the method (figure 3.7) involved deducing the volume of a unit cube of the crystal by measuring the crystal density and combining this with the Avogadro constant deduced from other measurements of fundamental physical constants. Thus

$$N_A = u_0 M f / d^3 \rho \tag{3.4.2}$$

where M is the molecular weight, f is the number of molecules per unit cube of side d, ρ is the density and $u_0 = 10^{-3}$ kg mol^{-1}. This essentially involved the charge on the electron and the faraday. Since 1925, x-ray wavelengths have been expressed in terms of their own unit, the x unit (xu), introduced by Siegbahn, and was intended to correspond to 10^{-13} m (0.001 Å). Unfortunately, it differed from this value by $\sim 0.2\%$, which was due in large measure to the unsuspected error in the assumed viscosity of air in the famous Millikan oil-drop experiment. The ratio (xu): 10^{-13} m is assigned the symbol Λ. The x unit was defined by assigning the value 3 029.04 xu for the effective spacing for first-order diffraction ($n = 1$) by calcite crystals at a temperature of 18 °C.

Several types of crystal were measured over a number of years. These included calcite (Bearden 1931, 1938), sodium chloride, diamond and potassium chloride by Tu (1932), lithium fluoride (Straumanis *et al* 1939 and Hutchinson and Johnston 1940), sodium chloride (Johnston and Hutchinson 1942) and

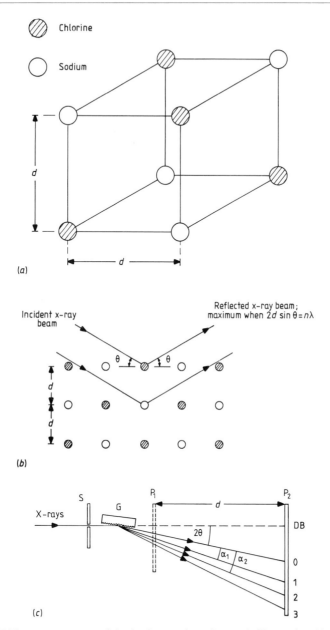

Figure 3.7 The measurement of the lattice spacing of crystals illustrating (*a*) the cubic structure of a rock-salt crystal, (*b*) Sir William Bragg's method of estimating the lattice spacing *d* (using x-rays at grazing incidence and the density and molecular weight of the crystal) and (*c*) the experimental arrangement for observing x-ray diffraction by a single crystal: S, slit; P₁, P₂, photographic plates; G, grating at grazing angle of incidence; DB, direct beam; 0, zero-order diffraction; 1, 2,..., first and second orders of x-ray diffraction.

potassium chloride (Hutchinson 1942). It is interesting that the results of Bearden (1931), after revisions by DuMond (1959), with those of Henins and Bearden (1964) were included in the 1973 evaluation of Cohen and Taylor (1973), some forty years after the original work. However, variations were found to occur with typical sources of materials, even within a particular crystal (Smakala and Kalnajs 1957). As a result, it became the practice to use x-ray emission lines as wavelength standards.

It was very difficult to find really suitable x-ray lines to use as standards, for even the narrowest were some 300 parts per million wide and, as for example Sauder *et al* (1977) have shown, the line shape is rather complicated and not expressible in terms of a simple asymmetry parameter. Several of these lines were used over the years to re-define the x unit and it was found that the units differed from one another by a few parts in 10^5.

Bearden (1965) proposed a new unit in order to obtain an x-ray scale which was more nearly based on the metre. This unit was termed the 'angstrom star' and was defined by $\Lambda(\mathrm{WK}_1) = 0.209\,010\,0\,\text{Å}^*$. This scale still differed from the metric length scale by an amount which was barely acceptable for the assignment of x-ray wavelengths. γ-ray wavelengths, as measured by double-crystal grazing-incidence spectrometers, were similarly affected. Moreover, the measurements of the Avogadro constant by crystal density techniques were also limited in precision by this lack of knowledge of the x unit. It was therefore increasingly desirable that an improved value be obtained for the x unit.

3.4.1 The lattice spacing of silicon

The rapid rise of the semiconductor industry led to considerable advances in the purity of germanium and silicon and it became possible to make nearly perfect cylindrically shaped 'crystals' which were some 5 cm in diameter. Conventional interferometry is very difficult with x-rays, since the sources are not very monochromatic and it is not easy to fabricate any precise optical components, such as collimators, in order to generate parallel beams from a point source. New techniques were developed in x-ray interferometry, by Bonse and Hart (1965) and others, and these led to a new method of measuring the lattice spacing of silicon directly in terms of the metre. The method adopted to measure the lattice spacing is illustrated in figure 3.8. When x-rays pass through the first crystal they are diffracted, and this happens again with the second and third crystals, which are mounted with their crystal planes very accurately parallel to one another.

The diffracted x-rays produced by the first two crystals produce a region of stationary wavefronts near the surface of the third crystal. This crystal also diffracts the x-rays and one of the diffracted beams is detected. As the third crystal is moved sideways with respect to the first two the intensity of the diffracted waves which are incident on the detector goes through cycles, passing from one cycle to the next as the crystal is displaced by one lattice spacing. The principle of the measurements of the lattice parameter is therefore

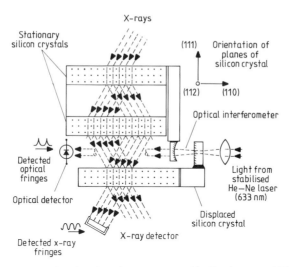

X-rays

Stationary
silicon crystals

(111) Orientation of
planes of
silicon crystal

(112) (110)

Optical interferometer

Detected
optical
fringes

Light from
stabilised
He–Ne laser
(633 nm)

Optical detector

Displaced
silicon crystal

Detected x-ray
fringes

X-ray detector

Figure 3.8 The x-ray and optical interferometers used by Deslattes and his colleagues at the National Bureau of Standards, USA, to measure the lattice spacing of silicon as part of their project to measure the Avogadro constant.

to displace the crystal very slowly through a known number of lattice spacings (known by counting the number of maxima or minima in the diffracted beam) and to measure this displacement in some type of optical interferometer.

Thus, if the Bragg spacing of the crystal in the direction of motion is d, the transmitted x-ray intensity becomes light and dark periodically for each $\frac{1}{2}d$ of translation. If the number of x-ray fringes is an integer p, plus a fraction ε, in displacing the crystal by one optical fringe, then

$$d = \tfrac{1}{2}\lambda_0/(p + \varepsilon)$$

where λ_0 is the wavelength of the standard source. For a 633 nm helium–neon laser, $p = 1648$, i.e. there is an integral number of 1648 x-ray fringes per optical fringe and the fractional part is ~ 0.3.

The first requirement for the measurement of d is to align the three silicon crystals very accurately with respect to one another. This is best achieved by fabricating them from a single crystal of silicon and mounting them on their respective supports before sawing them apart. The next requirement is to have some method for displacing the crystals sideways at a very slow rate. For this, Deslattes (1970) modified a device which was originally designed by Benioff for calibrating a geodetic strain gauge. This was machined from a solid block and, by means of an arrangement of cuts and holes, it was equivalent to a series of levers and pivots. In this way, the motion of the micrometer screw was reduced by a further factor of about 300 (materials such as brass are unsatisfactory since they give a rather jerky motion). This arrangement permitted displacement rates of as little as 0.1 nm per minute. The scan speed was conditioned in part by the necessity to observe the x-ray fringes with adequate precision. The experiment placed formidable demands on both the temperature stability and the uniformity of the temperature distribution over the volume of the crystals.

The distance through which the crystals had been displaced was measured in terms of interference fringes by means of a Fabry–Perot interferometer, one mirror of which was attached to the displaced crystal. This interferometer had a mirror separation of 0.3 mm. One mirror was plane and the other had a radius of curvature of about 1 m. Once the experiment was operational there was no need to count every fringe, for it was merely necessary to displace the crystal by a sufficient number of optical fringes so that the integral number could be calculated (there were 1648.... x-ray fringes per optical fringe) and the fractional part of a fringe (~ 0.3) measured with sufficient accuracy for the integral number of the next stage to be predicted. The reference wavelength standard and optical source for the interferometer was an iodine ($^{129}I_2(B)$) stabilised helium–neon laser operated with first derivative lock. The lattice parameter was finally obtained by displacing the crystal through about 150 000 x-ray fringes (or ~ 0.03 mm) and measuring the optical displacement with an accuracy of a few thousandths of a fringe, giving an overall uncertainty of little more than a tenth of a part per million.

It was, of course, necessary to ensure that the crystals were well aligned, otherwise the x-ray fringes would not have been seen, and also to ensure that the displacement was accurately perpendicular to the desired lattice spacing. Fortunately, a narrow feature was found at the centre of the rocking curve which enabled a very high alignment sensitivity to be achieved. This sharp feature occurred in the use of the second and third crystals in the single bounce mode. The usual angular half-width of the transmission curve as a function of the alignment was about 0.9″, but the central peak was only about 0.02″ wide. Consequently, by setting on the side of this peak it was possible to check that the displacement of the crystal was in a straight line to $\pm 0.000\,06''$. In fact, there was a small curvature of 0.0017″ which was taken into account and applied as a correction of 0.27 ppm to the lattice spacing.

Deslattes *et al* (1974) measured the d (220) lattice spacing of a vacuum float-zone refined specimen of silicon at temperatures between 22 °C and 23 °C. A value of 2.56×10^{-6} K^{-1} was used for the appropriate coefficient of expansion of silicon and the final result (Deslattes 1980) was

$$d(220) = 0.192\,017\,07 \text{ nm at } 25 \text{ °C}$$

with an overall assigned uncertainty of 0.15 ppm.

The above value applies to the particular sample used by Deslattes *et al* (1974) and there are, of course, possible differences due to different isotopic abundances, different amounts of impurity and lattice defects. Most of these parameters may be estimated by comparing one crystal sample with another. A slightly more difficult problem concerns the long-term stability of the lattice parameter for a particular crystal, but it is believed that at the present precision such effects are not important. However, the lattice spacing of one crystal may be very accurately compared with that of another and in this way the crystal measured by Deslattes *et al* can be transferred to other crystals. Different types of crystal may be calibrated. Thus the lattice spacing of germanium has also been determined relative to the $d(220)$ spacing of silicon.

Several groups are similarly engaged on a measurement of the lattice parameter. One of these, operating at the Physikalische Technische Bundesanstalt (PTB), has obtained a value which differs from the Deslattes (1980) results (Siegfried *et al* 1981). The difference may in part be accounted for by differences between the crystal samples; differences of 0.5 parts per million are not unexpected. The method used by Siegfried *et al* differs from that used by Deslattes and in particular uses the polarisation type of interferometer described by Curtis *et al* (1974). Their crystal was displaced through some 200 optical fringes and the precision and repeatability of the method corresponded to a few parts in 10^5 of an optical fringe. The temperature of the silicon was 22.5 °C and was the value required in order to keep the optical and x-ray fringes nearly harmonically related, so that it was necessary to measure only a small fraction of a fringe. The work used a binary build-up process. It is expected that these values will be supported shortly by other results and that the cause of the differences ($\sim 1.5 \times 10^{-6}$) will be resolved soon.

Given the lattice spacing, the possible applications diverge considerably. One is to obtain a value for the Avogadro constant, which is considered in §3.5. Another application is to use the crystals in spectrometers and make much more accurate measurements of x-ray and γ-ray wavelengths. This latter has led to some important results which have an impact on QED and also on the fundamental constants. We therefore consider the double crystal spectrometer, developed by Deslattes and others, in the next section.

3.4.2 The double crystal spectrometer

The general scheme of a double crystal spectrometer used by Kessler *et al* (1979) to measure γ-ray wavelengths is shown in figure 3.9. The collimated source is diffracted by the first rotatable crystal and then by the second rotatable crystal and the diffracted radiation falls on the two photo-detectors. The intensity profiles are recorded as a function of angle for two positions of the second rotatable crystal. These are the parallel position, which renders the emergent x-ray beam parallel to the original direction, and the antiparallel position. The angle of rotation of the crystal between the two conditions is then just twice the Bragg angle, θ. This angle is then used with the equation:

$$\sin \theta = \lambda/2D$$

in order to obtain the γ-ray wavelength.

In their spectrometer, the two crystals were separated by about 0.5 m and three slit collimators were used to prevent the detector from seeing the source

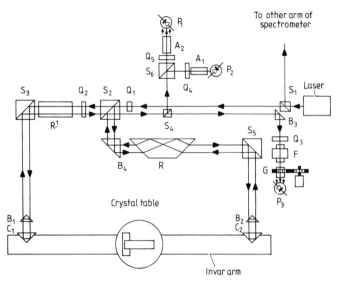

Figure 3.9 The polarisation-sensitive angle interferometer developed by Deslattes and others at the NBS as part of a double crystal γ-ray and x-ray spectrometer. The angle sensing sensitivity was ± 50 μrad for swing $\pm 2.5°$. A, analyser; B, prisms; C, cube-corner reflectors; F, Faraday rotation polarisation modulator; G, angle-encoded Glan–Thompson prism; P, photodiodes; Q, quarter wave plates; R, Dove prism and compensator (R'); S, beam splitters.

directly. Both x-rays and an autocollimator were used to align the crystals (which were 5 cm × 2.5 cm × few mm thickness) while fixing them with an epoxy resin to Invar mounts. The angular rotation was sensed by means of Michelson interferometers. These comprised cube-corner reflectors which were mounted at the ends of Invar arms. These rotated with the crystals and enabled angles of rotation of up to $\pm 5°$ to be measured with a precision of about 50 µrad. This extremely high sensitivity was accomplished by arranging that the plane of polarisation of the output laser light also rotated as the crystals were rotated. This permitted interpolation of the measurements to a small fraction of an optical fringe. The apparatus was evacuated and mounted on a vibration isolated table whose mass was some 5000 kg. The temperature of the room was stabilised to about 0.2 °C and the temperature of the thermally insulated apparatus changed very slowly, typically being stable to 10^{-2} °C over an interval of a few hours.

The calibration of the angular displacement was effected by a build-up process using a 72-sided polygon in place of the silicon crystals. Thus the table was rotated so that light from first one face and then an adjacent face of the polygon was reflected into a high resolution autocollimator. This process was repeated until all of the faces of the polygon had been observed. The sum of the angles was then accurately 360°.

The precision of the angle measurements in this spectrometer surpasses that normally achieved for angle measurement in metrology—indeed, polygon standards cannot normally be measured with such precision. Thus the faces cannot be lapped to be flat to much better than one tenth of an optical fringe and this limits the precision with which angles can be defined by the normal to the surfaces of the polygon. However, in this case, 2π was the cumulative calibration angle since the final setting was to the same face as the first. It is of interest that over a period of a few months a small systematic drift was found in the measurements. This drift was traced to small changes in the lengths of the Invar arms which carried the cube-corner reflectors, their length changing by about 0.7 ppm per annum. This change was consistent with that obtained for Invar by Berthold et al (1976). It is salutary to be reminded that in modern metrology materials such as Invar cannot be relied upon to be dimensionally stable when used for long periods—which is after all the reason why the use of material line standards to define the metre was abandoned.

Kessler et al (1979) used three sets of calibrated crystals in their spectrometer: Si(220), Si(111) and Ge(400), the figures in parentheses referring to the diffracting planes. The transfer measurements were made using methods based on those described by Hart (1969) and Ando et al (1978), and could be made with an accuracy of between 0.05 and 0.1 ppm.

The double crystal spectrometer has been used to measure the wavelengths of a number of γ-ray reference lines including the ^{198}Au 675 keV line and several ^{192}Ir lines between 205 keV and 612 keV. A number of x-ray energies have also been re-measured, including the tungsten K-α line and some Tm x-ray lines. The use of the new assignments to the γ-ray lines has been of particular interest in the use of muonic x-ray wavelengths to test the predictions of QED theory. The results of the earlier measurements (Dixit et al 1971) were in disagreement with theory by some 50–100 ppm. Since that time, the predictions of theory have been refined and these, together with the shift of 25 ppm in the ^{198}Au reference line, have largely removed the earlier disagreement with theory for measurements over the energy range from 157.5 keV (on calcium 3d–2p transitions) to 437.6 keV (barium 4f–3d transitions) (Kessler et al 1979).

Helmer et al (1979) have combined all of the available measurements to establish γ-ray standards between 45 keV and 1300 keV for use with germanium and lithium spectrometers and detectors. It should be pointed out that all of these measurements are made with respect to the Deslattes et al (1980) value for the silicon lattice parameter, and that in order to convert them to electron-volts, it is necessary to use the voltage wavelength conversion factor derived from the least squares evaluation of the fundamental physical constants. The Cohen and Taylor (1973) value was $1.239\,852\,0(32) \times 10^{-6}$ eV m and the readjusted value of 1983 is (Cohen and Wapstra 1983) $1.239\,851\,851 \times 10^{-6}$ e V_{BI} m, a change of -5 ppm. It is very probable that there will be some further advances in this area leading to new measurements of fundamental physical constants during the next decade. These include measurements of the Compton wavelengths of the electron, proton and neutron and the mass–energy conversion factor. It will be necessary in the future to make measurements of the lattice constant of silicon with rather greater precision than has been achieved so far if such determinations are to play a significant part in future evaluations of the physical constants. Deslattes (1980) has pointed out that such improvements are already under consideration and do not require a very major advance in the technology. The utility in the present work has been one of greatly extending the precision of the realisation of the electromagnetic energy or wavelength scale up to around 800 keV, with prospects of extension to the MeV range in the near future. If nuclear γ-ray measurements can be extended to sufficiently high energies, ~ 10 MeV, where mass defects can also be measured with high precision, then for an isomeric transition involving the emission of a γ-ray of wavelength λ and a mass changes ΔM^* nuclidic mass units, one would have

$$\Delta M^*(10^{-3}\,\mathrm{kg\,mol^{-1}})N_A{}^{-1} = hc/\lambda$$

enabling a new combination of fundamental constants, hN_A/c, to be measured.

3.4.3 Measurement of null angular displacements

The above impressively high sensitivity of 50 μrad for displacements of up to $\pm 2.5°$ can be bettered in null instruments (as with other branches of metrology). Thus, capacitance tiltmeters of the type described by Jones and Richards (1973) can surpass the above figure by some four orders of magnitude. In fact, the sensitivity of such tiltmeters is limited by microseisms, for at this level of precision the Earth is not a very stable platform. Capacitance transducers for observing displacements from 10^{-5}–10^{-14} m have shown stabilities of the order of 10^{-11}–10^{-12} m over a day. When made into a horizontal pendulum such devices also form sensitive gravimeters that are well able to show the tidal variations of gravity and have sensitivities of the order of $10^{-10}\,g$. Such devices are inherently quite small, the dimensions being of the order of a few centimetres. The principle of operation is that the displacement being sensed forms the central plate of a three plate capacitor which is arranged so that as one capacitance increases the other decreases. The electrode system is suitably connected to a transformer bridge circuit with associated amplifier and phase sensitive detector. Typically, a capacitance change of 1 fF (10^{-15} F) corresponds to a displacement of 10^{-9} m and a corresponding change in the output of the 16 kHz phase sensitive detector system of about 1 V.

3.5 The Avogadro constant

Following the formal introduction of the mole into the SI it became important to distinguish between the dimensionless Avogadro number and the dimensioned quantity which is now termed the Avogadro constant, symbol N_A or L. The terms Avogadro number and Loschmidt number are therefore rapidly falling into disuse.

The earlier measurements of the Avogadro constant (reviewed by Deslattes 1980), which were obtained by measuring the density and lattice spacing of a pure crystal, were inherently limited to about 70 ppm by problems of chemical purity and the necessity of measuring the lattice spacing in x units. The uncertainty in the conversion from x units to metres, which was discussed earlier, placed limits on the accuracy which could be achieved by these methods. However, the very high precision measurement of the lattice spacing of silicon achieved by Deslattes and his colleagues, as well as other technological advances, made it both possible and worthwhile to measure the other parameters involved in a determination of the Avogadro constant with higher precision than was hitherto possible. Consequently, they went on from their determination of the lattice spacing of silicon to derive the unit cell volume a_0^3 and also to make a determination of its density and atomic weight. Thus

$$N_A = u_0 nA/\rho a_0^3$$

where n atoms of average atomic weight A occupy a unit cell of volume a_0^3, ρ is the macroscopic density and $u_0 = 10^{-3}$ kg mol^{-1}.

3.5.1 The density measurement

The measurement of the density of silicon was achieved by means of an immersion method. Four highly spherical steel artefacts, 6.3 cm diameter, were constructed for use as transfer standards. Their masses were established by comparison with the NBS standard kilogram (replica number 20). The volumes of the spheres were calculated from a number of measurements of their diameters. These were measured by comparison with a hollow Fabry–Perot etalon (figure 3.10). The fringes were formed by radiation from an iodine stabilised helium–neon laser whose wavelength was known with respect to that of the orange radiation of a krypton-86 lamp used to define the metre. Johnson (1974) showed that the true volumes of the spheres differed insignificantly from the volumes calculated from the average diameters.

The density of four 200 g silicon disc samples was established by first weighing them in air and then while immersed in a fluorocarbon fluid, for which purpose they were suspended from fine wires (figure 3.11). The density of the fluorocarbon fluid was established by weighing the steel spheres in air and when totally immersed in the fluid. The change in mass, together with the known volume of the steel spheres, enables the density of the fluid to be calculated. The silicon samples were similarly weighed in air and when totally immersed. These measurements, when combined with the density of the fluid, provided sufficient information for their volume and density to be established. Corrections were of course applied for the buoyancy effect of weighing in air. (There were intended to be five samples, totalling 1 kg, but one met an untimely end!)

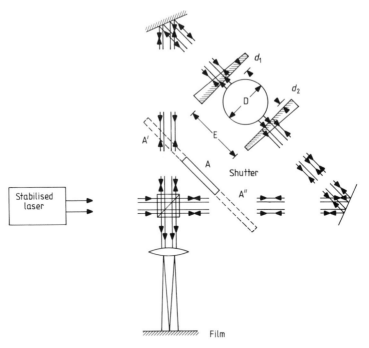

Figure 3.10 The interferometer used to measure the diameter of the steel spheres in the NBS Avogadro constant determination (after Saunders 1972).

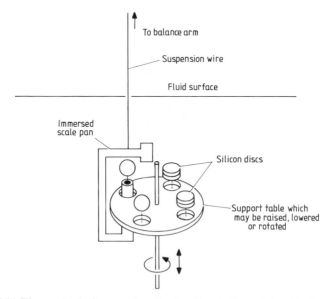

Figure 3.11 The method of measuring the density of silicon (after Deslattes 1980).

3.5.2 Calculation of the Avogadro constant

Once the density was known it was necessary to determine the relative isotopic abundances and impurity levels of the silicon crystal. The isotopic abundances were established by taking samples of the pure isotopes of silicon and mixing them together until the same signal was obtained in a mass spectrometer as that with the crystal sample. In this way it was possible to eliminate systematic effects in the mass spectrometer. The use of the 15 cm 60° magnetic sector mass spectrometer to make such determinations has been described by Smith *et al* (1956) and by Rodden (1972), and the procedure involved for the Avogadro constant work was described by Barnes *et al* (1975). The isotopic contents were approximately ^{28}Si:92.23%, ^{29}Si:4.67% and ^{30}Si:3.10% respectively. Three crystal samples were measured, and although the ranges of the individual densities and atomic weights were some 5 ppm, the quantity A/ρ was, as one would hope from equation (3.4.2), a formally invariant quantity, for the spread in these values was 0.3 ppm. These were consistent with a unique value within the estimated experimental uncertainties. The weighted mean value for A/ρ was 12.059 027 4 (0.94 ppm) which, when combined with the lattice spacing measurements, led to a final result of

$$N_A = 6.022\,097\,8 \times 10^{23}\ \text{mol}^{-1}\ (1.01\ \text{ppm}).$$

This value is about 0.6 ppm different from that reported earlier and is the revised result given by Deslattes (1980). It is interesting that since the weighings were performed at atmospheric pressure and the lattice measurements *in vacuo*, it was necessary to make a correction of 1.02 ppm for the change in the volume of the crystal unit cell due to the compression effect of atmospheric pressure. Thus the change in volume of the crystal as a function of pressure is given in terms of its elastic constants (Deslattes 1980) by

$$-V\frac{\mathrm{d}p}{\mathrm{d}V} = \frac{C_{11} + 2C_{12}}{3}$$

where typical values of C are $C_{11} = 0.1657$ and $C_{12} = 0.0639$, in units of 10^{-12} Nm^{-2} (McSkimin 1953).

It is expected that part or all of the Deslattes *et al* determinations of both the lattice spacing and Avogadro constant will shortly be confirmed by work at several standards laboratories throughout the world. In principle it should be possible to refine the precision of the measurements by at least an order of magnitude. This would be desirable from the point of view of making gross checks on the stability of the kilogram.

3.6 Can we replace the prototype kilogram?

It will be apparent from the discussion in other parts of this book that the fundamental constants are intimately involved in our realisation and maintenance of the second, the metre, the ampere and the volt. The kilogram now remains uniquely defined as a prototype unit (that is the mass of a specified platinum–iridium artefact), and it is natural to enquire whether it may be replaced. The present situation is that we do not know exactly how stable the kilogram is in an absolute sense, for there are no experiments which may be performed with adequate precision. The precision of measurements of the relevant fundamental constants, or combinations of several constants, is approaching the point where the limitations of the mass measurements are beginning to become apparent. One problem is that at present the precision of mass measurements decreases considerably as one departs from the kilogram (figure 3.12). It is apparent that considerable advances will be required in the precision of the subdivision of the kilogram if some atomic method comes along which is at the one gram level (figure 3.12).

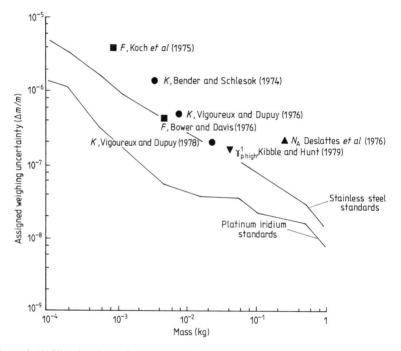

Figure 3.12 Showing how the accuracy of mass realisation decreases as one moves away from the kilogram, together with the weighing uncertainties assigned in some representative measurements of fundamental constants. The lower line is for the transfer to platinum–iridium standards and the upper line for the transfer to stainless steel standards where buoyancy effects must be allowed for as well.

We can begin to see the possibilities of replacing the kilogram by some atomic method, although it is not clear what the method might be. The Avogadro constant involves the mole, the definition of which invokes the prior definition of the kilogram, so that in any case some adaptations of the definitions would be required before this constant was used to derive the kilogram. Changes can take place in a comparatively short time. Thus, in the evaluation of DuMond and Cohen (1965), the conversion from the maintained to the absolute ampere was an auxiliary constant. Only four years later, in the Taylor *et al* (1969) evaluation its status had changed to a constant which was evaluated from the fundamental constants measurements as well as from the direct measurements. Only a further three years elapsed before the volt (and hence the ampere) were fixed in an absolute sense by the CCE-72 agreed value for the Josephson effect voltage to frequency conversion factor.

In the short term, we can expect that all of the fundamental constants which involve mass may be used to derive a 'best' value for the kilogram, in much the same way as the above mentioned electrically related constants are used to derive the absolute value of the ampere. In the longer term, the fundamental problem is that of ensuring that (i) the electrical and mechanical units of force and energy are the same, together with (ii) the allied problem of keeping μ_0 an exactly defined constant. Beyond that, it matters little whether the constant used to realise the unit of mass is primarily an electrical one involving force or energy, or a mechanical one. Clearly, much will depend on the direction in which metrology evolves during the next few years. Generally speaking, it is preferred practice to retain an existing method of defining a unit until a superior method is developed. Consequently, much will depend on the level of precision at which the prototype kilogram is demonstrated to be unstable in an absolute reference frame.

We can therefore begin to see possibilities for replacing the prototype kilogram by some atomic definition, possibly involving the Avogadro constant, but it is still rather early to make firm predictions as to which method will be used ultimately. There may be other quantum effects involving mass and frequency, which have yet to be discovered, which will change the situation overnight. One example might be the true demonstration of the Josephson effects in superfluid liquid helium, while another might be the combination of a Semiconductor Quantum Interference Device (SQUID), together with other techniques, for an atomic measurement of the Faraday constant. The one firm prediction is that there are likely to be some exciting times in the years to come.

3.6.1 The atomic mass unit

In discussing the measurements of the Avogadro constant it is important to state which mass scale is being used to specify the molecular weights. Some years ago there were two mass scales. The chemical scale took the atomic weight of oxygen as being 16 exactly, but this scale was arrived at before it was appreciated that naturally occuring oxygen contained three stable isotopes. The physical scale took the mass of ^{16}O as being exactly 16 atomic mass units. As the precision of the measurements advanced it became an increasing nuisance to have the two rival scales for atomic and molecular weights differing by 275(10) ppm. A very neat compromise was developed which was chosen so as to minimise the changes for practising scientists outside the highest level of precision. This compromise was to adopt the nuclidic mass unit or atomic mass unit, amu, whereby the mass of ^{12}C was exactly 12. This meant that the extensive data expressed in terms of the chemists' scale needed changing by only 40 parts per million and it was generally agreed at the time that, provided the change was less than about ± 50 ppm, the changed mass unit would not necessitate extensive changes of the published data. This change to the $^{12}C = 12$ scale, which was adopted by the appropriate international bodies, IUPAC and IUPAP, took effect from 1 January 1962. It

Table 3.4 Relationship between the various scales for expressing atomic weights.

Scale	Reference nuclide	Exact value on scale	Difference between the given scale and the chemical scale (ppm)
Chemical scale	$\Sigma(^{16}O, {}^{17}O, {}^{18}O)$	16	0
Physical scale	^{16}O	16	$+275$
Present atomic mass scale, adopted 1962	^{12}C	12	-43
Rival possibilities	^{17}O	17	$+8$
(considered, but	^{18}O	18	-4
rejected)	^{19}F	19	$+42$
	^{15}N	15	-50

The $^{12}C = 12$ scale is particularly convenient in mass spectrometry since carbon provides a convenient reference scale of ions which have charge to mass ratios ranging from 3 to 120, e.g. from $^{12}C^{4+}$ to $^{12}C^{+}_{10}$ while hydrocarbons of the form $C_n h_m$ can yield ions of intermediate mass values.

avoided the need to express such fundamental constants as the Avogadro constant, Faraday constant and the gas constant by values differing by 275(10) ppm in the chemical and physical scales (the uncertainty was the result of variations in the relative abundances). According to Rossini (1974) the change to the carbon-12 scale was suggested independently by Olander (1957) and Nier (1957). There were other possible candidates which also came within the desired ± 50 ppm range, which included ^{17}O, ^{18}O, ^{19}F and ^{15}N. The differences between these scales and the former chemical scales are shown in table 3.4.

3.7 The molar gas constant R and the molar volume of an ideal gas

The methods of measuring the gas constant are closely involved with the different methods which are used to establish the thermodynamic scale of temperature and the accuracy with which the fixed points of the scale have been fixed has an important impact on the work —the studies will undoubtedly lead to a better established temperature scale. Thus gas thermometry, acoustic thermometry, radiation pyrometry and noise thermometry may all be used with varying degrees of precision to measure the gas constant. They have been critically reviewed by Colclough (1981) and we consider only the main determinations here. Storm (1980) has made an analysis of the noise thermometry measurement which uses the Nyquist formula for the voltage noise V_n across a resistor r in a bandwidth Δf:

$$V_n^2 = 4kTr\Delta f. \qquad (3.7.1)$$

A measurement at liquid helium temperatures by the noise correlation method to several parts per million, even with a 20 kHz bandwidth, would require several million seconds of averaging time. However, Crovini and Actis (1978) have made noise thermometry measurements at higher temperatures, 630–962°C, which suggest that a 25–50 ppm measurement may be possible by this method.

3.7.1 The limiting density method

For an ideal gas the equation of state is given by

$$pV = RT(1 + B(T)/V + C(T)/V^2 + \cdots) \qquad (3.7.2)$$

where $B(T), C(T)\ldots$ are known as the virial coefficients. The molar gas constant R may be obtained from an extrapolation of the product pV of the molar volume at different pressures to zero pressure (or infinite gas volume). This is done by the method of limiting density, whereby a quantity

$$L(p) = (P_0/p)\rho(p) \qquad (3.7.3)$$

is measured by weighing the mass $M(p)$ of a constant volume V of the gas at a standard temperature T_0 at successively lower pressures p. Extrapolating to zero pressures yields

$$L(0) = M\rho_0/P_0T_0 = M/V_m$$

where T_0 is the thermodynamic temperature which corresponds to 0 °C (273.15 K), V_m is the molar volume of a perfect gas at standard conditions ($T = 0$ °C and $P_0 = 1$ atm (that is 101 325 Pa)) and M is the molecular weight of the gas.

Measurements of the absolute densities of gases between 1924 and 1941 primarily used oxygen as the gas, since the molecular weight of oxygen was exactly 32 on the old chemical scale of atomic weights. However, nitrogen has been used as well. The measurements have been reviewed by Batuecas (1964, 1972), and the value obtained from these, corrected for the changed conversion factor from litres to cubic metres (12th CGPM) becomes (Cohen and Taylor 1973)

$$V_m = 22.413\,83(70) \times 10^{-3}\ \text{mol}^{-1}$$

and

$$R = 8.314\,41(26) \text{ J mol}^{-1}\,\text{K}^{-1}$$

i.e. the uncertainty is 31 ppm. The method has also been reviewed by Colclough (1981). The reduction of this uncertainty would require taking into account recent data on the compressibility of gases which yields information about the virial coefficients (Dymond and Smith 1965). Although this would increase the accuracy of the extrapolation to zero pressure, the measurement of the gas pressure to better than 10 ppm requires that the heights of the mercury in the manometers are read to better than 10 μm while its purity and density must also be known to better than 10 ppm. In addition, the isotopic composition of oxygen (or nitrogen) must be known with an accuracy of 0.01% for the molecular weight to be evaluated to better than 10 ppm, and there is also the problem of allowing for the gas absorbed by the container. Thus the above uncertainty of 31 ppm is probably close to the limit achievable by the classical determinations.

Table 3.5 Measurements of the gas constant by the method of limiting density, corrected for standard gravity, the physical scale of molecular weights and from litres to dm^3 (random uncertainties only).

Authors	Gas	$V_m(\text{dm}^3\,\text{mol}^{-1})$	$R_0(\text{J K}^{-1}\,\text{mol}^{-1})$
Baxter and Starkweather (1924)	O_2	22.4133 (2)	8.31421 (7)
Baxter and Starkweather (1926)	O_2	22.4126 (6)	8.31395 (26)
Moles *et al* (1934–7)	O_2	22.4135 (2)	8.31429 (7)
Batuecas and Malde (1939–49)	O_2	22.4142 (9)	8.31455 (33)
Pereira (1978)	N_2	22.4132 (7)	8.31417 (26)

3.7.2 The acoustic interferometer measurement

Present day measurements, which are probably the first of the next generation of measurements of the gas constant, have made use of the fact that the velocity of sound in a perfect gas is given (at zero pressure) by the equation

$$v_0^2 = \gamma RT/M$$

where γ is the ratio of the specific heats and M is the molecular weight of the gas. This method has the particular advantage that, although relative pressure measurements are required with high accuracy, there is no need to know the pressure absolutely. However, in consequence of the virial coefficients being non-zero, the velocity of sound varies with pressure, and so the value at zero pressure must be obtained by extrapolation from measurements which are made at higher pressures. The acoustic interferometer method of measuring this constant was developed by Quinn *et al* (1976) at the NPL, as a refinement of the acoustic thermometry technique (figure 3.13).

A first requirement of the measurement is the production of a very precisely known temperature. Since the triple point of water is the most accurately realisable temperature (being defined as 273.16 K exactly), it was decided to construct a large triple point cell around the apparatus. The triple point of pure water corresponds to the temperature at which the liquid, solid and vapour phases of water are in thermodynamic equilibrium. The cell used for the experiment was much larger than normal triple point cells and the temperature was found to vary by a few mK over the cell volume so that the average

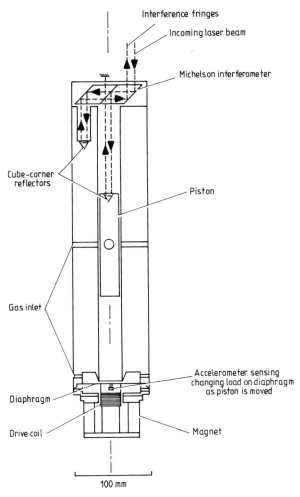

Figure 3.13 Scheme of the acoustic interferometer used by Quinn *et al* (1974) at the NPL to measure the universal gas constant.

temperature differed slightly from the triple point temperature and a small correction to the temperature was required. The inert gas argon was chosen since it approximates closely to an ideal gas. The ratio of its specific heats is $\frac{5}{3}$ and the molecular weight is very precisely known: $0.039\,947\,6$ kg mol^{-1}.

The velocity of sound in argon was measured by constructing an acoustic interferometer, and the velocity was deduced from the product of the frequency f of the acoustic waves and their measured wavelength. The interferometer is shown schematically in figure 3.13. The acoustic waves were induced in the argon by the electromagnetic transducer, which had an accelerometer attached to the diaphragm. The length of the cavity was varied by moving the plunger, and the distance moved by the plunger was measured by attaching the cube-corner to the opposite face. This cube-corner formed part of an optical interferometer and the number of stabilised helium–neon fringes by which the plunger was displaced could be measured by means of a photoelectric fringe counter system. The velocity of sound was measured as a function of pressure and a relationship of the form

$$v^2 = A_0(T) + A_1(T)p + A_2(T)p^2 + \cdots$$

where $A_0(T)\,(=\gamma RT/M), A_1(T), A_2(T)\ldots$ are termed the first, second, third ... acoustic virial coefficients, by analogy with the terms for the virial coefficients in the expansion of the expression for the product of pV given above. The acoustic virial coefficients could be expressed in terms of these and could, in principle, be used to extrapolate to zero pressure. However, they would involve the first and second derivatives with respect to temperature of both the second and third virial coefficients and these have not yet been determined to the required accuracy so that the extrapolation to zero pressure must be based on measurements at different pressures. The wavelength of the acoustic waves was measured with the interferometer and the method was to input constant energy into the diaphragm and observe the loading on it by means of the accelerometer. As the plunger moved through one acoustic wavelength, the acoustic impedance, i.e. the loading of the diaphragm by the gas in the cavity, followed an impedance circle (just as would the impedance of an electrical tuned circuit as it passed through the resonance condition). This variation in impedance was plotted on a pen recorder by deriving the in-phase and quadrature signals from phase sensitive detectors. The absorption of the acoustic waves in the gas and the walls of the apparatus led to the circles decreasing in diameter as the distance from the diaphragm increased.

Only plane waves could propagate because the acoustic cavity was sufficiently small in diameter to be below cut-off for the first and higher modes of oscillation. An important correction arose from the effects of the thermal conductivity and viscosity of the gas, at the pressures used in their experiment, for these reduced the velocity of sound at the boundary layer by around 0.1% to 0.3% (the effect varies with the gas density ρ as $\rho^{-1/2}$). There was also a change in the absorption coefficient associated with this velocity change. This increased the absorption by more than 100–400 times the amount expected from the normal 'classical' absorption in the gas. The magnitude of the effect depended on the square root of the density of the gas. Although the theory of Lee *et al* (1965) was used to modify some of the corrections, greater reliance was placed on the actual measurement of the absorption coefficients since these also enabled the corrections to the velocity to be made with adequate precision. Some absorption coefficient measurements were made at pressures below 30 kPa. In the initial work the majority of the velocity measurements were made between 30 kPa and 200 kPa. In this work it was found that the variation of v^2 as a function of pressure was better represented by a parabola than a straight line. The zero pressure value gave a value for the gas constant which differed by a significant amount from the Batuecas (1972) value. However, the fitting of the quadratic curve to the data was criticised by Rowlinson and Tildesley (1977) on theoretical grounds. Further work at higher pressures showed that the curvature was a result of non-linearities in the loading of the diaphragm at low pressures. There were also some distortions in the plotting of the impedance circles. The causes of the non-linearities were resolved by making measurements at higher gas pressures, up to 1.3 MPa, and when these values were combined with the corrected values for lower pressures, the value obtained for the square of the zero pressure velocity of sound in argon was

$$v_0^2 = 94\,756.8(17)\,\mathrm{m^2\,s^{-2}}.$$

The resulting value obtained for the gas constant R was

$$R_0 = 8.314\,48(15)\,\mathrm{J\,K^{-1}\,mol^{-1}}.$$

This revised value (Colclough *et al* 1979) is in good agreement with the earlier values. Further confirmation of the above results, at lower precision, comes from the work of Colclough (1979). He performed acoustic thermometry on ^{4}He gas between 2 and 20 K, and the value of R which he obtained from the acoustic isotherms corresponded to $8.314\,47\,\mathrm{J\,mol^{-1}\,K^{-1}}$, with an estimated uncertainty of 27 ppm.

It can be expected that there will be a number of measurements of the gas constant by these and other methods in the near future, together with a more detailed study of possible systematic effects, and work is already in hand to do so at several national standards laboratories throughout the world.

3.8 The Boltzmann constant

The universal gas constant is not regarded as being one of the fundamental physical constants, but it is closely related to the Boltzmann constant k, which is:

$$k = R/N_A.$$

Following the measurement of the Avogadro constant with part per million precision by Deslattes *et al* (1975), it may be regarded as an auxiliary constant when its uncertainty is compared with the 25 ppm precision of the above measurement of R by Colclough *et al* (1979). Therefore the Boltzmann constant is obtained directly from the above equation as

$$1.380\,662(44) \times 10^{-23} \text{ J K}^{-1}.$$

3.8.1 The radiation constants and the Stefan–Boltzmann constant

Planck's distribution law for the radiant energy $E(\lambda, T)$ of wavelengths between λ and $\lambda + \delta\lambda$, from unit surface area of a black body at an absolute temperature T is

$$E(\lambda, T)\delta\lambda = C_1 \delta\lambda / [\exp(C_2/\lambda T) - 1] \text{ Js}^{-1}$$

where $C_1 = 2\pi h c^2$ and $C_2 = hc/k$. Both of these constants are evaluated from the measurements which involve h, c and k and so the radiation constants are not measured directly with as much precision as they may be evaluated indirectly. Radiation pyrometry plays a very important part in temperature measurement, particularly at high temperatures. The distribution law has therefore been the subject of very careful studies. The colour temperatures of lamps are measured very carefully by this means.

The total energy radiated by a black body per unit area, that is the integral of the above equation over all wavelengths, involves another physical constant, which is known as Stefan's constant. This is more usually known as the Stefan–Boltzmann constant, with the symbol σ. Thus the radiant flux per unit area emitted by a black body at a temperature T over all wavelengths is given by

$$E = \sigma T^4.$$

In terms of the fundamental constants σ is given by

$$\sigma = 2\pi^5 k^4 / 15 h^3 c^2$$
$$= 5.670\,32(71) \times 10^{-8} \text{ W m}^{-2} \text{ K}^{-4}.$$

This constant has usually been obtained from the evaluation of the fundamental physical constants for it is very difficult to measure the radiant energy of a black body with adequate precision. However, since σ involves the gas constant to the fourth power, a 25 ppm measurement of R leads to a 100 ppm uncertainty in σ and vice versa. Therefore, an experimental determination of the value of σ, if it could be measured precisely enough, might be used to deduce R. Until recently this would have been thought unlikely, but Quinn and Martin (1981) have been making an absolute measurement of σ at the NPL by a primary radiometric method which it is hoped will lead to a check on R having a comparable precision to that of the acoustic interferometer measurement.

The principle of the method (figure 3.14) is to have an energy detector (a germanium thermometer of μK sensitivity) which is at ~ 1K surrounded by a bath of superfluid helium at temperatures of ~ 2K. This detector receives the

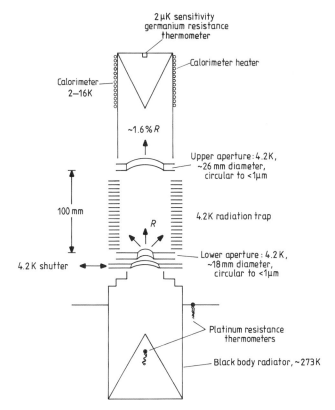

Figure 3.14 The absolute radiometer developed by Quinn and Martin (1983) to measure the Stefan–Boltzmann constant.

black body radiation from a thermal radiation source. As has been mentioned earlier, the most accurately realisable temperature is the triple point of water and hence their primary radiant source is a triple point cell. Other sources have been used and provided a check on the IPTS-68 temperatures scale as well. The radiation from this cell is incident on the detector after passing through a very precisely machined circular aperture. This aperture is then closed and an equivalent amount of electrical power substituted in the cryostat for the energy received from the radiant source. The overall target uncertainty for the value of the Stefan–Boltzmann constant is in the region of 100 ppm, or 25 ppm for the gas constant.

3.9 Conclusion

We have seen in this chapter how the measurement of the fundamental constants with the highest possible precision is intimately involved with the problem of realising the SI units with adequate precision. In some cases, the measurement of the atomic constant surpasses the precision with which the unit is realisable, in which case, as with the velocity of electromagnetic radiation and the metre, we can consider redefining the operational definition which is required to realise the appropriate base unit.

We move on in the next chapter to a consideration of the fundamental constants which are involved with the electrical units.

4

The electrically related fundamental constants

4.1 Introduction

In this chapter we consider the constants whose dimensions include one or other of the electrical units. As the precision with which the constants can be measured has advanced, it has gradually caught up on the precision with which the definition of the SI ampere may be realised. As a result, the last few evaluations of the fundamental constants have involved the conversion factor K from maintained to absolute volts and amperes, and this factor has been evaluated from the available experimental data along with the fundamental constants. The direct measurements are still included, but they have come to play an increasingly weaker part in the evaluations.

Information is provided from a number of measurements, including the Faraday constant of electrolysis, the gyromagnetic ratio of the proton in weak and strong magnetic fields, the magnetic moment of the proton in terms of the nuclear magneton and, interwoven with these, there is the measurement of $2e/h$ by the Josephson effects. The Klitzing et al (1980) quantised Hall conductance method for measuring the fine structure constant is still a comparative newcomer and in years to come it may well provide a natural quantum unit of impedance which is competitive in precision with the calculable capacitance method of realising the ohm.

It must be emphasised that the electrical units are defined through the definition of the ampere and the consequent definitions of μ_0 and ε_0. These definitions must be made in such a way that the electrical and mechanical units of both force and energy are exactly the same. Therefore, although it is tempting to set up a system of electrical units involving fundamental constants, for example, the Josephson effect for the volt and the quantised Hall conductance for the siemens, or ohm, such systems must be set up with care. If they are not, one finds that μ_0 has become a constant which must be evaluated experimentally, as it did during the 1930s for a while.

We begin with a discussion of the early measurements of the electronic charge and of h/e, before proceeding with the modern measurements of the latter using the AC Josephson effect.

4.2 The charge on the electron

The existence of the electron was foreshadowed in the work of Johnstone-Stoney (1881) who estimated its magnitude as $\sim 10^{-20}$ C. Following the confirmation of the existence of the electron in 1897, attempts were made by Townsend, Wilson and Thomson to measure the charge on the electron, but it was not until Millikan (1916) devised his famous oil-drop experiment that the order of magnitude experiments were replaced by measurements of high precision.

The principle of the oil-drop experiment was quite simple (figure 4.1), but it was no easy task to use it to obtain high precision results, as those students who have tried to use it in undergraduate experiments will confirm. The method was to produce fine oil drops from a spray, some of which became charged in the process. These were allowed to fall into the field of view of a microscope and their further fall was largely arrested by the application of an opposing electric field which was adjusted to balance the gravitational force. The terminal velocity of the drop was then measured under various combinations of gravitational and

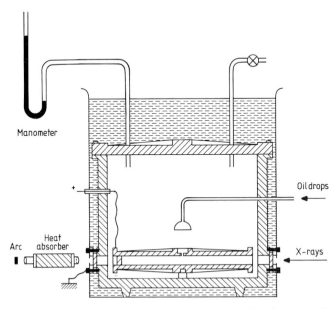

Figure 4.1 The Millikan oil-drop experiment to determine the charge on the electron. The apparatus has its modern counterpart in the experiments which search for fractionally charged particles, i.e. quarks.

electrical forces. The former variation was effected by studying drops having different masses. The charge on the drops was shown to be consistent with their each being an integral multiple of a fundamental unit of electric charge—the charge on an electron. The experimental uncertainty of Millikan's work was estimated to be 0.17%. His result was initially accepted by the scientific community, but there gradually appeared a growing discrepancy between the Millikan value and the indirect values. This was subsequently traced to the use of an incorrect value for the viscosity of air in the Millikan work. Modern counterparts of this experiment are found in the levitation experiments in search of the quark which are discussed in chapter 8.

4.2.1 The consequences of the error in the viscosity of air

In evaluating his results, Millikan had assumed that the measurement of the viscosity of air by one of his colleagues, Harrington (1916) was the most accurate. The latter used the method which required measuring the torque transmitted across the air gap between two concentric cylinders when the outer cylinder was rotated. Houston (1937) and others later showed that Harrington had over-looked, or rather underestimated, two important sources of error, one due to an end effect and the other due to air being dragged round by the rotating cylinder, which increased its moment of inertia. These led to Harrington's value being too low by 0.4% and Millikan's value for the elementary charge being too low by 0.6%. This systematic error remained undetected for about fifteen years (1915–31) and, because of the great importance of the elementary charge e, this systematic error had many far reaching effects. One useful result of the error was that all of the experiments concerned with the charge on the electron were performed a little more carefully, and the sources of error examined a little more closely, than they might otherwise have been. Consequently Harrington appears to be destined to be remembered as one who got it wrong rather than one who nearly got it right—*pour encourager les autres*!

There is a further interesting point to be made concerning these results, for they illustrate the dilemma of the reviewer. In 1944, Birge reviewed the experimental values for the electron charge by oil-drop experiments. These comprised the (corrected) results of Millikan (1924), Hopper and Laby (1944), Blocklin and Flemberg (1936) and Ishida *et al* (1937). The assigned uncertainties were as shown in table 4.1. (They have been converted here to standard deviation uncertainties and the values given in coulombs, whereas Birge considered probable errors and gave the value in esu.) Birge considered that the Hopper and Laby work was probably comparable with Millikan's determination and had reservations concerning the Ishida *et al* value as well. He therefore assigned the results the relative weights shown in the table and obtained the weighted mean. He found that the internal consistency of the results was better without the Ishida *et al* result, giving a weighted mean of $1.6033(20) \times 10^{-19}$C. (It should be noted that the uncertainties in each case

Table 4.1 The oil-drop values for e considered by Birge (1944) taking η_{28} (viscosity of air) $= 1832.5(10) \times 10^{-7}$ cgs units.

Date	Author	Value of $e(10^{-19}$ C)		Assigned weight
1924	Millikan	1.603 5	(19)	2
1940	Hopper and Laby	1.605 7	(15)	2
1936	Blocklin and Flemberg	1.598 1	(56)	1
1937	Ishida et al	1.616 2	(15)	1
	Weighted average	1.605 4	(32)	—
	Final value taken in evaluation	1.605 4	(37)	—
1973	Recommended value	1.602 189 2	(46)	

were the random uncertainties and the uncertainty did not include an allowance of the uncertainty in the value of the viscosity of air. Thus the first uncertainty assigned by Birge to the oil-drop experiment was considerably larger, being the value given in the table.) It is interesting that this 1944 value was only one standard deviation higher than the 1973 value of Cohen and Taylor, showing that by then the earlier cause of discrepancies had been largely removed.

4.3 The ratio of h/e

Planck's constant appears as a fundamental unit in spectroscopy, relating the frequency of photons to a particular energy, and has a deeper role in quantum physics. It also appears throughout physics as a fundamental unit of angular momentum with sufficient frequency to be assigned the symbol $\hbar(= h/2\pi)$. Despite its universality, neither h nor $\hbar$ appear in situations where high precision measurements may be made directly. Effects such as the Einstein–de Haas effect, although representing the required gyromagnetic phenomena, are not so far usable. The combination h/e occurs in a number of experiments since they represent effects in which the energy is measured as an electromagnetic energy and the frequency is that associated with a photon. The value of h/e has therefore been obtained in the past from such experiments as:

(i) photoelectric measurements in the visible region of the spectrum
(ii) the use of x-rays to produce photoionisation of atoms
(iii) from measurements of the excitation or ionisation potentials which involve the valence electrons
(iv) measurement of the upper frequency limit of the continuous x-ray spectrum.

The first method, (i), made use of the Einstein equation

$$eV = h\nu - e\phi$$

where ϕ was the workfunction of the atom and eV the observed kinetic energy of the photo-electron. The value of the retarding voltage required to stop the photoelectric current was observed for different values of the frequency v. The potentials were normally in the region of a volt for photons in the visible region. There were difficulties in the experiment, due to the Fermi energy distribution and band structure in the electron energies at non-zero temperatures, which affected the above equation, and it was also difficult to ensure that the workfunction ϕ remained constant during the experiment. It was for the latter reason that Millikan (1916) devised his famous 'lathe in a vacuum' in order to prepare clean surfaces of sodium and lithium. Later measurements were made by Lukirsky and Prilezaev (1928) and by Olpin (1930). The accuracy of these experiments is not much better than 0.2%, as evidenced by the lack of agreement between the different experiments.

Method (ii) made use of x-rays of energy eV_e electron-volts, which were just sufficient to ionise an atom by removing one of the electrons from the filled shells. The onset of photoionisation also caused a sharp edge in the x-ray absorption spectrum, at a wavelength λ_e. The constant hc/e was then derived from

$$hc/e = \lambda_e V_e.$$

The disadvantage of the method was that the x-ray wavelengths were measured by crystal diffraction techniques and hence in terms of x-units, so that the Siegbahn conversion factor also entered into the above equation. The method was used by Bearden and Schwarz (1941), who investigated the emission of K-shell electrons from copper, gallium, nickel and zinc and the L-shell electrons from tungsten, and also by Nilsson (1953), who investigated the K-shell electrons from cobalt, copper, iron and nickel. These experiments yielded h/e to an accuracy of about 100 ppm.

In method (iii), the pioneering work of Frank and Hertz (1916) was followed by many other measurements of the minimum energy V_{min} required for an electron to excite an atom from its ground state to emit a certain wavelength λ_{min}. Thus the determining equation was

$$hc/e = \lambda_{min} V_{min}.$$

Dunnington (1933) used the work on the ionisation potentials of argon, helium, mercury and neon by Lawrence (1926), Van Atta (1932) and Whiddington and Woodroofe (1935) to obtain a value for h/e which had an uncertainty of about 0.3%. Later, Dunnington et al (1954) made a precise measurement of the ${}^1S_0 - {}^1P_1$ excitation energy in helium by bombarding the gas with electrons of a known energy. The electrons were produced in a similar apparatus to that used by Dunnington to measure e/m_e and hence there were uncertainties in the electron energies as a result of contaminating films and space charge, despite care being taken to minimise or eliminate the effects of these. Dunnington *et al* obtained a result having an uncertainty of about 150 ppm.

Method (iv) yielded the most precise measurement of h/e and consequently it was analysed in rather greater detail. The principle was that x-rays produced by electrons which fell on a target were produced in the form of a continuous spectrum from zero frequency up to a maximum frequency which corresponded to all of the electron energy being converted into that of a single photon. Once again, the x-ray wavelength was measured in x-units and the relation used was

$$h/e = \lambda_{min} V/c.$$

The potential V was the sum of the potential difference between the electron source and the target metal and the workfunction of the cathode surface. The estimation of this limiting wavelength was made difficult by the limited resolving power of the spectrometer, which smoothed out the limit. A further complication was that the electrons were incident on a thick target (in order to achieve sufficient x-ray intensity) and so there could sometimes be more than one bremsstrahlung per electron, which blurred the limit. The situation was further complicated by the presence of small irregularities in the shape of the spectrum near the limit, discovered by Ohlin (1940, 1942, 1944, 1946). These prevented the accurate computation of the limit, even if the instrument function was accurately known. Over the years, the method was improved considerably from that of the initial work by Duane *et al* (1921) to the 100 ppm obtained by Bearden *et al* in 1951. In comparing the values, it is necessary to take account of the changes in the recommended values for both the speed of light and the Siegbahn conversion constant.

The ratio h/e has the dimensions of magnetic flux, and indeed was the value expected for the magnetic flux quantum. The more recent measurements of $h/2e$, the magnetic flux quantum associated with the (paired) electrons in superconductivity, making use of the Josephson effects, have completely replaced the methods which have just been discussed.

4.4 The measurement of $2e/h$ by the Josephson effects

As is discussed in chapter 5, at the time of 1963 evaluation by Cohen and DuMond there was difficulty in deciding which of two sets of measurement of the fine structure constant were correct. With hindsight, we can see that, faced with a difficult choice, the reviewers rejected the wrong set of experimental determinations but, although suspected earlier by DuMond and Cohen, this did not become really clear until the advent of the Josephson effects. These effects have had a far reaching effect on measurements of the fundamental constants and also on other areas of electrical metrology.

The Josephson effects were predicted by Brian Josephson (1962) who was at that time working for his PhD at Cambridge. They serve as a source of inspiration to the budding young scientist, for, although he shared the Nobel prize for this work in 1972, it formed only a part of his thesis—the remainder concerned his work (partly as an undergraduate) on the effects of gravitational acceleration on photons. The Josephson effects occur in superconductivity, and are a result of the long range phase ordering consequent on the Cooper pairing of the superconducting electrons, which was so brilliantly incorporated into the Bardeen, Cooper and Schrieffer theory of superconductivity of 1961. As a consequence of this ordering, the waves associated with the superconducting electrons have a coherence which, in the case of a superconducting solenoid, extends for distances of the order of kilometres. Outside the superconductors, the waves decay in a distance which is characteristically about 1 nm. Josephson considered the situation in which there were two superconductors which were separated by a distance such that the pair waves could interact weakly with one another. He found that it should be possible for a superconducting current to pass between them by a tunnelling process, and that as this current increased the phase difference between the pair waves increased until it reached the value of $\phi = \pi/2$, beyond which the direct supercurrent could no longer be sustained. This supercurrent was also predicted to vary with the magnetic flux through the area of the tunnelling region, and led to some remarkable experiments which clearly demonstrated the quantisation of magnetic flux.

Once there is a potential difference between the two superconductors, the pair waves no longer have the same frequency associated with them and they beat together (if the coupling is close enough). Josephson showed that under these conditions it was still possible for a superconducting current to make a net transfer across the insulating barrier as long as the potential difference satisfied the simple equation

$$2eV = nhf$$

where e was the electronic charge, h Planck's constant, n an integer, and the frequency f was either that appropriate to the self-resonance frequencies of the barrier (in the microwave region), or was the frequency of microwave radiation which was applied externally to the barrier (figure 4.2). His predictions were rapidly verified experimentally and the group of scientists at the University of Pennsylvania were particularly interested to discover whether the apparent simplicity of the above equation was sustained by experiment. In a series of classic measurements of increasingly high precision they improved the accuracy of their work until the uncertainty in the value of $2e/h$ which they measured was comparable with that achieved from the rest of physics as represented by the Cohen and DuMond (1963) least squares evaluation of the constants.

As indicated above, there were at that time discrepant values for the fine structure constant and the value deduced for $2e/h$ depended critically on these. Aside from the Josephson effect results, other measurements of the fine structure

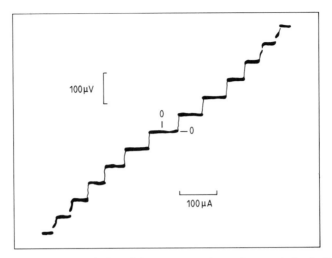

Figure 4.2 An oscilloscope display of the current–voltage characteristic of a Josephson junction (the barrier between two superconductors), when irradiated with 36 GHz microwaves. The precision measurement of the microwave frequency together with the consequent voltage separation of an integral number of supercurrent steps enables the constant $2e/h$ to be deduced.

constant (which involved quantum electrodynamics in varying degrees), also appeared to be discrepant with the review values. This provided Taylor *et al* (1969) with the motivation to perform their own evaluation of the fundamental constants, which evaluation they performed with considerable skill. Meanwhile, the Josephson effects had also been shown to apply in slightly modified form in other situations than that envisaged by Josephson. Indeed, it became apparent that in order to observe the Josephson effects it was only necessary to ensure that the two superconductors were weakly connected electrically. The author, working with Morris at the NPL, and using the solder-drop type of junction, was the first to confirm the Pennsylvania values (Petley and Morris 1969). The two sets of results agreed to well within the estimated uncertainties. The error estimates may with hindsight have been a little pessimistic and this was partly because in 1969, measurement of millivolt potentials with even one part per million precision was still a new and rapidly developing area of metrology.

4.4.1 The CCE-72 value for 2e/h and consequences

Following the above measurements there were a number of confirmatory measurements by other standards laboratories which have been well documented (see e.g. Petley 1982). These results showed excellent agreement, but after a period it became apparent that they were changing with time. This change was rapidly attributed to the drifts in the units of potential maintained in the various countries by means of Weston-cadmium standard cells.

As a result of the drift detected in an absolute reference frame by the Josephson effect work, it was decided to assign the value of 483 594.0 GHz to the volt as maintained at the BIPM in 1969. Many countries have adopted this value as a method of keeping their maintained unit of potential stable. Others have adopted the value of $2e/h$ appropriate to their maintained voltage unit at the time that they changed to using the Josephson effects. The adoption of a fixed value ensures that the maintained volt remains stable with a precision of a few parts in 10^8, but it does not ensure that it is exactly one volt as required by the SI definition of the electrical units, which is made in terms of the ampere. However, it is possible to make a hindsight adjustment to a measurement if it is made with respect to a quantity that is known to be stable, and such an adjustment will be

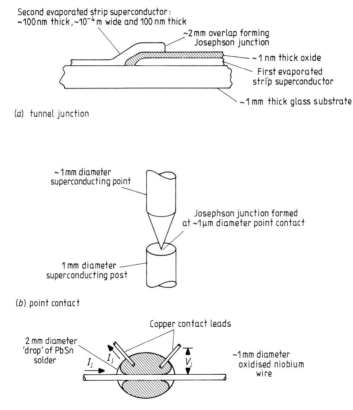

Figure 4.3 The different types of Josephson junction that have been used to measure $2e/h$. Today, measurements mostly concentrate on using tunnel junctions since these may be connected in series on the same substrate and also give measurable steps to rather higher voltages: $\lesssim 10\,\mathrm{mV}$.

made as soon as the accuracy of the absolute measurements of the ampere (or volt) have reached the desired level.

As far as evaluations of the fundamental constants are concerned the Josephson effect measurements of $2e/h$ now have the status of an auxiliary constant. Associated with this there is the assumption that one is indeed measuring the true value of $2e/h$ by this method and no further corrections are required. Here, one may make two points. The first is the practical one that the precision is so far in advance of that of alternative methods of deducing $2e/h$ that any experimental conclusion to the contrary is unlikely for some time to come. Indeed, the evidence from internal measurements, whereby searches have been made for dependence on the material or other experimental parameters such as current, voltage, frequency or temperature, have all been consistent with the same value for $2e/h$ at precisions ranging from parts in 10^{16} to parts in 10^8. The second point is that all attempts at a theoretical overturning of the simplicity of the equation, by finding small correction terms, have also failed—providing that the measurements are made as potential measurements between the terminals connected across the Josephson junction. All that one can say, therefore, is that if metrology is being led astray by assuming that it is truly the value of $2e/h$, then it is at a level which is undetectable for the present.

4.4.2 The present day methods of measuring $2e/h$

The detailed description of the methods which are used for measuring $2e/h$ in the many standards laboratories throughout the world is beyond the scope of this book and the techniques which are discussed in this section should be taken as being representative.

The type of junction employed has stabilised on a tunnel junction which comprises, for example, a strip of lead about 0.1 mm wide which is evaporated onto a fused silica substrate. Indium contacts for external corrections are often evaporated first. The lead is then oxidised under carefully controlled conditions in order to provide an insulating barrier about 1 nm thick and this is followed by the evaporation of a second lead strip which forms the second superconductor. The overlapping region has self-resonant frequencies somewhere in the microwave region. In earlier work, these were typically around 9 GHz (depending on the size of the junction) and, in order to achieve steps to a few millivolts, the applied microwave frequency had to be close to one of these self-resonant frequencies. The tunnel junction is normally mounted in a 9 GHz waveguide which is often tapered in order to reduce the waveguide impedance to give a better match to the impedance of the junction. Not all laboratories use 9 GHz; the PTB for example use 72 GHz. The power required from the microwave source, whose frequency must be highly stable, is up to a few tenths of a watt (depending on the coupling efficiency). For the working temperature one may choose the temperature of liquid helium at atmospheric pressure (4.2 K) or alternatively the pressure may be decreased so that operation is below the

λ-point of liquid helium at 2.2 K. The advantage of working below the latter temperature is that the thermal conductivity is greater and there is better thermal equilibrium.

The microwave power and current through the junction are adjusted, together with the microwave frequency, until the junction is operating on a high order step. Since the step voltage is about 20 μV it is necessary to operate on the 250th step if operation is required at 5 mV. It should be remarked at this point that although steps have been observed to beyond 10 mV, these voltages have proved difficult to achieve on a regular basis. It is therefore usual to cascade two junctions in series if steps to 10 mV are required. This potential must usually be decided in advance since the design of the high precision potentiometric system is rather inflexible and operation at a specific voltage is necessary.

4.4.3 Method of comparing the cell and junction voltages

The junction voltage must be compared with the EMF of a standard cell (figure 4.4). This has two consequences. The first is that it is necessary to step up the junction voltage for comparison with the 1.0186...V EMF of the standard cell. This means that some effective form of high precision current or potential divider is required and this divider must be calibrated, or be self calibrating, with high precision. The second consequence is that there must be leads from the cryogenic part of the apparatus up to room temperature (indeed the cell

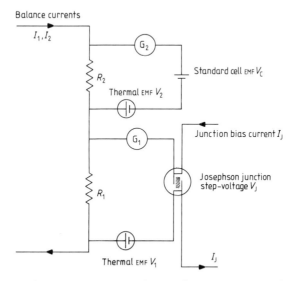

Figure 4.4 The basic potentiometer circuit used for the $2e/h$ determinations. (More recent methods have used the same resistor for the junction and standard cell and used very high precision current dividers to change the required potentiometer current from I_1 to I_2.)

may well be maintained at 35 °C). This results in thermal EMFs, for even if the leads are made from copper wire from consecutive parts of the same reel, the mechanical strains in the wires may be different, thereby producing differences in the thermoelectric properties of the wire and give rise to a net thermal EMF. A further disadvantage of the standard cell is that its internal impedance lies between $500\,\Omega$ and $1000\,\Omega$ so that the noise voltage from this must ultimately limit the precision with which the voltage comparison can be made. Some day, reliable room temperature voltage standards of lower internal impedance may be developed—Zener diodes usually show monotonic drifts with time and are not capable of much better than part per million precision. Meanwhile the all-cryogenic systems which have been designed have some precision in hand.

Many of the potential divider systems made use of an experimental realisation of a simple principle originally developed by Hamon at the NSL in Australia (Hamon 1954). The principle is simply that if n almost indentical resistors are switched from being connected in series to being connected in parallel, then the ratio of their series to their parallel resistance is n^2. If the resistors are identical to a part in 10^4 the ratio is n^2 to a part in 10^8, provided that the temperature or power coefficient of the resistors is small enough. In this way, by making pairs of Hamon resistors, each comprising ten resistors, ratios of 100:1 may be established quite readily. Initially Hamon resistors were built to operate at room temperature, but it rapidly became apparent that the performance was even better at cryogenic temperatures, where the leads could be made from superconducting materials. The first working cryogenic comparator system to be used to measure $2e/h$ was that of Petley and Gallop (1974). The system employed a ten-element Hamon resistor and balance between the junction voltage, and the potential drop across this resistor was detected with the superconducting type of galavanometer which is known as a SLUG (Superconducting Low inductance Undulating Galvanometer) or DC SQUID. There was a second three-element Hamon resistor in series with the first which was operated at room temperature and the standard cell EMF was balanced using a room temperature type of galvanometer amplifier (i.e. the galvano-meter spot fell on two differentially connected photocells whose difference current could be further amplified electrically).

In addition to Hamon resistors, an alternative method of making the voltage build-up has been to establish known current ratios across a resistor whose temperature coefficient and power coefficient was very small. It was found, for example, that resistors made from phosphor-bronze wire had a temperature coefficient of only one part per million per degree and also a low power coefficient. The current ratio was then established with very high precision with the aid of a DC current transformer. The type of transformer is simple in operation for it is essentially a null device whereby two coils are wound on the same core, one having n times the number of turns of the other. A SQUID is then used to sense the net flux in the core. Since the detection sensitivity of the SQUID is typically $\sim 10^{-4}\phi_0$ where

$$\phi_0 = h/2e = 2 \times 10^{-15} \text{ Wb}$$

current ratios of $n:1$ may be established with very high precision, $\sim 10^{-8}$ or better.

4.5 Quantised Hall conductance: the Ando–Klitzing method for the fine structure constant

Just as the Josephson effects provided a new method of measuring $2e/h$ with the application as an absolute voltage standard, so too there has recently emerged a method of measuring e^2/h, which not only leads to a simple determination of the fine structure constant, but also promises to provide a natural conductance standard which is based on fundamental physical constants. The effect arises as a quantisation of the Hall resistance of a semiconductor when the degenerate electron gas in the inversion layer of a MOSFET (metal oxide semiconductor field-effect transitor) becomes fully quantised. These effects occur when the transistor is operated at liquid helium temperatures and in flux densities of the order of 15 T. The experimental observation of this effect was reported by Klitzing *et al* (1980).

The type of MOSFET device which has been used is shown in figure 4.5. The electric field perpendicular to the surface, i.e. the gate field, produces sub-bands in the motion of the carriers normal to the semiconductor–oxide interface, and the magnetic flux produces Landau quantisation in the perpendicular directions, i.e. parallel to the interface. Since only the lowest sub-band is fully occupied, a two-dimensional electron layer is formed. As a result, provided that the conditions $\omega_c \tau \ll 1$ and $\hbar\omega_c \gg kT$ are satisfied (where $\omega_c = eB/m^*$, ω_c is the cyclotron frequency, m^* the electron mass, τ the relaxation time of the electrons in the layer and B the flux density), then the energy of the electrons in the layer is fully quantised.

Thus the energy levels may be written

$$E_{ns v} = E_0 + (n + \tfrac{1}{2})\hbar\omega_c + sg\mu_B B + vE_v$$

where $s = \pm 1$ (spin quantum number), g is the Lande g-factor, $v = \pm\tfrac{1}{2}$ (valley degeneracy quantum number), E_v is the valley splitting and $\hbar\omega_c = eB_z/\sqrt{m_x m_y}$ is the energy difference between Landau levels with quantum numbers n and $n+1$ (m_x, m_y are the effective carrier masses in the x, y surface plane). The Landau splitting $\hbar\omega_c$ depends on the component B_z and the spin splitting B. Hence their energies may be changed independently by altering the direction of B. The density of states $D(E)$ comprises broadened δ-functions as described by Ando (1974) and if the magnetic field is strong enough these states have minimal overlap.

The number of states N_L within each Landau level is given by the simple

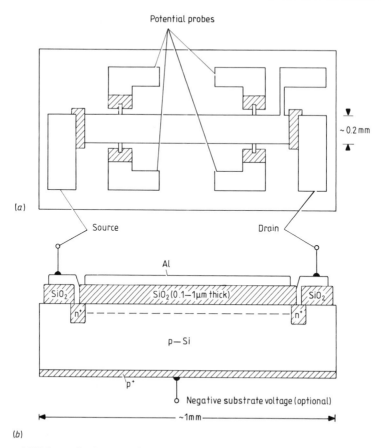

Figure 4.5 Schematic diagram of a quantised Hall effect MOSFET device: (a) top view and (b) section, as used in the high precision measurement by Klitzing et al (1980).

equation (excluding the spin and valley degeneracies)

$$N_L = D(E)\hbar\omega_c = eB/h$$

where the density of states $D(E)$ per unit area is given by $D(E) = m_x m_y/2\pi\hbar^2$.

If the density of states at the Fermi energy $N(F)$ is zero, the electrons in the inversion layer cannot be scattered and the centre of the cyclotron orbit drifts in the direction perpendicular to both the electric and magnetic fields. This lack of scattering also applies when $N(F)$ is finite, but still remains small. In consequence of this, when the Fermi level is set between the Landau levels, the device current is thermally activated and there is a minimum in the conductivity term σ_{xx}. This minimum can be as much as 10^7 times smaller than the maximum values. The ratio may be enhanced further by increasing the magnetic field and decreasing the temperature. Whereas the Hall conductivity tensor σ_{xy} is usually a complicated function of the scattering process, the absence of scattering under these conditions produces a major simplification, and the Hall conductivity is then given by

$$\sigma_{xy} = -Ne/B$$

where N is the carrier concentration. When the Fermi energy is set by the gate voltage to be between the Landau levels the correction term to this expression σ_{xy} is of the order of $\sigma_{xx}/\omega\tau$ and

$$\sigma_{xy}(\text{min}) \sim 10^{-7}\sigma_{xx}(\text{max}).$$

Subject to any error as a result of a non-zero value of σ_{xy}, under conditions when a Landau level is fully occupied one obtains

$$N = N_L i \qquad i = 1, 2, 3, 4, \ldots.$$

Hence, after eliminating N_L from the above equations, this leads to

$$-\sigma_{xy} = e^2 i/h. \qquad (4.5.1)$$

The Hall resistivity ρ_{xy}, which is given by

$$\rho_{xy} = -\sigma_{xy}/(\sigma_{xx}^2 + \sigma_{xy}^2) \approx \sigma_{xy}^{-1}$$

is defined by the ratio of the Hall field E_H to the current density j, and this may be rewritten as the ratio of the Hall voltage to the current flowing, which is just the Hall resistance R_H. Thus the expression $-\sigma_{xy} = e^2 i/h$ becomes

$$R_H = \alpha^{-1}\mu_0 c/2i$$

where μ_0 is the permeability of vacuum, c the velocity of light and α is the fine structure constant.

Although solid-state physicists had earlier used the above relationships in order to compare the performance of their devices with theory, it was not until Klitzing et al (1980) had a sufficiently pure sample and a sufficiently strong magnetic field that the possibilities of using the effect to measure the fine structure constant became a reality. They operated at a temperature of 1.5 K in a flux density of 18 T with a source–drain current of 1 μA. The device was some 400 μm long and 50 μm wide and the distance between the potential probes was 130 μm. A record was taken of both the Hall voltage and the voltage drop between the potential probes as a function of the gate voltage. These curves are shown in figure 4.6. It is apparent that the potential drops to zero at certain gate voltages, i.e. $\sigma_{xx} = 0$, and that when this occurs the Hall voltage shows a constant voltage step. Data on several steps and samples were analysed by Klitzing et al and all were consistent with the same value for α in equation (4.5.1) within the experimental uncertainties. The value obtained by Klitzing et al (1980) was

$$h/4e^2 = 6453.17(2) \, \Omega$$

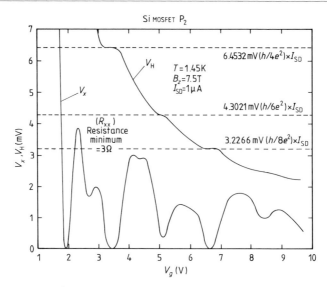

Figure 4.6 Example of the voltage V_x along and the Hall voltage V_H transverse to the current as a function of the gate voltage V_G, for a silicon MOSFET device (after Hartland NPL).

which corresponds to a value of

$$\alpha^{-1} = 137.0353(4).$$

This value may be compared with that from the work on the gyromagnetic ratio of the proton by Williams and Olsen (1979) of 137.035 963(15). Braun *et al* (1980) later reported a value of

$$h/4e^2 = 6453.198(9)\ \Omega$$

which corresponds to $\alpha^{-1} = 137.035\,92(18)$. Thus, although the earlier work was 5 ppm lower than the 1973 value recommended by Cohen and Taylor, the later work was in excellent agreement with their recommended value of 137.036 04(11). The initial work was performed on silicon MOSFET devices, but gallium arsenide provides higher electron mobilities and so gives an enhancement of the effects at a given magnetic field, even allowing weaker fields to be used. Tsui *et al* (1982) have reported a measurement using gallium arsenide heterostructures which yielded

$$\alpha^{-1} = 137.035\,968(23).$$

This result may also be combined with the Williams and Olsen (1979) value of γ_p' (low) to eliminate the realisation of the absolute ohm to yield

$$\alpha^{-1} = 137.035\,965(2)\ (0.089\ \text{ppm}).$$

One could combine them differently to derive a value for the ohm conversion factor $\bar{R}$. This serves to illustrate that the method of measuring γ'_p (low) essentially incorporates a realisation of the ohm (Petley 1976).

The work to date, therefore, has been very encouraging and will no doubt continue to lead to a considerable amount of theoretical activity to evaluate any possible corrections, and experimental work to check for dependence on the physical variables. Dissipative non-linear behaviour at high currents in the quantum Hall regime has been reported by Cage *et al* (1983). This is shown by structure in the resistance steps. The definitive intercomparison of the results in Japan, West Germany, the USA and elsewhere involves the knowledge of the relationships between the maintained ohms in these countries and, in turn, requires better realisations of the absolute ohm. A possible disadvantage of the method at present is the requirement of a high magnetic field, for this demands the very best superconducting magnets, operating close to the limit of present technology. However, it may well be that other high-purity materials will become available which have a higher electron mobility (as work with gallium arsenide has shown already). If so it will be possible to use lower magnetic fields which are more reliably available commercially.

The work illustrates strongly the difficulties facing those involved in fundamental metrology. Early in 1980 it was considered that the absolute derivations of the ohm were adequate. However, by June 1981 the quantised Hall effect measurements were already being made in West Germany, the USA and Japan. In order to intercompare the measurements it was necessary to know the relationships between the various nationally maintained ohms to better than 0.1 ppm. This relationship needed to be established with care since the maintained ohms are all drifting at different rates.

4.6 The gyromagnetic ratio of the proton

The proton has a magnetic moment μ_p and an angular momentum $\frac{1}{2}\hbar$, and consequently, in a magnetic flux density B, protons, in changing from being aligned parallel to antiparallel to the magnetic flux, absorb a quantum of energy of frequency f_s given by the simple equation

$$\Delta E = hf_s = 2\mu_p B \qquad (4.6.1a)$$

or

$$f_s = (2\mu_p/h)B \qquad (4.6.1b)$$

The quantity $2\mu_p/h$ is termed the gyromagnetic ratio of the proton and the angular frequency ω_s is given by $\omega_s = (2\mu_p/\hbar)B$. The symbols f_s or ω_s derive from the fact that in classical terms they would correspond to the proton spin precession frequency. It might have been more logical to call it the magnetogyric ratio of the proton. Evidently, to measure the gyromagnetic ratio of the proton γ_p one must measure a frequency and a magnetic flux density. The former is effected quite simply using a digital frequency counter. The magnetic flux B may either be produced by passing a constant direct current through a precisely wound solenoid of known dimensions, or alternatively be established with a powerful electromagnet. There is a limit to the flux density that may be obtained with a solenoid, and so measurements of this type are termed low or weak field determinations. The magnetic flux density with permanent or iron-cored electromagnets can be a thousand times greater than that achieved with an air-cored solenoid and consequently it is termed the high or strong field method.

The net magnetisation of a sample containing N spins in a magnetic flux B is

$$M_0 = \chi_0 B_0 \tag{4.6.2}$$

where the static nuclear susceptibility χ_0 is given by

$$\chi_0 = N\gamma_p^2 \hbar^2 I(I+1)/3kT. \tag{4.6.3}$$

The motion of a group of nuclei is governed by the Bloch differential equation

$$\frac{d\boldsymbol{M}}{dt} = \gamma_p \boldsymbol{M} \wedge \boldsymbol{B} - (M_x/T_2)\boldsymbol{i} - \frac{M_y}{T_2}\boldsymbol{j} - \left(\frac{M_0 - M_2}{T_1}\right)\boldsymbol{k} \tag{4.6.4}$$

where T_1 is the longitudinal or spin–lattice relaxation time, and T_2 is the transverse or spin–spin relaxation time: $T_1 \sim T_2$. These lead to damped oscillation at frequency $\omega_0 = \gamma B_0$ and the magnetisations M_x, M_y decay as $\exp(-t/T_2)$. If there is an external rotating magnetic field of the form

$$B_x = B_1 \cos \omega t \qquad B_y = -B_1 \sin \omega t \tag{4.6.5}$$

then in the rotating frame the solutions are of the form

$$M'_x = \chi'(\omega)2B_1$$
$$M'_y = \chi''(\omega)2B_1 \tag{4.6.6}$$

with the frequency dependent susceptibilities

$$\chi''(\omega) = \tfrac{1}{2}\omega_0 T_2\chi_0/[1 + (T_2\Delta\omega)^2 + \gamma_p^2 B_1^2 T_1 T_2] \tag{4.6.7}$$

and

$$\chi'(\omega) = \chi'' T_2\Delta\omega.$$

The optimum amplitude B_1^{opt} is given by

$$B_1^{\text{opt}} = 1/\gamma_p\sqrt{T_1 T_2}. \tag{4.6.8}$$

The signal may be observed as an absorption curve $\chi''(\omega)$ or as a dispersion curve. It is important not to observe a mixture of the two since the absorption is symmetric and the dispersion is antisymmetric about the centre

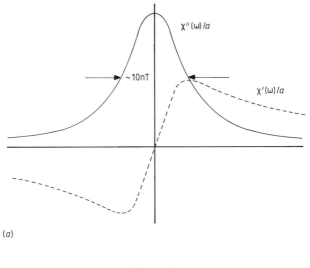

$\chi''(\omega)/a$

$\sim 10nT$

$\chi'(\omega)/a$

(a)

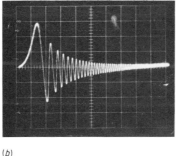

(b)

Figure 4.7 (a) The nuclear magnetic resonance absorption $[\chi''(w)/a]$ and dispersion curves $[\chi'(w)/a]$ and (b) the 'wiggles' observed with too rapid a passage through resonance (by the author at NPL).

(figure 4.7(a)) and so the combined value, their sum, would not be symmetric. Equally, if the absorption curve is traversed too rapidly the characteristic wiggles are observed (figure 4.7(b)) and the curve is distorted. This problem may be overcome by doping the sample to reduce the relaxation time by the addition of a paramagnetic salt such as ferric chloride. A possible disadvantage in high precision work is that such a salt produces a chemical shift of the resonance centre and this must be corrected for.

4.6.1 The involvement with the ampere conversion constant

In the weak field method, the magnetic flux density is normally produced by a solenoid of n turns per metre, carrying a current I, so that the magnetic flux at the centre is $\mu_0 nI$. Consequently, the metrological equation in the low field determination of γ_p is

$$2\pi f_s = \gamma_p' \mu_0 nI. \tag{4.6.9}$$

This expression only applies for a uniformly wound solenoid which is infinitely long. In practice of course this condition cannot be completely achieved and for shorter solenoids the flux density which is produced depends on the diameter, the length, and also on the number of turns wound on the solenoid. The measurement of the effective diameter of the coil is complicated by the fact that the solenoid is usually wound from copper wire which must, of course, be under some tension if it is to follow the shape of the cylindrical former exactly. As a result, the current density varies over the cross section of the wire, necessitating a correction. This correction requires a knowledge of the tension in the wire and, since copper wire is used, such tension may well relax with time. Thus, with winding tensions of 30–100 N the strain correction corresponds to a decrease in the effective diameter of the solenoid of about 0.3 μm. A further problem arises in measuring the diameter of the wire. Because the measuring device is contacting a cylindrical surface, the pressure must exceed the yield point of the metal locally, so that even with the lightest contact force there must be some distortion of the surface. A correction, using a formula originally derived by C V Boys, is made in order to allow for this effect.

The current which is required in equation (4.6.9) is of course normally measured in terms of maintained amperes and this must be multiplied by the (unknown) dimensionless factor K, which is within a few parts per million of unity, in order to convert it to absolute amperes. Thus the equation becomes

$$\omega_s / \mu_0 nI_{B169} = K\gamma_p' = (\gamma_p')_{weak} \tag{4.6.10}$$

where we have used $(\gamma_p')_{weak}$ to denote the value of γ_p' measured by the weak field method.

With the strong field method the magnetic flux density in equation (4.6.1b) is produced by a permanent magnet, an electromagnet or a superconducting magnet. The flux density is measured by using a long current-carrying coil of wire of rectangular shape and suspending it from the arms of a sensitive balance (known as a Cotton balance) such that the lower end of the coil is at the centre of the magnet. If the coil consists of n turns of width a and the current flowing is I (absolute) amperes $(= K^{-1}I_{B169})$ the force equation is

$$F = mg = BIa \tag{4.6.11}$$

where g is the local acceleration due to gravity at the scale pan and m is the mass required on the pan to balance the force due to the current in the coils. Hence, eliminating B between equations (4.6.10) and (4.6.11), we have

$$\gamma_p' / K = \omega_s / (I_{B169} / mg)$$
$$= (\gamma_p')_{strong}. \tag{4.6.12}$$

Since we do not know the currents in absolute amperes we have to use the values obtained using maintained amperes. However, it is apparent that by combining the two measurements in two different ways we may obtain either γ'_p or K; thus

$$\gamma'_p = [(\gamma'_p)_s(\gamma'_p)_w]^{1/2} \tag{4.6.13}$$

and

$$K = [(\gamma'_p)_w/(\gamma'_p)_s]^{1/2}. \tag{4.6.14}$$

In the early measurements of γ_p it was not necessary to distinguish between the two methods, since the value of K was thought to be known with adequate precision. However, the precision of present day work is such that the error in K is the dominant one and consequently the two methods of measuring γ'_p provide important information about the value of K.

It is apparent from equation (4.6.12) that the dimensions of γ_p, in terms of the base SI units, are $kg\,s^{-2}A^{-1}$. However, in both the weak and strong field determinations, it is necessary to make very precise measurements of length — almost as if there is a realisation of the ohm built into the measurements. It may be, therefore, that it will be possible to devise a method which does not require high precision measurements of length (the dimensions are required with the present methods because μ_0 (with dimension $H\,m^{-1}$) is involved in the weak field equation and the acceleration due to gravity (with dimension $m\,s^{-2}$) in the strong field measurements).

The gyromagnetic ratio is measured for protons which are in a very pure sample of water. There is a small correction of about 26 ppm which is necessary because of the diamagnetic shielding due to the electrons in the water molecule. This correction was first calculated by Ramsey (1950) and, although experiment and theory are in good agreement, it is usual to give the results both without and with this correction, the former case being denoted by γ'_p. The techniques developed in the USA around 1946 by Bloch and others at Stanford and by Purcell, Torry, Bloembergen and Pound and others at Harvard rapidly led to the development of the powerful analytical tool of nuclear magnetic resonance which is so useful to chemists today.

To understand how the signal is detected, it is convenient to remember the classical picture where the proton spin is inclined initially at some angle θ to the magnetic field direction and precesses about the field direction until finally it is aligned with the field. Provided that the field is sufficiently uniform the signal decays in about two or three seconds in water, or up to 15 seconds in benzene. Evidently, then, the detection coil should be perpendicular to the magnetic field direction. Sometimes two mutually perpendicular coils are used. One of them is supplied with an RF signal of the appropriate resonance frequency and the precessing protons cause a signal to be induced in the second coil, thereby indicating the resonant condition. As indicated by equation (4.6.1b), the strength of the signal depends on magnetic flux density. The excess of protons in the higher energy state is small, and the ratio of those in the upper to lower energy state depends on $\exp(2\mu_p B/kT)$, where $2\mu_p B/kT$ is typically 10^{-8}–10^{-4}. Since the energy absorbed or emitted depends on the flux density as well (equation (4.6.1a)), the signal to noise ratio varies roughly as the square of the magnetic flux density.

A very popular method of detecting the nuclear resonance signal is to use a single RF coil which forms part of the tuned circuit of a marginal oscillator, that is an oscillator which is only just oscillating. The change in the quality factor (or Q) of the tuned circuit due to the absorption of energy by the precessing protons makes a correspondingly larger change in the amplitude of the oscillations and the level of the oscillations provides a sensitive indication of the resonance condition. In order to use phase sensitive detection techniques and thereby make detection easier it is usual to modulate either the magnetic flux, using small coils wound over the water sample (which is typically a sphere of a few millimetres in diameter), or alternatively to modulate the frequency. This latter method has been made much easier following the introduction of low-noise variable capacitance diodes which were not available to the earlier experimenters.

4.6.2 The sample shape correction

The flux density inside a sample is affected by a demagnetisation effect which depends on the susceptibility of the sample and also its shape. Thus

$$(B_i - B_o)/B_o = (\tfrac{4}{3}\pi - \varepsilon)\chi$$

where B_i is the flux density inside and B_o the flux density outside the sample and χ is the susceptibility. For a spherical sample the value of the correction ε is $4\pi/3$ so that the shape correction is zero and hence this shape of sample is the one preferred for the measurements of γ'_p. It is, however, rather difficult to make a small spherical sample which is completely filled with water and hence many experimenters have preferred to use a cylindrical sample and apply a correction. Although the shape factor for a cylindrical sample is very difficult to work out analytically, the value for an ellipsoid of revolution may be obtained exactly and hence a value for a cylindrical sample may be obtained by assuming an ellipsoid of infinite length, for which the value of ε is 2π. This is an adequate assumption as long as the cylindrical sample length is more than about five times the diameter. The overall correction amounts to about -0.56 ppm.

The variation of the demagnetisation coefficient with the length to diameter ratio of the ellipsoid is shown in figure 4.8 and depends on whether the axis of the ellipsoid is parallel or perpendicular to the magnetic field. It is apparent that the correction varies rapidly when the shape is nearly spherical, which has implications in situations where one is concerned with small departures from exact sphericity.

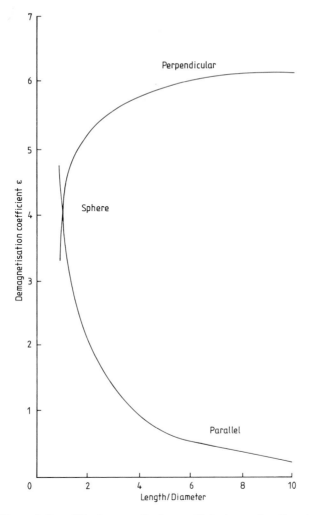

Figure 4.8 The variation of the demagnetisation coefficient ε as a function of the sample ellipticities for ellipticity parallel and perpendicular to the direction of the applied magnetic flux.

4.6.3 The susceptibility of water

The susceptibility of water is, in SI units, $-0.9 \times 10^{-8}\,\mathrm{kg}^{-1}$. Its value is required for the demagnetisation term and also to correct for the change in the field experienced by the protons due to the susceptibility of the spherical water sample. Since all measurements are referred to a spherical sample this correction is not required for many purposes. However, the susceptibility of water varies with temperature (as a result of changes in the density and degree of molecular association). For very precise measurements, it has become essential to state the temperature at which the measurements were made in order that a temperature correction ($\sim 10^{-8}\,°C^{-1}$) may be applied. Phillips et al (1977) fitted a quadratic to the measurements of Hindman (1966) and obtained:

$$[\mu_p'(T) - \mu_p'(0)]/\mu_p'(0) = 0.2(15) - 11.88(8)T - 0.020(9)T^2$$

where the value is that of μ_p' at a temperature T °C compared with that at 0 °C, and the correction is in parts per billion. In most cases, the temperature is between 20 °C and 35 °C and the above expression allows the results to be reduced to a standard temperature of 20 °C with a precision which is adequate for present day results. Petley and Donaldson (1984) obtained $-10.36(30) \times 10^{-9}\,°C^{-1}$ and Taylor obtained a similar value from Hindman's original data.

4.6.4 Diamagnetic shielding

The correction for the diamagnetic shielding of the protons in the water by the electrons in the molecule compared with the value for the free proton was calculated by Ramsey (1950). The value for protons in molecular hydrogen was found by Newell (1959) who combined experimental data with the work of Ramsey. Thus they obtained

$$\sigma(H_2O) = 26.0(3) \text{ ppm}$$

and

$$\sigma(H_2) = 26.6(3) \text{ ppm}$$

respectively. Liebes and Franken (1959) quote Hardy as obtaining

$$\sigma(H_2O) - \sigma(H_2) = -0.60(15) \text{ ppm}$$

while Gutowsky and McClure (1951) gave

$$-0.3(3) \text{ ppm.}$$

The precision of the above measurements has been overtaken by that obtainable from more recent experimental work as follows. The value of μ_p/μ_e for the free proton may be obtained by taking the hydrogen maser value of the ground-state splitting $g_j(H)/g_p(H)$ of Winkler et al (1972), and their value must then be corrected in order to give that for the free proton in terms of the electron magnetic moment μ_p/μ_e by using the theory of Grotch and Hegstrom (1971). (Their theory has since been confirmed by other workers.) Finally, μ_e is converted to μ_B using the results for the anomalous magnetic moment of the electron obtained from the g-2 experiments of Wesley and Rich (1972) and Dehmelt et al (1979). In this way the three results may be combined to give a value of μ_p/μ_B for the free proton.

The value of μ'/μ_B for protons in a water sample has been measured by the

Lambe–Dicke (1969) and the Phillips *et al* (1977) experiments (see § 6.1). Combining the results for μ_p/μ_B with those for μ_p'/μ_B allows a value of 25.637(67) ppm to be obtained for $\sigma(H_2O)$ (revised to 25.6(67) ppm).

4.6.5 The strong field measurement of γ_p'

The first high accuracy measurement of γ_p' was made at the National Bureau of Standards, Washington DC, by Thomas *et al* (1950). They used an electro-magnet, having pole pieces some 30 cm in diameter with a pole gap of 5 cm. This produced a flux density of 0.477 T, with a corresponding nuclear magnetic resonance frequency of 20 MHz. The latter frequency could be readily compared with that of a standard crystal oscillator. A second 20 MHz NMR probe was used to servo the magnet power supply current in order to maintain the resonance condition. The magnetic flux was measured by weighing the force on a nine-turn coil wound on a precisely constructed rectangular former, some 100 mm wide and 700 mm high (figure 4.9). It was convenient to effectively double the force by reversing the current through the coil and this procedure has been followed in subsequent experiments. Thus the change in force was given by

$$F = mg = 2aN(B - B')E_s/R$$

where the coil and N turns, the separation was a, m was the change in mass on the scale pan and B and B' were the flux densities at the ends of the coil which were in the magnet and remote from it respectively. The current through the coil was adjusted until the EMF E_s across a series resistance R balanced the EMF of a standard cell.

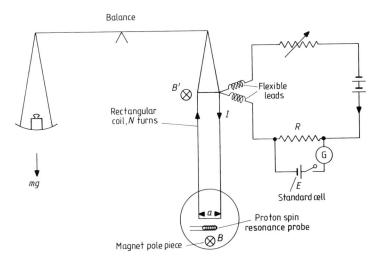

Figure 4.9 The Cotton-balance method of determining the gyromagnetic ratio of the proton in a strong magnetic field, ~ 0.47 T (Kibble and Hunt 1979).

Other measurements, by essentially the same technique, were made subsequently by Capptuller (1964), Yagola *et al* (1966) and Kibble and Hunt (1976) at the NPL. The latter improved on the original measurements by nearly a factor of ten in precision. They made use of the very precisely constructed rectangular coils which were made by Sim and Taylor (1968) at NML in Australia—a good example of the strong international cooperation which exists between the metrological laboratories. The coil was also rectangular in shape and comprised three turns of a gold-coated silver strip conductor which were fixed to a Pyrex former by a thin layer of an epoxy resin. The gold was only a few microns thick and it was considered that the current distribution in the strips would not be disturbed either by diffusion of the gold into the silver or by mechanical strain due to the bonding forces. The rectangular former was 800 mm long and 190 mm wide and was tapered in the horizontal plane from a 9 mm width at the edges to a 3 mm central web, thereby reducing the overall mass to about 1.5 kg. The turns could be energised separately or connected in series, which together with the use of various energising currents enabled the force on the balance to be varied. Both contacting and non-contacting methods were used to measure the separation a. Particular attention was paid (i) to reducing the field at the top of the coil, (ii) to measuring the field distribution in the magnet, (iii) to ensuring that the magnetic flux was truly horizontal, and (iv) that the coil was perpendicular to the magnetic field. Kibble and Hunt used a permanent magnet which produced a steady flux density of 0.47 T in a 5 cm pole gap. In order to measure the NMR frequency, a marginal oscillator was used together with a pure water cylindrical sample. The homogeneity of their magnetic flux was such that the resonance was scanned in a time which was comparable with the relaxation time, and so 'wiggles' were observed (see figure 4.7). These caused a distortion of the resonance, with a corresponding shift in the proton resonance frequency. Such effects are well understood and readily corrected. Measurements have also been described by Chiao *et al* (1980) and by Forkert and Schlesok (1980), who used broadly similar techniques.

The magnetic flux densities in all of the methods have been similar, so that the NMR frequency has been around 20 MHz. The force on the coils has been such that the mass required on the scale pans to counterbalance the force change when the current direction is reversed has been between 10 and 50 g. It is apparent that the achievement of part per million precision of the force, in the presence of the 1.5 kg mass of the coil former, imposes formidable requirements on the balance, for it must be used to weigh to parts in 10^9 of the mass on the fulcrum. This is comparable to the precision achieved in the balances used to intercompare standard kilogram masses, where there are fewer problems with convection currents or buoyancy corrections.

Although a permanent magnet produces a more stable field than an electromagnet, the improvement is rather less than one might at first expect. Thus electric currents may be stabilised to rather better than a part in 10^5, so

Table 4.2(a) Low field γ'_p measurements since 1970. (As far as possible all values have been expressed in terms of the BIPM 1976 maintained unit of current.)

Date	Authors and laboratory	γ'_p (low) $(10^8\,\mathrm{s}^{-1}\,\mathrm{T}^{-1}_{\mathrm{BI}-76})$		Uncertainty (ppm)
1972	Olsen and Driscoll (NBS)	2.675 138 4	(54)	2.0
1975	Olsen and Williams (NBS)	2.675 135 4	(11)	0.4
1978	Vigoureux and Dupuy (NPL)	2.675 117 8	(13)	0.5
1979	Williams and Olsen (NBS)	2.675 136 25	(57)	0.2
1980	Chiao and Shen (NIM)	2.675 139 1	(21)	0.8
1980	Forkert and Schlesok (ASMW)	2.675 155	(13)	5.0
1981	Tarbeyev (VNIIM)	2.675 122 8	(16)	0.6

Table 4.2(b) High field γ'_p measurements.

Date	Authors and laboratory	γ'_p (high) $(10^8\,\mathrm{s}^{-1}\,\mathrm{T}^{-1}_{\mathrm{BI}-76})$		Uncertainty (ppm)
1950	Thomas et al (NBS)	2.675 256	(26)	9.7
1964	Capptuller (PTB)	2.676 275	(100)	37
1966	Yagola et al (NVIIM)	2.675 130	(20)	7.4
1979	Kibble and Hunt (NPL)	2.675 168 9	(27)	1.0
1980	Chiao et al (NIM)	2.675 157 2	(95)	3.5
1980	Forkert and Schlesok (ASMW)	2.675 132	(41)	16.0

that with additional NMR stabilisation, electromagnets may produce flux densities which are stable to considerably better than a part in a million. On the other hand, with a permanent magnet, changes in the permeability of the iron and of the gap as a result of changes in the temperature cause the magnetic flux to change, despite the advantage of the large thermal capacity. Thus, very good temperature control is required for sub-ppm use. In addition, the movement of apparatus which contains iron causes changes in the magnetic path with consequent changes in the flux density. Smaller effects are observed with superconducting solenoids which are used in the persistent mode.

4.6.6 The weak field measurements of the gyromagnetic ratio of the proton

The earliest measurements of γ'_p were made before the signals had been detected in weak magnetic fields, and so the first weak field measurements were not made until some seven years after the initial high field measurement by Thomas et al (1950) at the NBS. The first weak field measurement was by Kirchner and Wilhelmy (1957) at Cologne university and this was followed by the measurements of Driscoll and Bender (1958) at the NBS. Wilhelmy placed the proton resonance sample in a solenoid which was wound on a 1 m long former having a diameter of 60 mm and produced a flux density of about 0.01 T. The coil former was made from brass and was water cooled. It is thought that there was either magnetic contamination of the apparatus or problems with values assigned to the electrical standards, which may account for the discrepancy between their value and the other values.

Driscoll and Bender adopted an ingenious way of enhancing the proton signal by pre-polarising the protons in a much higher flux density of about 0.5 T, thereby enhancing the Boltzmann factor, $\exp(2\mu_p B/kT)$, by many times the value in the weak field of the solenoid. The polarising field was some 12 m away from the solenoid and the water or benzene proton sample was transferred rapidly along a pneumatic tube to the centre of the solenoid. The direction of the polarisation followed that of the net field, and so a short pulse at roughly the precession frequency was applied in a direction perpendicular to the solenoid field in order to produce a $\pi/2$ change of the polarisation direction. The protons were then left to precess freely, thereby inducing a synchronous signal in a detector coil which was wound around the sample, and the frequency of this signal, ~ 52.5 kHz, was then measured. With water the signal lasted for about three seconds and with benzene for about fifteen seconds. It was necessary to locate the experiment at an isolated hut at the Fredericksburg Magnetic Observatory. This site was relatively free from man-made disturbances, but even so, diurnal fluctuations in the Earth's magnetic field were troublesome at certain times of day. The components of the Earth's field perpendicular to the solenoid field were largely cancelled by means of large Helmholtz coils wound around the apparatus (figure 4.10). The residual component parallel to the axis of the solenoid was eliminated by reversing the direction of the current through the solenoid and taking the mean of the precession frequencies for the two current directions.

Yagola and Zingerman (1966, 1967) similarly used a pneumatic tube in their measurements near Leningrad. Vigoureux (1965) used an *in situ* method by winding the polarising coil around the sample using carefully selected copper wire to avoid magnetic effects. He initially used a Helmholtz coil from the Smith (1914) Lorentz apparatus which was used for measuring the ohm absolutely, but for his later work he used a long solenoid which was mounted vertically. The fluctuations in the Earth's magnetic field are smaller in a vertical direction at remote localities. These variations were also servoed out in his experiment by sensing the Earth's field a short distance from the solenoid with a fluxgate magnetometer.

The Leningrad group have reported a measurement using a polarising coil (Tarbeyev 1981) and the group at IMM, Peking, have reported a value (Chiao *et al* 1980) using a vertical Helmholtz coil system.

The above measurements were essentially pulsed, but the most precise measurements to date are those of Olsen and Williams (1979). They used a cw

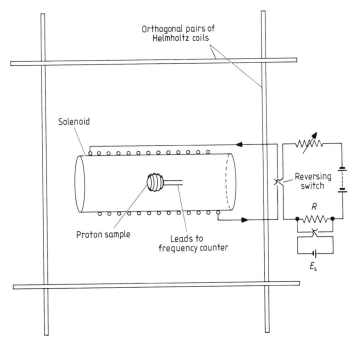

Figure 4.10 The air-cored solenoid method used to determine the gyromagnetic ratio of the proton in a weak magnetic field.

method utilising nuclear induction by means of two orthogonally wound coils. These were wound on an accurately made hollow quartz sphere which contained their pure water sample. Their solenoid was 1 m long and 0.28 mm in diameter and the 1000 turns of copper wire were wound on a helical thread of 1 mm pitch. This thread was accurately machined on a fused silica former and the precision was such that no turn deviated in position by more than 2 μm from that expected for a perfect solenoid. The solenoid was, of course, of finite length and hence the magnetic field depended on the diameter as well as on the pitch of the helix. However, in their determination this dependence on diameter was considerably reduced by the use of compensation windings (by a factor of about 8.6), albeit at the expense of reducing the flux density from the usual 1.2 mT to 0.8 mT. Their method of compensation was to apply the main current, together with four other currents, into selected turns. These reduced the second-, fourth- and sixth-order gradients. Five separate one ampere sources were required to energise these windings. This method resulted in a flux density which was uniform to 0.1 ppm over the water sample.

A new non-contacting technique was devised on order to measure the pitch of the coil, whereby a current was injected into selected turns. This current produced a magnetic flux which could be sensed by sets of coils mounted on the moving carriage of a laser interferometer along the axis of the solenoid. Two of the coils sensed the axial position of the injected current and a further three acted as a diameter to magnetic field transducer. The position of the carriage which carried the contacts for injecting current into the selected turns was changed as the interferometer carriage was moved. In this way both the pitch and diameter variations of the solenoid could be accurately measured. The tube containing the interferometer could be evacuated, thereby avoiding the correction for the refractive index of air. A further feature of their determination was the use of an inert fluorocarbon oil to cool the solenoid, which gave a greater dimensional stability. The temperature gradients along the solenoid were particularly reduced by this technique.

Although the sensitivity to the diameter of the solenoid was greatly reduced, some of the other uncertainties were increased considerably. Thus, the uncertainty in the pitch variation was trebled and the effects of the return leads doubled over the corresponding uncertainties when the solenoid was used in the normal way. The overall uncertainty was 0.29 ppm for the conventional method, and 0.23 ppm for the multi-current method. A new determination is in progress using much longer coils, which will include a large solenoid into which the main solenoid may be inserted while an NMR signal is observed, thereby allowing the effect of the susceptibility of the coil former and winding to be determined experimentally.

Particular attention was paid to ensuring that the nuclear induction signal was symmetric and an example of their line shapes and a pulsed signal observed by Petley is shown in figure 4.7. As will become apparent in chapter 5, the advent of the sub-ppm determinations of low field γ_p' has for the present considerably reduced the impact of the direct determinations of the fine structure constant. The reason for this is that the low field value may be combined with other more precisely measured constants in order to give a value for the fine structure constant. Thus:

$$\alpha^{-2} = \frac{c}{4R_\infty} \frac{1}{(\Omega_{\mathrm{NBS}}/\Omega)} \frac{\mu_p'}{\mu_B} \frac{(2e/h)_{\mathrm{BI}}}{\gamma_p'(\mathrm{low})_{\mathrm{BI}}}.$$

It will be noted that the ampere conversion constant K does not enter into this equation. The reason for this is that it is involved equally in the denominator (with γ_p') and numerator (with the CCE-72 value of $2e/h$). The involvement with other constants will be discussed in chapter 6, which considers the experiments having a bearing on the fine structure constant.

4.7 The faraday

The faraday, next to the velocity of light and the universal gravitational constant, is one of the oldest of the fundamental physical constants. The broad phenomena of electrolysis were known even before Michael Faraday began the researches which led to the enunciation of the laws of electrolysis which bear his name. The two laws are (i) that the amount of chemical decomposition produced by an electric current (that is the mass of substance deposited or dissolved at an electrode) is proportional to the quantity of electricity passed, and (ii) the amounts of different substances released or dissolved at electrodes by the same quantity of electricity are proportional to their relative atomic masses divided by the valencies of their ions. It was not until after Faraday's death that the significance of his laws for atomic theory was realised, for in 1881 von Helmholtz pointed out that if elementary substances were composed of atoms it followed from Faraday's laws that electricity would be composed of elementary portions which behave like atoms of electricity. This observation led to the study of the conduction of electricity through gases and then to the discovery of the electron at the end of the century.

The study of the silver coulometer by Rayleigh and Sidgwick (1884) and Kohlrausch, and also by many other scientists, showed that it possessed a high degree of reproducibility when used under carefully controlled conditions. Washburn *et al* (1914) at the NBS studied the iodine coulometer as well. Faraday's appreciation that electrolysis could be used to measure electric current reached fruition with the adoption of the silver coulometer as the official primary standard of current at the Chicago meeting of the International Electrical Congress in 1893. Their decision was endorsed at the London meeting of the International conference in 1908. The international ampere was defined as the steady current causing silver to be deposited at the cathode of a silver coulometer at the rate of $1.118\,00\ \mathrm{mg\,s^{-1}}$, thereby fixing the electrochemical equivalent of silver. (Note that the use of international electrical units was finally abandoned, by international agreement, in 1948.)

This coulometer realisation of the unit of current enabled a known voltage to be produced by passing the current through a standard resistor, which voltage could then be used to establish the EMF of a Weston-cadmium standard cell. Such cells were used to maintain the volt on a day-to-day basis and indeed they came to be used so extensively that the coulometer definition effectively fell into disuse.

It is apparent from the above that the faraday was historically the first of the atomic metrological standards. The fact that it was abandoned after some years ought to temper our enthusiasm for the use of the velocity of light to define the metre, or the Josephson effect value of $2e/h$ to maintain the volt or the quantised Hall effect for the ohm, for, surprising as it may seem at present, these too may be abandoned at some future date after they have played their part in the evolution of the metrological arts.

The faraday has also played a varying role in the evaluation of the fundamental physical constants. Thus it was used by R T Birge in his classic 1929 review of the 'best values' of the fundamental physical constants. In that review he used the Faraday as an auxiliary constant which he combined with the Millikan oil-drop value of the electronic charge in order to arrive at a value for the Avogadro constant (note that until the advent of the mole the dimensionless Avogadro number was used rather than the Avogadro constant which has dimensions mol^{-1}). In later evaluations Birge combined the faraday with an x-ray determination of the Avogadro constant in order to derive a value for the electronic charge. This value differed by some 0.6% from that derived from the oil-drop method and helped to demonstrate the existence of a systematic error in the oil-drop method which was later shown to be due to the use of an incorrect value for the viscosity of air.

The measurement of the faraday constant, while intrinsically simple in concept, poses formidable experimental problems. Although silver was the favoured material to use in a coulometer, especially at the turn of the century, other materials have been investigated. One problem is that silver is not isotopically pure and the relative isotopic abundances must be carefully measured. For this reason, iodine, which has only one natural isotope, has some very attractive features. Oxalic acid and amino-pyridine have also been used. The results for the electrical–chemical equivalent of silver and the values obtained for the faraday are given in tables 4.3 and 4.4, together with the 'best' value obtained by Cohen and Taylor (1973).

Perhaps one impressive conclusion from these values is how little they have

Table 4.3 Values obtained for the electrochemical equivalent (ECE) of silver.

Date	Author(s)	ECE of silver $(mg\,C^{-1})$	
1884	Rayleigh and Sidgwick	1.117 94	
1884	Kohlrausch	1.118 3	
1884	Mascart	1.115 6	
1890	Pellat and Potier	1.119 2	
1893	Chicago Congress	1.118 00	
1898	Kahle	1.118 3	
1898	Patterson and Guthe	1.119 2	
1903	Pellat and Leduc	1.119 5	
1904	Van Dijk and Kunst	1.118 2	
1906	Guthe	1.118 2	
1907	Smith *et al*	1.118 27	
1908	London Conference	1.118 00	
1912	Rosa *et al*	1.118 04	
1929	Birge (adjusted)	1.118 05	(7)
1956–1960	Craig *et al*	1.117 971	(11)
1969	Taylor *et al*	1.117 972 2	(72)
1980	Bower and Davis	1.117 963 9	(9.5)

Lord Rayleigh and Mrs Sidgwick would surely have been pleased by the result obtained by Bower and Davis about a hundred years after their own measurement!

Table 4.4 Some recent measurements of the faraday constant of electrolysis.

Date	Authors	Method	Faraday $(C_{BI69} \text{ mol}^{-1})$		Uncertainty (ppm)
1960	Craig *et al*	Silver	96 486.72	(66)	6.8
1968	Marinenko and Taylor	Benzoic acid	96 487.30	(112)	12
1968	Marinenko and Taylor	Oxalic acid	96 486.25	(157)	16
1971	Bower	Iodine	96 485.36	(148)	15
1979	Koch	Amino-pyridine	96 484.41	(100)	11
1979	Davis	Silver	96 486.20	(20)	2
1973	Cohen and Taylor	Review value	96 484.63		2.8

changed over the years. This suggests that the claimed reproducibilities of around 20 ppm, achieved in the first decades, were remarkably accurate. Present day metrologists have the advantage that the purity of the chemicals is much higher and they also have a better knowledge of the isotopic abundances. However, the mass of material deposited or liberated is still around ten grams and the current densities that can be employed are much the same. Consequently, as so often happens in metrology, progress over the years has been represented by steady improvements in technique rather than by a spectacular breakthrough leading to a rapid advance in precision.

Weighing with the required accuracy also imposes formidable problems. First, the chemical balance is being pushed close to the limits with which a mass of a few grams may be measured in terms of the kilogram, and second because the chemical materials concerned must be completely dry.

4.7.1 Silver

As was indicated above, the mass of silver eroded from a silver anode, following the passage of a known quantity of electricity, is the quantity which must be determined. The electrolyte is perchloric acid and, because the silver tends to dissolve slightly in it, some silver perchlorate is added to the electrolyte. Since the amount of silver dissolved is only a few grams, the silver must be dried very carefully before the mass loss can be measured. This is complicated by the formation of a sediment and this is still as much of a problem in today's measurements as it was to Rayleigh and Sidgwick a century ago. Thus the amount of sediment which falls away can be some 1000 ppm. Bower and Davis have reported that making the silver into a single crystal by careful vacuum annealing reduces the amount of sediment, but the mechanism of sediment formation is still not very well understood. Some of the sediment may contain impurities, but the bulk is silver, and it is probably safe to assume that this has the same relative isotopic abundance of the silver isotopes as the anode. An alternative method is to measure the amount of silver deposited at the cathode, but the difficulty here is that some of the electrolyte may be buried in the cathode.

4.7.2 Iodine

For this method, the two electrodes are made from an inert material such as platinum. The coulometer comprises two electrode compartments which are joined by a tubular bridge which is filled with electrolyte. The coulometer contains a solution of potassium iodide. The electrolytic reaction is

$$I_3^- + 2e \rightleftharpoons 3I_3^- \qquad (4.7.1)$$

which proceeds in the left to right direction at the cathode and in the opposite direction at the anode. The tri-iodide ion is formed by combination of iodine with the mon-iodide ion as

$$I_1 + I^- \rightarrow I_3^- \qquad (4.7.2)$$

and consequently the molecular iodine which would form at the anode remains in solution as the tri-iodide ion. The reaction therefore involves the transfer of iodine from the cathode to the anode compartment. When the electrolysis is finished, the two compartments are separated and the amount of iodine transferred is measured by chemical analysis. A solution of pure arsenious acid is used and it is standardised by titration against weighed quantities of pure iodine. The reaction which occurs between iodine and arsenous acid is:

$$H_3AsO + I_3^- + H_2O \rightarrow H_3AsO_4 + 2H^+ + 3I^-. \qquad (4.7.3)$$

As a check on the method, the iodine lost from the cathode compartment should equal that gained by the anode compartment. A major difficulty with the method has proved to be that of ensuring that the iodine in the standard titration is free of any water of crystallisation and this has prevented the potential major advantage of using an isotopically pure material from being realised.

4.7.3 4-aminopyridine

The silver and iodine methods have existed in various forms since the turn of the century, but the 4-aminopyridine method has been a comparatively recent development. The material may be purified by sublimation, which removes occluded solvents (as also can iodine). There are more isotopes involved than is the case with iodine, although nowadays the relative isotopic abundances may be measured with the required precision. Koch and Diehl (1976) used a molar solution of sodium perchlorate as the electrolyte. Nitrogen was bubbled through their electrolyte in order to remove carbon dioxide, and a standard (commercially available) coulometer was employed. It was intended to pass a current for a sufficient time for a neutralisation titration to occur with the hydrogen ions evolved at the anode. However, a side reaction, amounting to about 4% of the main reaction, was identified at the anode which prevented this method being used. Instead, new techniques were developed using a 'hydrazine–platinum anode' and a further method which involved the addition of an excess of perchloric acid: both methods circumvented the side reaction at the anode. Measurements were made at both the anode and cathode. The method used at the cathode was similar to that in the anode compartment. Perchloric acid was added from a burette into the cathode compartment until the base generated by the electrolysis was neutralised. In both compartments the neutralisations could be monitored by measuring to the ± 0.02 pH uncertainty required for a ppm faraday determination. A weighed sample of 4-aminopyridine, which was contained in a platinum boat suspended from a platinum wire, was lowered into the electrolyte. One of the difficulties in this method is that of determining the molecular weight with the required accuracy, and this limits the precision at present.

4.8 Measurement of e/m_e

The measurements of the charge to mass ratio of the electron followed the initial experiments by J J Thomson (1897), demonstrating the existence of the electron. Throughout the 1930s it became apparent that there were discrepancies between the two methods of measuring e/m for the electron: the spectroscopic method and the deflection methods, the two methods being distinguished by whether the electrons were bound to an atom or were free electrons, respectively. The measurements on the free electron used either combinations of electric and magnetic fields (Thomson (1897, 1903), Kaufman (1897, 1898), Bucherer (1908, 1909), Classen (1908 a, b), Busch (1922), Wolf (1927), Kirchner (1929, 1932), Dunnington (1933, 1937) and Shaw (1934, 1937, 1938)), or alternatively the time of flight of electrons of known energy was measured over known distances (Weichert (1899), Hammer (1914), Kirchner (1924) and Perry and Chaffee (1930)).

It is now thought that all of these experiments were affected to a greater or lesser extent by a fundamental limitation of the experimental technique. This was a result of the difficulty of ensuring that the metal surfaces inside the vacuum system formed equipotential surfaces whose contact potentials remained unaffected by contamination. Oil films are a particular source of difficulty, since they are used in both roughing and diffusion pumps, and the air in most laboratories probably contains oil vapours from one source or another—machine shops are a particular source of contamination. As a result of these films, even with the best vacuum techniques used today, it is very difficult to define the potential differences inside a vacuum system to much better than a few tenths of a volt (a later development has been to make the surfaces infinitely dirty by covering everything with a layer of colloidal graphite!). The existence of these contaminating films was first recognised by Shaw and he relied on focusing of the electrons in crossed electric and magnetic fields in order to obtain a result which was independent of the electron velocity. Dunnington also tried to eliminate the effects of unknown electric fields by varying the energy of his electron beam and extrapolating to infinite energy.

The spectroscopic methods were of two kinds. In the first, the Zeeman splitting of the spectral lines emitted by a source operated in a strong magnetic field was used to obtain e/m_e. Thus the energy levels of an atom in a magnetic flux density B are displaced by an amount $MgB\mu_B$ where M is the magnetic quantum number, μ_B the Bohr magneton ($eh/4\pi m_e$) and g is the Lande g-factor. Thus the shift in frequency is given by

$$\delta\omega_c = MBe/(4\pi mc).$$

where $g = 1$ for the normal Zeeman effect. Insler and Houston (1936) obtained a precision of about 400 ppm by this method. In earlier spectroscopy it was conventional to measure frequency shifts in terms of wavenumber $\bar{\nu}$, since the speed of light was not known with adequate precision. As we saw in chapter 3, this distinction is no longer strictly necessary, although spectroscopists are likely to continue to express their results in terms of wavenumbers, rather than Hz, for some time to come.

The alternative spectroscopic method strictly involved the electrochemical measurement of the faraday as well, for it involved combining a measurement of the faraday with a measurement of $N_A m_e$ and also used the values of the Rydberg constant for hydrogen and deuterium in order to obtain m_p/m_e. This could then be combined with a measurement of the mass of the proton in terms of the atomic mass unit (which at that time was based on the ^{16}O scale) and hence, remembering that the reciprocal of the Avogadro constant is essentially the atomic mass unit, it is possible to combine all of the measurements to obtain a value for e/m_e for the electron. It should be noted that the exact relationship between the Avogadro constant and the atomic mass unit depends on whether one is referring to the value before the $^{12}C = 12$ scale and the mole were introduced—of course the answer for e/m_e is the same.

The above methods are no longer used, although the principles involved are still used today in different types of mass spectrometer. Following the 1939–45 war, the techniques of radio-frequency and microwave spectroscopy led to a period of rapid change which affected particularly the experiments having a bearing on the charge to mass ratio of the electron and proton. The particular change came as a result of the development of the techniques of nuclear magnetic resonance which allowed the magnetic fields in the e/m_e experiments to be measured in terms of the proton spin precession frequency. This in turn led to the experiments being regarded as measuring a different combination of the fundamental constants, so that instead of measuring e/m_e for the electron or e/m_p for the proton, the corresponding measurements today would be of the proton magnetic moment μ'_p (for protons in a pure water sample) in terms of the Bohr and nuclear magnetons, i.e. μ'_p/μ_B and μ'_p/μ_N which are discussed in the next chapter.

4.9 Direct measurements of the ampere conversion constant K

It is apparent that the imprecision of the absolute determinations of the ampere has become a limiting factor in the evaluation of the 'best values' of the fundamental constants and considerable thought has been devoted to finding ways in which modern measurement techniques might be applied to improve the precision. Two lines of attack have resulted, one at room temperature and the other at cryogenic temperatures.

The room temperature measurements follow a suggestion by Kibble (1976). If a coil carrying a current i is suspended in a magnetic flux Φ, the force in a vertical direction z is given by

$$F = -i(\partial \Phi_z/\partial z) \tag{4.9.1}$$

and if the coil is suspended from a balance and the force is balanced by a mass M, then

$$Mg = -i(\partial \Phi_z/\partial z). \tag{4.9.2}$$

If the coil is now moved vertically there will be an induced EMF V' given by

$$V' = -\partial \Phi/\partial t = -i(\partial \Phi/\partial z)(dz/dt). \tag{4.9.3}$$

If the experiment is designed such that $\partial \Phi/\partial z$ is the same in the dynamic and static case (we may suppose that any correction may be made to the desired precision) then

$$Mg(dz/dt) = iV'. \tag{4.9.4}$$

If the current i is measured by passing it through a resistor R giving a voltage V, then we may convert all of the measured electrical quantities to maintained units using

$$V V'/R = K^2(\Omega/\Omega_{SI})(V V'/R)_{\text{maintained}} \tag{4.9.5}$$

and hence

$$K^2 = \frac{Mg(R/V V')_{\text{maintained}}}{(\Omega/\Omega_{SI})(dz/dt)}. \tag{4.9.6}$$

The parameters which must be measured experimentally are therefore the two voltages, the mass on the scale pan and the uniform velocity dz/dt.

In the NPL apparatus (Kibble 1982) an astatically wound coil of 3362 turns is suspended in a flux density of about 0.7 T and, when servoed to fall vertically with a velocity of about 2 mm s^{-1}, gives an induced EMF of about 1 V. The velocity of the coil system is measured by attaching a cube corner to the coil and this cube is part of a laser fringe-counting interferometer system. The coil velocity is uniform to about one part in a hundred thousand over a distance of about 4 cm, the total displacement being about 5 cm. The current in the coil to produce a force change requiring the addition or subtraction of a mass of 1 kg to the balance pan is only 10 mA, so that the heating effects are quite small.

A second approach has been tried at the NBS using a modified Ayrton–Jones arrangement of two opposed coils to produce a uniform rate of change of the mutual inductance dM/dz. Some preliminary work at room temperature using a modified Pellat balance has also been reported by Olsen *et al* (1981). In their apparatus, the rotation rate of the coil has been made uniform over a small displacement and the angular velocity measured by means of a cube-corner reflector attached to the balance pan. This method allows the distance involved in the torque measurement to be eliminated. This system has been built to enable the appropriate measurement techniques to be developed. It is expected that the final NBS system will embody cryogenic fixed coils and a room temperature moving coil. The superconducting coils will have a much greater number of turns than in the fixed coil of a conventional current balance since there is no need to measure the dimensions of the coils. The vertical force on the suspended coil will therefore be at the kilogram level and, to first order, will be independent of the coil diameter for a displacement of about 6 cm on either side of the mid-plane of the coils. The motion of the coil will be rendered exactly vertical by appropriately servoing the horizontal position of the fulcrum of the balance.

The other cryogenic approach follows a suggestion by Sullivan and Frederick (1978). A superconductor behaves as a perfect diamagnet and hence may be levitated by a static magnetic field in a stable manner. The levitated mass is typically about 1 kg and conical in shape. The method is to levitate the mass by passing a current through a coil and to change the current by a known amount thereby increasing the field and displacing the levitated mass vertically through a distance which may be measured with a laser interferometer.

Prototype systems have been described by Hara *et al* (1981) (figure 4.11) and Tarbeyev (1981). The change in current is to be monitored by balancing the EMF across the coil against the voltage across a Josephson junction or alternatively by passing the current through a SQUID of known sensitivity.

The energy conservation relation is

$$E(z) = E_m(z) + mgz + m\dot{z}^2/2 = \text{constant}$$

where $E_m(z)$ is the magnetic energy of the system and m is the mass of the cone.

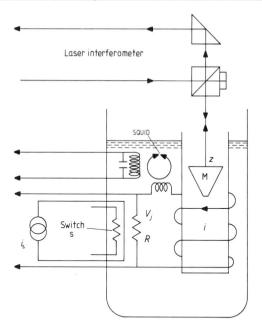

Figure 4.11 A prototype levitated superconducting cone method of realising the ampere by Hara *et al* at ETL, Japan.

When the switch S is opened for a time interval 0 to T the flux change is

$$\Phi = \int_0^T V_j \, dt.$$

Hence if the current in the changing and persistent modes is monitored with a SQUID and the count rate is an integer n (the separately measured SQUID calibration constant being K) the current is given by

$$i = Kn\Phi_0.$$

where $\Phi_0 = h/2e$. We have

$$\Phi = \int_0^T V_j(t) \, dt = N\phi_0$$

where an exact number N of flux quanta link the area of the levitating coil. The coil self inductance L is then

$$L = \Phi/i = NK/n$$

and the magnetic energy

$$E_m = \Phi i/2 = KnN\Phi_0^2/2$$

In the prototype system described by Hara *et al* (1981) the mass of the cone was 25.24 g and was made of aluminium coated with lead. The persistent current was ~ 1A and the coil comprised 18 layers of 720 turns per layer, $\frac{1}{4}$ mm diameter niobium wire giving a self inductance, without the cone, of 48.6 mH. The vertical vibration of the cone was about 1 μm or ~ 100 ppm. Kibble (1982) has suggested that the method could be modified to give the cryogenic equivalent of his room temperature method, thereby avoiding the need to perform the integration.

4.10 The realisation of the ohm

For many years the SI unit of resistance, the ohm, was derived electromagnetically by using the method of Campbell (1907). This method took the form of a precisely calculable mutual inductance and it was necessary to produce a very uniformly wound helical solenoid for the primary and a slightly less precisely wound secondary winding mounted in the position where small displacements lead to negligible change in the mutual inductance. This method was used by Rayner (1967) with an estimated uncertainty of a part in a million which represented the limit of the precision which could be achieved by this method. Fortunately however, an alternative method derived from electrostatics was discovered by Thompson and Lampard (1956) at the National Measurement Laboratory in Australia. This method relied on their new theorem in electrostatics concerning the cross capacitance between right cylindrical conductors which were of arbitrary shape in two dimensions but were infinitely long in the other. This cross capacitance, for an arbitrary shape divided into four segments, was found by Lampard (1957) and van der Pauw (1958) to satisfy the relationship

$$\exp(C_1 \pi/\varepsilon) + \exp(-C_2 \pi/\varepsilon) = 1 \qquad (4.10.1)$$

where C_1 and C_2 were the two pairs of cross capacitances per unit length, (figure 4.12(*a*)) and ε was the permittivity of the medium. The theory was further extended by Lampard and Cutkosky (1960). If the segments were chosen to make C_1 and C_2 approximately equal, then the capacitance per unit length became

$$C = (\varepsilon/\pi) \ln 2 \text{ F m}^{-1}$$

or

$$C = (10^7/4\pi c^2) \ln 2 \text{ F m}^{-1}$$

in vacuum where c is the speed of light in vacuum. Inserting the CCDM value for c this becomes 1.953 594 pF m^{-1} in vacuum and is independent of the shape of the segments. In practice, of course, C_1 and C_2 are not exactly equal,

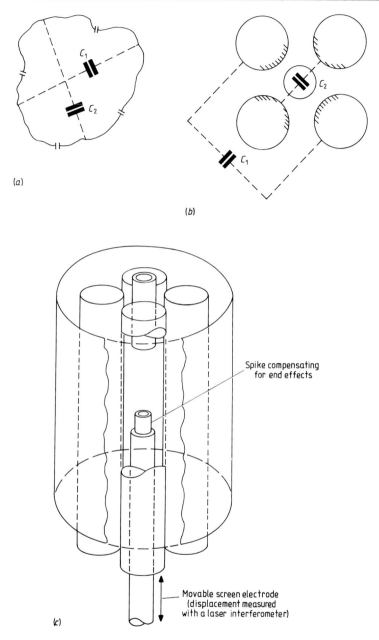

(a)

(b)

(c)

Spike compensating
for end effects

Movable screen electrode
(displacement measured
with a laser interferometer)

Figure 4.12 The calculable capacitor method of realising the absolute ohm: (a) the theorem; which applies to the cross capacitance between opposite segments of an infinite cylinder of arbitrary cross section; (b) the more convenient cylindrical electrode system; and (c) the experimental realisation (see Petley (1980), for example, for a description of the method).

but if we write

$$C_m = (C_1 + C_2)/2 \qquad \text{and} \qquad \alpha = (C_1 - C_2)/(C_1 + C_2)$$

then

$$C_m = C[1 + (\alpha^2/2)\ln 2 - \alpha^4(\ln 2)^3/48 + \alpha^6(\ln 2)^5/45 + ...]$$

which shows that the error made by taking C_m for C is

$$(C_m - C)/C < (C_1 - C_2)^2/11C^2 \qquad (\lesssim 10^9 \text{ for } C_1 - C_2 \lesssim 10^{-4}C).$$

It is apparent, therefore, that the requirement of equality for the cross capacitances is not unduly stringent. A useful form of equation (4.10.1) for evaluating corrections $\delta C \ll C_2$ is

$$\delta C = (\varepsilon_0/\pi)\exp(-\pi C/\varepsilon_0)$$

so that if C_2 is the capacitance between adjacent electrodes, then δC, the leakage capacitance between the electrodes and the screen, is 10^{-20} F m^{-1}, or less than a part in 10^8 of the capacitance.

The simplest experimental configuration which may be used to realise the theorem comprises an array of four parallel cylinders, and the capacitance between them may be changed by inserting a fifth cylinder inbetween them as illustrated in figure 4.12(b). The theorem applies to infinitely long parallel cylinders and so there are some end effects from the finite length which is used in practice. There are small additional corrections if the cylinders are not parallel, and these corrections are position dependent. However, it was shown by Thompson and Lampard that if the movable guard electrode had a spike on the end it could be so dimensioned as to compensate for these effects. Further corrections are required if the four cylinders do not lie on a square and these are of the form

$$\delta C/C = \beta \delta d/d$$

where d is the displacement from the square position and the value of β varies between 3.4 and 4.6, depending on the direction of the displacement. The capacitors at the NPL, NML and NBS all operate with a capacitance change of between 0.1 and 0.7 pF. It is a tribute to the accuracy of modern AC bridge techniques that this small capacitance may be measured to approaching a part in 10^8. The capacitance is first scaled up to a 10 pF standard capacitor by means of a transformer bridge operated at a frequency of 10^4 rad s^{-1} and built up to the 500 or 1000 pF required for a transfer to a 10 kΩ resistor using a quad bridge which was devised independently by Cutkosky at the NBS and Thompson at the NML. The subsequent transfer to the standard 1 Ω resistor may be effected either at very low frequencies or DC.

Despite the length of the comparison chain the overall uncertainty is today in the region of a few parts in 10^8. This precision makes it necessary to know the velocity of light with high precision for the uncertainty associated with the

Froome result would have dominated the other uncertainties. It is possible to devise a five-element standard capacitor (Elnekave 1965). Although this design has not been generally used and has not yet achieved the highest precision, it does have the advantage that it has five sets of capacitance relationships which are only identical if the capacitor is properly set up.

The natural standard of impedance, from the Maxwell electromagnetic theory, is the impedance of free space, which is simply $\mu_0 c$ Ω (or $(\mu_0/\varepsilon_0)^{1/2}$ Ω). It has not been possible to make a direct high precision realisation of this impedance, but of course the calculable capacitor is entirely compatible with this relationship. The quantised Hall effect which we have discussed earlier in this chapter is a more recent natural impedance standard and the ratio $\bar{R}$ of the maintained to SI ohms is now evaluated along with K and the fundamental constants.

4.11 The absolute realisation of the volt

It is quite possible to realise the volt absolutely. All that is required is to make use of an expression, derived from electromagnetic theory, which is consistent with the SI definition of the ampere. One usable expression results from the energy W stored in a capacitor, which is $\frac{1}{2}CV^2$. If this capacitor is of the parallel plate type and the plates are in the horizontal (x, y) plane, then the force in a direction perpendicular to the plates is

$$F = \mathrm{d}W/\mathrm{d}z = \tfrac{1}{2}V^2\,\mathrm{d}C/\mathrm{d}z.$$

This is the principle on which the attracted disc electrometer was based, which was first described by Sir William Snow Harris in 1834, and was greatly improved by Lord Kelvin in 1880 by the addition of a guard ring electrode. Several attempts were made to improve this design, including the work of Brooks et al (1939) at the NBS who used it to measure high voltages to $\sim 0.01\%$.

One of the difficulties is that for precision measurements the voltages must be quite high in order to achieve a sufficiently large force and, as with the ampere balance, one ends by having a large mass suspended from the beams of the balance. There are corona discharge problems as well, and difficulties with the effects of insulating films on the electrode surfaces. Clothier (1965) has been working at the National Measurement Laboratory on a design which will use mercury surfaces as the electrodes and Slogget et al (1981) described progress on the measurements. The vertical displacement of these will be measured by making the system into a laser interferometer. The system is illustrated in figure 4.13. The force equation is

$$\varepsilon_0 \varepsilon_r V^2/2(S - z)^2 = z\rho g$$

or

$$V^2 = k^2(S - z)^2 z$$

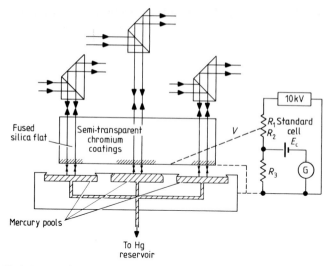

Figure 4.13 Scheme of the attracted mercury surface method of realising the volt absolutely devised by Clothier and others at the NML in Australia. The applied voltage V is expressed in terms of the standard cell EMF E_C by $V = E_C R_2/(R_2 + R_3)$.

where the gap for zero voltage is S, $k^2 = \rho g/\varepsilon_0 \varepsilon_r$, ρ is the density of mercury, g the acceleration due to gravity and ε_r is the relative permittivity of the medium. Under certain conditions the system may become unstable and this condition may be derived by differentiating the above equation

$$dV/dz = k^2(S - 3z)/2z^{1/2}$$

from which

$$dz/z = [2(S - z)/(S - 3z)] \, dV/V.$$

If the gap with the voltage applied is denoted by $l \, (= S - z)$ then the expression becomes

$$dz/z = [2l/(l - 2z)] \, dV/V.$$

From this, it is apparent that the system becomes unstable for $z \gtrsim l/2$. For large gaps the sensitivity is approximately $2 \, dV/V$. There is obviously some advantage in working as close as possible to the condition $z = l/2$, but great care must be taken to avoid vibration from external sources — mercury pools were often used as vibration detectors in the past. Taking an initial gap of 4 mm and a potential of 22 kV gives an elevation of 1 mm, which gives a sensitivity of 6 nm per ppm change in the applied voltage. Increasing the separation to 6 mm only halves the sensitivity, which means that the experiment may be performed for a number of different separations. The upper plate is made from fused silica and transparent conducting electrodes are evaporated into it, thereby allowing the separations to be measured by laser interferometric techniques. The plate must be flat to $\lambda/300$, the diameter of the mercury pool must exceed about 60 mm for surface tension effects at the centre to be negligible, and if the plate diameter is too large one must remember that the curvature of the Earth gives a change from flatness of 1 nm over a 22 cm long arc.

4.11.1 Attracted cylinder electrometer

Harris (1972) has been working with others at the NBS on a rival determination which involves the attraction between coaxial cylinders one partly inside the other (radii a and b respectively). The capacitance per unit length is given by

$$\mathrm{d}C/\mathrm{d}z = (4\pi\varepsilon_0/2)/\ln(b/a)\ \mathrm{F\,m^{-1}} = 5.56 \times 10^{-11}/\ln(b/a)\ \mathrm{F\,m^{-1}}.$$

The inner cylinder is suspended from the beam of a balance and a balancing mass of about 5 g is required to offset the force due to an applied potential of about 5 kV, the total load on the beam being about 1.5 kg. The target precision is about one ppm.

Elnekave (1965, 1980) has described a flat plate type of attracted disc electrometer which achieved a precision of about 7 ppm and Yamazaki *et al* (1972) have described a hybrid type in which two sets of four vertically suspended rectangular plates of 140 mm width are attracted into the gap between five plates of 220 mm width and 10 mm gap. The capacitance changes from 50 to 70 pF for an insertion of 20 mm and the force for an applied potential of 5.3 kV is equivalent to a mass of about 1.5 g.

As with the ampere, fundamentally new techniques are required if the absolute volt determinations are to achieve sub-ppm accuracy. However, given the realisations of the ampere and the ohm, there is no need to realise the absolute volt as well.

5

Dimensionless and calculable fundamental constants and tests of quantum electrodynamics

5.1 Introduction

In this chapter we discuss some of the dimensionless and calculable constants. The Sommerfeld fine structure constant is deeply involved throughout and consequently many of the experiments which are discussed have contributed towards the derivation of the 'best value' of the fine structure constant. In many respects the process has been and continues to be an oscillatory one, for we can either

    (i)   rely on the theory and use the experiments to derive a value for α, or

    (ii) assume a value for α, from other results, and use the experiments to test the theory.

For the most part, the calculations have a common theoretical framework lying in the subject known as quantum electrodynamics, which is discussed in the next section. Comparisons of the theory and experiment are therefore often referred to as 'tests of quantum electrodynamics'. These tests stretch both the theorist and experimentalist to the limit. Although this book concentrates for the most part on experiment, some of the theoretical expressions have been given so that one may glimpse both the skill and dedication of the theorist and also appreciate better the deep underlying purpose behind theory and experiment at the nth decimal place.

After a discussion of quantum electrodynamics, we move on through the spectroscopic measurement of the fine structure constant and the Lamb shifts, through to the hyperfine splittings in hydrogen, positronium and muonium and the g-2 determinations for the electron and muon. In order to further illustrate the unexpected spin-offs from such work we discuss how the g-2 experiments have been used to provide a very high precision test of special relativity.

5.2 Quantum electrodynamics (QED)

Originally, quantum mechanics was formulated non-relativistically and it was not until 1929 that the method of field quantisation incorporated relativity. Some twenty years later QED came into prominence with the failure of the old Dirac theory to accommodate satisfactorily the discoveries of both the Lamb shift and the anomalous magnetic moment of the electron. Quantum electrodynamics is the term applied to the quantised theory that describes the interactions between leptons (electrons, positrons and muons) and radiation which is based on the quantised form of the Maxwell equations, together with the Dirac electron theory. The QED theory has been characterised by its remarkably accurate predictions and successfully described the interactions of electrons and photons at the low-order perturbation theory. However, the higher-order correction terms were always divergent and this destroyed the good agreement between low-order theory and experiment. These problems arise as a result of the divergent integrals that appear when the theory is developed by perturbation techniques in terms of an expansion in the form of a power series involving the fine structure constant α. The physical source of these infinities may be pictured by recalling that in quantum field theory a vacuum is a complicated concept having fluctuating zero-point electric fields and electron pairs present. It is the interaction of the lepton with the vacuum fluctuations in the electromagnetic field that leads to the mass renormalisation. Similarly, the presence of the charge disturbs the electron pair distribution by repelling the like charge and attracting the electron of unlike charge. The resulting polarisation of the vacuum produces a change of the apparent charge.

The difficulties produced by the above infinities were overcome by the renormalisation techniques that were introduced by Bethe, who calculated the Lamb shift by including the divergence as part of the electron mass. Further renormalisation by Tomonaga, Schwinger and Feynman soon followed in 1948. The difficulties were avoided, but not removed, by the relativistically covariant development of their theory through the use of particular terms which were to be understood as electrodynamic contributions to the charge and mass of the particle and were invariant under a Lorentz transformation. The experimentally observed quantity is the sum of the original term and this extra one. Thus the self-charge is the extra contribution to the electric charge from the vacuum polarisation arising from the field produced by the original charge. The self-charge and original charge cannot be identified separately and the process of identifying their sum with the observed charge is termed renormalisation. A possible way in which these renormalisations might arise naturally would be if our ideas of space and time require modification for interactions at very short distances and times (high energies). In this case, the renormalisations or truncations of the integrals would arise as a cut-off at high energies that consequently made the integrals finite when applied to the lower energy situations that involve quantum electrodynamics.

Quantum electrodynamics is the most developed of all fields of elementary particle physics. It is far from being dormant and important progress continues to be made both experimentally and theoretically. On the experimental side there have recently been high precision measurements of:

the anomalous magnetic moment of the electron,
the anomalous magnetic moment of the muon,
the hyperfine structure of muonium,
the fine structure of positronium,
the Lamb shift in hydrogen,
the hyperfine structure in the ground state of hydrogen
the decay rate of orthopositronium,
high-energy colliding beam reactions of the type $e^+e^- \rightarrow e^+e^-, \mu^+\mu^-$.

The theoretical studies of QED can be classified into two groups. First, there is the study of the structure of QED from a general field theoretical viewpoint, such as large orders of perturbation theory, and second, work that is directed towards improving the precision of various predictions of QED. The latter include calculations of the anomalous magnetic moments of the electron and muon (which CPT conservation requires are charge independent), muonium and positronium hyperfine structure and the decay rate of orthopositronium. The latter three types of calculation require a deep understanding of relativistic bound state equations, such as the Bethe–Salpeter and Gross equations. It should not be thought that QED is a totally isolated discipline, and indeed tests of QED must be a part of the test of a unified field theory such as the SU(2) × U(1) model of Weinberg and Salam. However, at the present levels of precision, most electromagnetic phenomena are still insensitive to non-electromagnetic interactions.

The high-energy tests of QED involving colliding beam experiments are a fruitful source of investigation. This is a very active area of research which lies outside the scope of this book, but basically the results for such processes as $e^+e^- \rightarrow e^+e^-$ or $\mu^+\mu^-$ are expressed in terms of a parameter Λ which is a measure of possible deviation of the photon propagator k from the standard theory:

$$\left(\frac{1}{k}\right)^2 \rightarrow \left(\frac{1}{k}\right)^2 \left(1 \pm \frac{k^2}{k^2 - \Lambda_\pm^2}\right).$$

The discussion that follows is therefore restricted to the low-energy, high-precision tests of QED. These are tests of higher-order radiative corrections and/or theories of relativistic bound states.

QED is considered to be part of a broader theory in which the electromagnetic and weak interactions (possibly the strong interactions too) form part of a single unified interaction. There is therefore considerable interest in testing the predictions of theory against experiment each time that both advance in precision by another decimal place.

5.3 The spectroscopic measurements of the fine structure constant

The Lamb shifts, discussed in § 5.4, may be used to derive a value for the fine structure constant, but more accurate values may be derived from a measurement of the fine structure interval ΔE (figure 5.1). Here, the $n = 2$ state of hydrogen has proved a very fruitful source of experiments, although higher energy levels have also been explored (table 5.1). Hydrogen-like atoms such as deuterium and H_e^+ have also been studied.

Table 5.1(a) Measurements of ΔE and $\Delta E - \mathscr{L}$.

Atom	n	Interval	Experimental value (MHz)		Author(s)
H	2	ΔE	10 969.6	(7)	Wing (1968)
H	2	ΔE	10 969.13	(10)	Baird et al (1973)
H	2	$\Delta E - \mathscr{L}$	9 911.377	(26)	Kaufman et al (1971)
H	2	$\Delta E - \mathscr{L}$	9 911.250	(63)	Shyn et al (1971)
H	2	$\Delta E - \mathscr{L}$	9 911.173	(42)	Cosens and Vorberger (1970)
H	2	$\Delta E - \mathscr{L}$	9 911.117	(41)	Safinya et al (1980)
D	2	$\Delta E - \mathscr{L}$	9 912.59	(10)	Treibwasser et al (1953)
H	3	$\Delta E - \mathscr{L}$	2 933.5	(12)	Fabjan et al (1971)
D	3	$\Delta E - \mathscr{L}$	2 935.2	(8)	Wilcox and Lamb (1960)
H	4	$\Delta E - \mathscr{L}$	1 235.9	(13)	Fabjan et al (1971)
H	4	$\Delta E - \mathscr{L}$	1 237.79	$\binom{+23}{-27}$	Braun and Pipkin (1971) (also PMFC1)
H	5	$\Delta E - \mathscr{L}$	622.4	(101)	Fabjan et al (1971)
$^4He^+$	3	$\Delta E - \mathscr{L}$	47 844.05	(48)	Mader et al (1971)
$^4He^+$	4	$\Delta E - \mathscr{L}$	20 179.7	(12)	Jacobs et al (1971)

Table 5.1(b) Post 1969 values for the fine structure interval $\Delta E - \mathscr{L}$ in hydrogen.

Date	Authors	$(2\,^2P_{3/2} - 2\,^2S_{1/2})$ (MHz)	
1970	Cosens et al	9911.173	(42)
1971	Shyn et al	9911.250	(63)
1971	Kaufman et al	9911.377	(26)
1980	Safinya et al	9911.117	(41)
	Theory (see text)	9911.167	(13)

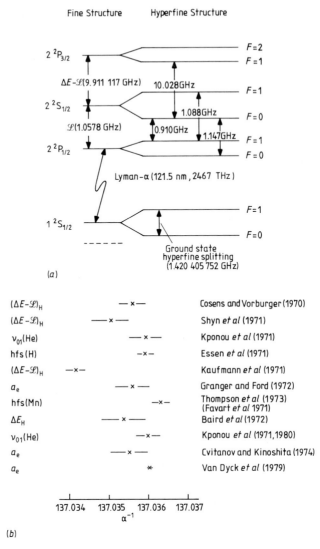

Figure 5.1 (a) The fine structure and hyperfine structure of the $n = 1$ and $n = 2$ levels in atomic hydrogen, illustrating the Lamb shifts $\mathscr{L}$ and energy separation ΔE (not to scale); (b) the principal QED-dependent determinations of α^{-1} published between 1970 and 1980.

The direct measurements of the fine structure constant by spectroscopic methods played an important part in the evaluation of the fundamental constants for many years. Their role is decreasing rapidly today, for the indirect methods have overtaken the spectroscopic methods in precision. In addition to the Lamb shift $\mathscr{L}$ of the levels, which gives rise to the splitting of the $2\,^2S_{1/2}$ and $2\,^2P_{1/2}$ states and from which the value of the fine structure constant may be deduced where the theory is sufficiently reliable, the fine structure constant may be obtained from two other quantities. These are the separations ΔE and $\Delta E - \mathscr{L}$ of the $P_{3/2}$ and $P_{1/2}$ levels and of the $P_{3/2}$ and $S_{1/2}$ levels, i.e.

$$\Delta E = n\,^2P_{3/2} - n\,^2P_{1/2}$$

and

$$\Delta E - \mathscr{L} = n\,^2P_{3/2} - n\,^2S_{1/2}$$

where $n = 2, 3, \ldots$. The system rapidly becomes more complex at higher values of n, but values up to $n = 5$ have been studied.

Following Taylor $et\ al$ (1969), the expression for ΔE for the nth level of a hydrogenic atom of charge Z and nuclear mass M_i may be written

$$\Delta E_n = \frac{Z^2 R_\infty (Z\alpha)^2 c}{2n^3}\left[[1 + F_n(z\alpha)^2]\left(1 + \frac{m_e}{M_i}\right)^{-1} - \left(\frac{m_e}{M_i}\right)^2\left(1 + \frac{m_e}{M_i}\right)^{-3}\right.$$
$$\left. + 2a_e\left(1 + \frac{m_e}{M_i}\right)^{-2} - G_n\frac{4\alpha}{3\pi}(\alpha Z)^2 \ln(Z\alpha)^{-2}\right] \qquad (5.3.1)$$

where $F_n = (7n^2 + 18n - 24)/16n^2$ and $G_n = (1 - 1/n^2)$. The first term in this equation comes from the Dirac solution together with the reduced mass term $(1 + m_e/M_i)^{-1}$ obtained by Grotch and Yennie (1969). The second term is the contribution from the normal Dirac moments of the electron and the nucleus (Barker and Glover 1955), and the third term is the contribution from the anomalous magnetic moment of the electron and this term and the final term, which is the radiation correction first calculated by Yennie (1960), are the QED contributions. This equation is complex enough, but it must be further modified to include more subtle effects in order to be used to the precision required today. As will be seen in § 5.4, the expressions for the Lamb shift can be equally fearsome. It is clear, therefore, that the spectroscopic measurements of the fine structure constant make formidable demands on the theoretician as well as the experimenter.

5.3.1 The $n = 2$ state

For many years the most accurate values for ΔE were derived from the measurements of the $2\,^2P_{1/2} - 2\,^2P_{3/2}$ separation in deuterium by Treibwasser $et\ al$ (1953). Their work led to a number of other experiments which were all on broadly similar principles. The source of atomic hydrogen was a beam of hydrogen atoms which were produced by dissociation of molecular hydrogen in a tungsten oven. The beam of atoms then passed through a cross-beam of electrons of optimum energy $\sim 6\,eV$ which excited the atomic beam into the $n = 2$ state. The lifetime of the $2\,^2P$ states is in the region of $10^{-8}\,s$ and hence such atoms rapidly decay into the ground state. The beam then comprised atoms in the metastable $2\,^2S_{1/2}$ state and this state should have a lifetime of $\sim \frac{1}{8}$ s. (However, very careful experimental techniques are required to achieve such a lifetime, for quite small dynamic electric fields can cause mixing of the $2\,^2S_{1/2}$ and $2\,^2P_{1/2}$ states and the latter decays back to the ground state.) The

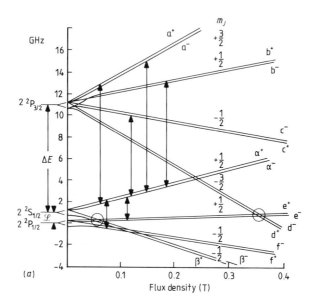

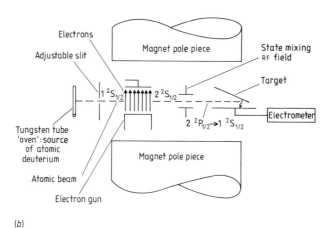

Figure 5.2 (a) The Zeeman splitting of the $n = 2$ levels of atomic hydrogen, showing the crossings of the levels which have been utilised to measure ΔE and $\mathscr{L}$; (b) the apparatus used by Lamb et al to measure the fine structure of deuterium by an atomic beam method.

beam then passed through a region in which it was exposed to an RF field, either in a single cavity or a separated double cavity (by the Ramsey technique). In many of the experiments the beam was also in a uniform magnetic field at this point, and this had the value required to cause certain of the Zeeman-split energy levels to be at their exact crossing point (figure 5.2(a)) so that state mixing could occur. A measurement of the frequency of the applied radiation which was required to excite the desired $2\,^2S_{1/2}$–$2\,^2P_{1/2}$ transition, together with the appropriate extrapolation of the Zeeman levels to yield the value at zero magnetic field, enabled a value of the Lamb shift $\mathscr{L}$, ΔE and $\Delta E - \mathscr{L}$ to be deduced.

The peak effort in this area probably came at around 1970 and this was a result of attempts to resolve discrepancies between the value obtained for the fine structure constant from the work of Lamb and that from the work of other experimenters (for example as discussed by Cohen and Dumond (1965)). The resolution of these discrepancies partly provided the motivation for the Taylor *et al* (1969) review of the best values. Overall, the review showed that the discrepancies among the values of the fine structure constant could be accounted for, partly by improved theoretical calculation and partly by closer attention to the many factors which could affect the experiments. The shape of the resonance curves was of particular importance, since their widths were

Table 5.2 Table of spectroscopic values for the fine structure constant used in the 1973 evaluation.

Date	Authors	Quantity measured	Value (MHz)		Uncertainty (ppm)	Value derived for α^{-1}		Uncertainty (ppm)
1972	Baird *et al*	ΔE_H	10 969.127	(87)	7.9	137.035 44	(54)	3.9
1970	Cosens and Vorburger	$(\Delta E - \mathscr{L})_H$	9 911.173	(42)	4.2	137.035 63	(31)	2.3
1971	Kaufman *et al*	$(\Delta E - \mathscr{L})_H$	9 911.377	(26)	2.6	137.034 16	(20)	1.5
1971	Shyn *et al*	$(\Delta E - \mathscr{L})_H$	9 911.250	(63)	6.4	137.035 08	(46)	3.3
1971	Kponou *et al*	$\nu_{01}(^4\text{He})$	29 616.864	(36)	1.2	137.035 95	(42)	3.1[†]
						137.036 113	(110)	0.8[‡]

[†] 1973 value
[‡] 1982 value with revised theory

largely determined by the 10^{-8} s lifetimes of the 2P states. These experiments were discussed at the Gaithersberg 1970 Conference on Precision Measurements and the Fundamental Constants, and the reader is referred for further details to the proceedings of this conference (Taylor and Langenberg 1970; see also Taylor 1970).

The experiments which were included in the Cohen and Taylor (1973) review of the best values are shown in table 5.2. The review of DuMond and Cohen (1963) gave considerable weight to the Lamb *et al* determination of the fine structure constant. However, when they re-examined the deuterium work, Taylor *et al* (1969) found that the value rested on comparatively few results (~ 6); indeed one run carried most of the weight in the final result.

Treibwasser *et al* measured the α-a and α-c transitions in deuterium. Kaufman *et al* (1971) studied the α-a transition at 0.1465 T, the α-b transition at 0.1860 T and the α-c transition at 0.1090 T, for which the applied microwave frequencies were about 11.970, 9.170 and 7.430 GHz respectively. In their experiment, the dissociation of the molecular hydrogen and excitation to the $2\,^2S_{1/2}$ state were carried out in one step by a beam of 25 eV electrons and the microwave field was applied over the same region. The metastable atoms were excited to the $2\,^2P_{3/2}$ state by the microwave field and Lyman radiation from their decay gave a measure of the number of metastable atoms. The resonance was scanned by varying the magnetic field, whose intensity was measured with an NMR water sample. The α-c transition was affected by the nearby β-d resonance and these measurements were given a low weight. The rather messy environment meant that it was necessary to extrapolate the results to zero pressure and electric field.

The measurement of Metcalf *et al* (1971) was of the level crossing at 0.3484 T of the e to d transitions. In their apparatus Lyman-α radiation was incident on a cloud of H atoms, $\sim 10^9$ atoms cm^{-3}, in the magnetic field and the scattered Lyman-α radiation was used to detect the crossing condition. The resonance was scanned by monitoring the detected scattered signal as the magnetic field was scanned through the Lorentzian resonance, which had a halfwidth of about 7 mT. A ten part per million measurement of $\Delta E - \mathcal{L}$ required estimating the centre of the resonance to about 1/2000 of the halfwidth, or about 3 μT, and also that the theory of the lineshape was understood to about 0.2%.

The Shyn *et al* (1971) measurements were similar to the Dayhoff *et al* arrangement in that a microwave electric field was applied to a beam of metastable $2^2S_{1/2}$ state atoms and the 5.46 mT wide β^+-b transition at 0.779 T was studied. The hydrogen from the tungsten oven at 2400 K was bombarded with 14 eV electrons in a flux density of 0.545 T and, since the motional electric fields quenched the β states, it was necessary to regenerate them by passing the beam through a flopper field in order to generate the single hyperfine β^+ state. There were about 3×10^4 atoms s^{-1} at the detector in the metastable state.

The Vorberger and Cosens (1971) apparatus was similar to that of Shyn *et al* except that there was a further flopper field region between the microwave field and the detector. Adjustment of the first flopper flux density ($0 - 1.5$ mT) and the flopper frequency produced atoms in the β^+ or β^- state. The cavity frequency was about 10.844 GHz for the β-b$^-$ and β-b$^+$ states and 9.4 GHz for the β-d$^-$ and β-d$^+$ states. The second quencher contained a further magnetic field at the β-e transition crossing point and, when a 3 V cm^{-1} electric field was turned on, the β-state atoms were quenched by transition to the e-state. With the quencher off the observed signal was a mixture of β^--b$^-$ and α-c states, but with it on only the α-c state remained so that the difference of these signals yield the pure β^--b$^-$ signal. It is interesting that the original Dayhoff et al measurement in deuterium has not been repeated.

Since the 1973 review of Cohen and Taylor there has been further measurement of the Lamb shift (§ 5.3) by Safinya et al (1980) using a fast beam method with separated RF cavities. They measured the $2^2P_{3/2}$–$2^2S_{1/2}$ fine structure interval in atomic hydrogen in zero magnetic field and obtained a value of $\mathscr{L} = 9911.17(41)$ MHz.

This value may be compared with the value of

$$\mathscr{L} = 9911.167(13) \text{ MHz}$$

which is obtained using the 1973 value for α^{-1} of

$$\alpha^{-1} = 137.035\,963(15)$$

together with Mohr's (1975a, b) recalculation of the Lamb shift and taking a value of $0.84(1) \times 10^{-15}$ m for the radius of the proton. The above results are, to some extent, of a preliminary nature and the experiment is thought to be capable of a further order of magnitude improvement in precision. If so, it should be possible to obtain a value for the fine structure constant by this method which has a precision which rivals that obtained through the low field measurements of the gyromagnetic ratio of the proton and the quantised Hall effect.

The atomic hydrogen beam in the work of Safinya et al was obtained from a 107.5 keV or 48.7 keV beam of protons. These produced atomic hydrogen in the H(2S) state by a charge exchange scattering interaction with gaseous nitrogen. The atoms then passed through a 1.1 GHz RF field which provided the state selection before entering the spectroscopy region. The number of atoms in the $2S_{1/2}$ state traversing the spectroscopy region was observed by quenching the atoms with a 910 MHz RF field and observing the Lyman-α photons, which were detected by a photomultiplier. The atoms passed through 6.34 mm diameter holes in the spectroscopy region; these were in the narrow sides of two microwave waveguides and were separated by 27.21(3) mm. The microwave frequency was varied slowly in the region of 10.0285 GHz and the excitation to the $2^2P_{3/2}$ state produced a diminution of the photomultiplier signal. The Doppler shift was eliminated by reversing the direction of propagation of the microwave signals. A correction of -118.386 MHz was required in order to allow for the hyperfine structure of the states. The separated oscillatory field led to a narrower linewidth, ~ 40 MHz, since this provided a preferential selection of the longer lived P state atoms. This gave an absorption curve which was narrower than the 100 MHz natural linewidth.

5.3.2 The fine structure constant from helium

The measurement of Kponou et al (1971) was the only spectroscopic one to be included in the least squares evaluation by Cohen and Taylor (1984). They studied the $2\,^3P$ state of atomic helium which had several advantages over the corresponding transition in atomic hydrogen. First, the lifetime of the helium $2\,^3P$ state is about 10^{-7} s compared with the 1.6×10^{-9} s lifetime of the hydrogen $2\,^2P$ state, and second the fine structure interval from $J = 0$ to $J = 1$ in helium is three times larger than the hydrogen $2\,^2P_{3/2} - 2\,^2P_{1/2}$ fine structure interval. Although a disadvantage of using helium over hydrogen is that the theory of two electron atoms is less well developed, considerable progress has been made, and the $2\,^3P$ state, being the lowest state of a particular symmetry in spin and orbital angular momentum, is the one to which variational theoretical calculations can be applied with greatest accuracy.

The principle of the experiment of Kponou et al (figure 5.3) was to allow the helium atoms in the $1\,^3S$ state to effuse into the vacuum region through a slit. They then traversed a beam of electrons which excited some of the atoms into the metastable $2\,^3S$ state (with an efficiency of about 2×10^{-4}). Next, they passed through an inhomogeneous 'A' magnetic field, followed by a uniform 'C' field and then a further inhomogeneous 'B' field before being detected. The

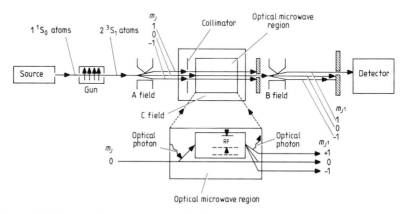

Figure 5.3 Scheme of the measurement by Kponou et al (1971) of the fine structure of helium, with the beam stops and detector arranged for the $2^3S_{m_J} 0 \rightarrow 1$ transition. The microwave/optical region (lower inset) involves (i) optical excitation, (ii) RF mixing and (iii) optical decay.

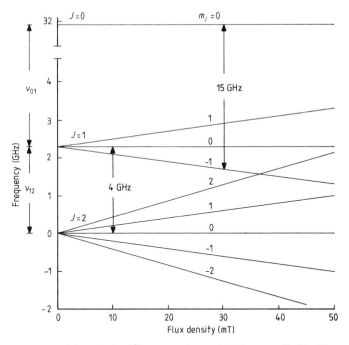

Figure 5.4 The transitions in the 2^3P_3 state of atomic helium studied by Kponou *et al* (1971).

inhomogeneous fields and the collimating slits served to allow selection of atoms with a particular m_j component of the $2\,^3S$ state. In the C field region there was also provision for radiating the atoms with optical radiation from a helium lamp to induce the $2\,^3S \rightarrow 2\,^3P$ optical transition, and a microwave field of about 0.14 mT to induce the transitions labelled 4 GHz and 15 GHz of figure 5.4. The resonances were observed by keeping the microwave frequency fixed and varying the applied C field, the signal being observed by the change in the normalised count rate of the detector as the microwave field was switched on and off.

The overall precision achieved was 36 kHz or 1.2 ppm, the value being

$$\nu_{01} = 29\,616.864(36)\ \text{MHz}.$$

This value could not be used in the 1973 evaluation, since at that time the theoretical value had an uncertainty of about 6 ppm and there was a possible 3.09 MHz discrepancy between their result and the theoretical value computed as far as some of the terms of order α^4 by using the evaluated value of the fine structure constant. However, in 1976, Lewis *et al* reported calculations of the $2\,^3P$ level which were later extended by Lewis and Serafino (1976) to order α^6 and the accuracy of the $J = 0 \rightarrow J = 1$ transition calculations is now 1.5 ppm. The latest calculations yield $\nu_{01}/\alpha^2 = 556\,171.73(83)\ \text{GHz}$;

hence using Kponou *et al* measurement of v_{01} leads to

$$\alpha^{-1} = 137.036\,113(110) \qquad (0.8\ \text{ppm}).$$

The uncertainty of the theory still remains larger than the measurement and an extension to higher precision, including the α^7 terms, would be useful and desirable.

It would appear that the measurements of the fine structure constant have nearly reached the limit in precision of which conventional spectroscopic techniques are capable. The major obstacle is the relatively short lifetime of the P states, for these lead to the broad resonances. However, in recent times we have had the advent of tunable dye lasers, polarisation spectroscopy and the excitation of double quantum transitions and these, when combined with experimental ingenuity, may yet lead to further advances. For the present, the direct spectroscopic determinations of the fine structure constant play a part in putting limits on the validity of QED theory, but have a decreasing role in the evaluation of the best value for the fine structure constant.

5.3.3 'Theoretical' values for the fine structure constant

As is discussed elsewhere in the book, it is of questionable value to discuss time variations of dimensioned fundamental constants, since such variations depend on the system of units that is employed. It is evident that dimensionless constants cannot vary with time if some exact theoretical expression can be found for them. Such theories are quite respectable in some cases. Thus one can cite calculations of the Lamb shift, and of the magnetic moments of the electron and muon in terms of their classically expected values. Thus the g_e value of the electron is given by

$$2(\mu_e/\mu_B - 1) = A(\alpha/\pi) + B(\alpha/\pi)^2 + C(\alpha/\pi)^3 + \cdots$$

where $A = \frac{1}{2}$, $B = -0.328\,478\ldots$, $C = 1.29(6)$ and the discrepancies between theory and experiment lie in the high-order terms involving the fine structure constant α. It is natural to enquire whether there have been any expressions calculated for the constants α or β ($= m_e/m_p$). The answer is that there have been several attempts over the last 50 years and some of these are more questionable, or respectable, than others (table 5.3). No claim is made that this is a complete list, but the earliest to be found in the textbooks is that of Lewis and Adams (1914) who derived an expression for α from considerations of the new quantum electrodynamics as

$$\alpha^{-1} = 8\pi(8\pi^5/15)^{1/3} = 137.348.$$

Another is a prediction by Perles (1928) of a relation between α and m_p/m_e, which is the reciprocal of our constant β

$$\alpha^{-1} = [2\pi(\pi - 1)]^{-1} m_p/m_e = 136.4557.$$

Table 5.3

(a) Some theoretical expressions that have been 'derived' for α^{-1}.

Date	Author(s)	Expression	Value	
1914	Lewis and Adams	$8\pi(8\pi^5/15)^{1/3}$	137.348	
1928	Perles	$[2\pi(\pi-1)]^{-1}m_p/m_e$	136.455 7	
1930	Eddington	$(16^2-16)/2+16+1$	137 (exactly)	
1931	Beck $et\ al$	$T_0 = -(2/\alpha-1)\,°C$	137.075	
1970	Wyler	$(8\pi^4/9)(2^45!/\pi^5)^{1/4}$	137.036 082	
1972	Aspden and Eagles	$108\pi(8/1843)^{1/6}$	137.035 915	
1973	Cohen and Taylor	Review value	137.036 04	(11)

(b) Some empirical ways of synthesising α^{-1}

Date	Authors	Expression	Value	
1971	Robertson	$2^{-19/4}3^{10/3}5^{17/4}\pi^{-2}$	137.035 94	
	(with Roskies and Prosen)	$2^{5/3}3^{-8/3}5^{5/2}\pi^{7/3}$	137.036 01	
	(Wyler's value)	$2^{19/3}3^{-7/4}5^{1/4}\pi^{11/4}$	137.036 08	
		$2^{2/3}3^{7/3}5^{11/3}\pi^{-7/2}$	137.036 12	
		$2^{-13/4}3^{17/4}5^{2/3}\pi^{5/4}$	137.036 16	
		$2^{8/3}3^{3/4}5^{-1/2}\pi^{8/3}$	137.036 29	
1973	Burger	$(137^2+\pi^2)^{1/2}$	137.036 015 7	
1973	Cohen and Taylor	Review value	137.036 04	(11)

(c) Values of m_p/m_e

Date	Authors	Expression	Value	
1951	Lenz (empirical)	$6\pi^5$	1836.118 108	
1970	Wyler (derived)			
1969	Taylor $et\ al$	—	1836.109	(11)
1973	Cohen and Taylor	—	1836.151 52	(70)

Probably the most famous is the prediction by Eddington, who derived α as a pure number:

$$\alpha^{-1} = \frac{16^2 - 16}{2} + 16 + 1.$$

This came from considerations of the number of independent elements in a symmetrical matrix in 16-dimensional space where $16 = 4 \times 4$ (4 being the number of dimensions in Minkowski's world). The theory lost respectability, partly perhaps because Eddington at first predicted the number as 136. Birge (1929), however, pointed out that the latest values were in good accord with a number closer to 137. Eddington later re-examined his theory and found that a further degree of freedom was required to allow for the forces between pairs of electrons. Somehow a disagreement of sorts seems to have developed between Eddington and Birge, for Eddington was moved to write a paper 'on the method of least squares' for evaluating the values of the fundamental constants. In particular, Eddington claimed that for someone to 'use the probable error in referring to his experimental results was equivalent to dealing with someone whose habit was to tell the truth only half the time'. Some twenty years later Birge was still quoting this paper and saying that it was full of mistakes!

It would appear that not all theoreticians took Eddington's paper seriously because in 1931, Beck, Bethe and Riezler wrote a paper which was published in *Naturwissenschaften* (in German):

Remark on the Quantum Theory of Zero Temperature

We consider a hexagonal crystal lattice. The absolute zero of this is characterised by the condition that all degrees of freedom of the system freeze, that is all internal movements of the lattice cease. An exception to this is, of course, the motion of the electron in its Bohr orbit. According to Eddington each electron possesses $1/\alpha$ degrees of freedom, where α is the Sommerfeld fine structure constant. Besides electrons, our crystal contains only protons, and the number of degrees of freedom for them is the same since, according to Dirac, a proton can be regarded as a hole in the electron gas. Thus, since one degree of freedom remains because of the orbital motion, in order to attain absolute zero we must remove from a substance $2/\alpha - 1$ degrees of freedom per neutron ($= 1$ electron $+ 1$ proton; since our crystal has to be electrically neutral overall). We obtain therefore for the zero temperature T_0

$$T_0 = -(2/\alpha - 1)\text{Degrees.}$$

Setting $T_0 = -273°$ we obtain for $1/\alpha$ the value 137, which, within the limits of error, agrees completely with the value obtained in an independent way. One can easily convince oneself that our result is independent of the special choice of crystal structure.
Cambridge, 10 December 1930
G Beck H Bethe W Riezler

This was intended to be a complete spoof of the Eddington approach, which neither Eddington nor the editor of the journal took very kindly when they found out after the paper had been published. This story had a later sequel, for in 1948, when the time came to publish the work discussed in Alpher's PhD thesis, Gamow waved the earlier paper in front of Alpher as evidence that Bethe had a keen sense of humour, and suggested that the authors of the paper were given as Alpher, Bethe and Gamow to obtain the obvious euphony, with *in absentia* after Bethe's name. However, the phrase *in absentia* was apparently deleted after it left their hands. Clearly, the story had a happy ending, for Bethe was one of Alpher's PhD examiners! This according to Alpher (1973), is the explanation of how the famous *Physical Review* paper by Alpher, Bethe and Gamow on cosmology came to be published. Physics would surely have been the poorer without the publication of that paper.

There have been other attempts since then to predict α^{-1}. Thus Wyler made a prediction as

$$\alpha^{-1} = (8\pi/9)(2^45/\pi^5)^{1/4} = 137.036\,082$$

with the associated prediction $m_p/m_e = 6\pi^5$ (as also Lenz (1951) had noted earlier), which caused something of a stir a few years ago. The latter paper (reproduced from *Physical Review*, by permission of the American Physical Society) must set a world record for brevity!

The Ratio of Proton and Electron Masses
Friedrich Lenz
Düsseldorf, Germany
(Received April 5, 1951)

The most exact value at present[1] for the ratio of proton to electron mass is 1836.12 ± 0.05. It may be of interest to note that this number coincides with $6\pi^5 = 1836.12$.

[1]Sommer, Thomas, and Hipple, Phys. Rev. 80, 487 (1950).

The expressions had values which were amazingly close to the value recommended in the 1969 adjustment of the atomic constants. However, the work by Mamyrin *et al* in Leningrad, and by Petley and Morris at NPL led to the 1973 value of 1836.151 52(70) which seems to be too far away from the Wyler value and, as is shown below, recent work confirms these values. More recently, a further attempt in this area has been made by Stansbury (1983) which has been commented on by Rosenberg, by Good and by Crawford (Rosenberg *et al* 1983). No doubt the theoretical attempts to calculate the values of α and β will continue—possibly with a Nobel prize winning success. Aspden and Eagles (1972) (authors of *Physics without Einstein*) obtained

$$\alpha^{-1} = 108\pi(8/1843)^{1/3}.$$

Incidentally, Burger (1978) pointed out that a good approximation for α^{-1} was

$$\alpha^{-1} = (137^2 + \pi^2)^{1/2} = 137.036\,01.$$

This is very close to the 1983 values for α^{-1} and is easily remembered and entered into a hand calculator.

It is not too difficult to synthesis an eight digit decimal number by combinations of the integers 2, 3 and 5, and the irrational quantity π, raised to various powers. Thus Robertson (1971) gives the values quoted at the end of the table as being derived by Roskies and Prosen independently and there were further values given in *Physics Today* 24 **1** 9(1971). At this point the subject passes somewhat into disrepute, for using a combination of 12 digits to synthesise an eight digit decimal number is an arbitary and rather unrewarding process unless it is soundly based theoretically. However, we have the success of Balmer, with the series of spectral lines that bear his name in hydrogen, at one end and the acknowledged hoax of Beck *et al* (1931) at the other: which only serves to underline the problems facing anyone who has to referee a paper in this subject area.

5.4 The Lamb shift

The Dirac theory predicts that, for hydrogen-like atoms, states with the same total quantum number n and angular momentum j are degenerate (aside from some small nuclear effects). Investigations of the hydrogen lines by conventional spectroscopic techniques showed that, for the fine structure doublet in the hydrogen Balmer-α line (656 nm), the components were separated by only about 96% of that predicted by the Dirac theory. This reduction was suggested to be due to the $2\,^2S_{1/2}$ and $2\,^2P_{1/2}$ levels, which according to the Dirac theory were degenerate, being in fact separated by a small amount. This was confirmed in the classic measurements of Lamb and Retherford (1950, 1951, 1952) who found that the $2\,^2S_{1/2}$ level was 1058 MHz above the $2\,^2P_{1/2}$ level. This frequency is in the microwave region and hence the Lamb shifts have usually been studied directly by microwave techniques. However, the advent of the techniques of saturable absorption and polarisation spectroscopy have enabled the Lamb shifts to be investigated by optical spectroscopic techniques, although the accuracy of the measurements has not yet surpassed the impressive work that is possible in the microwave region.

Other energy levels are also Lamb-shifted: thus Lamb and Skinner (1950) showed that the $2^2S_{1/2}$ level in He^+ is 14 021(60) MHz above the $2^2P_{1/2}$ level. The Lamb shift may be explained by taking into account the interaction of the electron with the radiation field and the effect of vacuum polarisations, i.e. the disturbance of the electron distribution in negative energy states in close proximity to a proton (many of the S states have a higher probability density close to the nucleus than the P states and hence they suffer greater Lamb shifts).

The Lamb shifts may be calculated very precisely in terms of fundamental constants, involving the fine structure constant in particular. The expressions for the Lamb shift are quite complex, for example Taylor *et al* (1969) gave the following expression for the $2\,^2S_{1/2}-2\,^2P_{1/2}$ splitting:

$$\mathscr{L} = \frac{8\alpha^3 R_\infty c Z^4}{3\pi n^3}\left[\left(1+\frac{m_e}{M_u}\right)^{-3}\ln\left(1+\frac{m_e}{M_u}\right)^{-3}\left(\ln\frac{(1+m_e/M_u)}{(Z\alpha)^2}\right)\right.$$

$$+\frac{19}{30}+\ln\frac{k_0(n,1)}{k_0(n,0)}+\frac{1}{8}\left(1+\frac{m_e}{M_u}\right)^{-2}+\frac{\alpha}{\pi}[\tfrac{3}{2}m-0.3285-\tfrac{82}{81}(\tfrac{3}{4})]$$

$$+\pi Z\alpha\left(\frac{427}{128}-\frac{3\ln 2}{2}\right)+(Z\alpha)^{-2}[-\tfrac{3}{4}\ln^2(Z\alpha)^{-2}+C_n\ln(Z\alpha)^{-2}$$

$$-(4\pi^2/3+4+4\ln^2 2)]+\frac{Zm_e}{M_u}\left(\frac{1}{4}\ln(Z\alpha)^{-2}+\ln\frac{k_0(n,1)}{k_0(n,0)}-\frac{1}{2}+D_n\right)\Bigg]$$

$$+\frac{4R_\infty c Z^4}{3n^3}\frac{r_n^2}{a_0^2} \qquad\qquad (5.4.1)$$

where $\ln k_0$ (n, 0 or 1) is the Bethe logarithmic excitation energy, m is the coefficient of the fourth-order radiation correction calculated by Soto (1966) as being numerically equal to 0.211 296 114, C_n and D_n are constants which depend on n

$$C_n = 7\ln 2 + 3\sum_{q=1}^{n}q^{-1}-\frac{757}{240}-\frac{4}{5n^2}-3\ln n \qquad (5.4.2)$$

$$D_n = \frac{91}{24}+\frac{7}{2}\left(\ln\frac{2}{n}+\sum_{q=1}^{n}q^{-1}\frac{-1}{2n}\right) \qquad (5.4.3)$$

and Erickson (1967) showed that the expression for D_n is correct for all n. The last term takes nuclear structure into account and r_n is the root mean square radius of the nuclear charge distribution and a_0 is the Bohr radius. Since these involve highly sophisticated quantum electrodynamic calculations, a precise comparison between the theoretical and experimental values provides a sensitive test of QED. In addition, the Lamb shifts are necessary in order to evaluate the determinations of the fine structure constant by spectroscopic methods using hydrogen, deuterium and singly ionised helium.

Although measurements of the Lamb shift in hydrogenic atoms provide sensitive low energy tests of QED, the sensitivity of the tests depends on both the accuracy of the theory and the experiments. The difficulty with hydrogen is that the uncertainty of the RMS charge radius of the proton ($\sim 1.4\%$) is a major limiting factor in the theoretical calculations. On the other hand, the uncertainty in the radius of $^4He^+$ is only $\sim 0.3\%$. This, together with the absence of hyperfine structures in $^4He^+$, which simplifies the experiments, means that measurements in $^4He^+$ are very attractive. Experiments are in progress at Harvard University to measure the $4\,^2S_{1/2}-4\,^2P_{1/2}$ Lamb shift in $^4He^+$ using the fast beam, separated oscillating field technique developed by Lundeen and Pipkin (1981).

The recent work has involved several of the lines in atomic hydrogen, including the ground state level. The measurement involving the ground state level is of particular interest not only for its accuracy (although in the context of the difficulties of the measurement this was quite impressive), but also for the method which was used, for it is an indication of the type of measurement which is likely to be made with increasing precision in the future.

5.4.1 The measurement of the ground state Lamb shift

The level in hydrogen which has the greatest Lamb shift is the $1^2S_{1/2}$ ground state and the Lamb shift of this level was first demonstrated experimentally by Herzberg from his measurement of the absolute wavelength of the Lyman-α line in deuterium (121.4 nm).

Further measurements were reported by Lee *et al* (1975). They were able to measure the isotope shift of the 1S–2S transition in atomic hydrogen and deuterium. Their technique (figure 5.5) was to use the Doppler-free two-photon spectroscopy with the aid of a frequency doubled pulsed dye laser. The fundamental dye laser output near 486 nm was used to observe the Balmer-β line (486 nm) by high resolution saturation laser spectroscopy (as is described

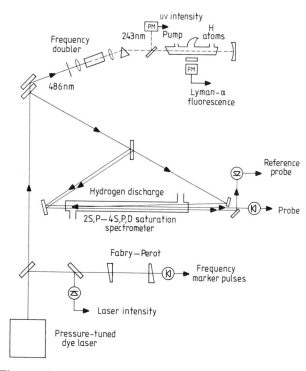

Figure 5.5 The experimental layout used by Lee *et al* (1975) to measure the ground state Lamb shift in atomic hydrogen.

for the measurement of the Rydberg constant). If Bohr's formula were correct, the $n = 1$ to 2 interval would be exactly four times the $n = 2$ to 4 interval, so that the two transitions would occur at exactly twice the frequency. The observed displacement is in part due to the Lamb shift and there are also some small nuclear structure effects. The $1\,^2S_{1/2}$–$2\,^2S_{1/2}$ transition would normally be a forbidden transition, but the transition is allowed if two photons may be absorbed simultaneously, the sum of their frequencies being equal to the transition frequency.

In the experiment of Lee *et al* the 486 nm radiation was first doubled in a lithium formate crystal and the 243 nm radiation focused to a 0.2 mm waist diameter inside a cell which contained hydrogen atoms (these were produced in a ~ 26 Pa gas discharge in a mixture of H_2 and D_2 and carried into the cell by gas flow). The 300 W peak power beam was then reflected back through the cell on a counter-propagating path in order to provide a standing wave field. The pressure-tuned dye laser system (Wallenstein and Hänsch 1975) provided laser pulses of about 15 kW peak power at 486 nm. These had a linewidth of about 120 MHz and a repetition rate of 17 pulses per second. The excitation of the $1\,^2S_{1/2}$–$2\,^2S_{1/2}$ transition was observed through a MgF_2 window by means of a solar blind photomultiplier and narrow band optical filter. These enabled the collision induced $2\,^2P_{1/2}$–$1\,^2S_{1/2}$ Lyman-α fluorescence radiation to be detected; the filter had a 15% transmission and a collection solid angle of about 2%. Apart from a Fabry–Perot etalon, which was used to give frequency markers, there was no need for any absolute wavelength measurements in their work since the Balmer line served as the wavelength reference.

The results of the measurements were 8.20(10) GHz and 8.25(11) GHz for the Lamb shifts of the 1S ground states for hydrogen and deuterium respectively. This precision may be compared with the only other experiment of its kind, which was that of Herzberg (1956). He determined the 1S shift of deuterium with 15% precision by comparing the absolute wavelength of the Lyman-α line with the value calculated from the Dirac theory.

The isotope shift is principally due to the nuclear mass difference between hydrogen and deuterium and it is possible that more precise measurements of the difference of the $1\,^2S_{1/2}$–$2\,^2S_{1/2}$ levels of hydrogen and deuterium shifts may in the future improve our knowledge of the ratio of the electron mass to the proton mass. Although the present work was performed with a pulsed dye laser it should be possible to perform CW experiments with much lower laser intensities, and ~ 1 mW should suffice. The Balmer-β line could be observed more precisely by using the following methods: a Doppler-free two-photon spectroscopy approach, using an infrared laser whose second harmonic is locked to the 486 nm laser, double photon absorption to excite $2S_{1/2}$–$4S_{1/2}$ transitions or it may be possible to refine the polarisation spectroscopy approach of Lee *et al* (1975) to give the required precision. It appears likely that the new techniques may in time provide us with very stringent tests of QED.

5.4.2 Lamb shifts in the n = 2 level

The Lamb shifts in the ground state level cannot be observed by microwave techniques because there is no nearby P level. However, the $n = 2$ state does possess suitable levels and these have been used to obtain measurements of the Lamb shifts (table 5.4). As with atomic clocks, a beam of atoms is often used and the transitions may either be observed by using a single microwave cavity, as in the measurements of Andrews and Newton (1976), or by using separated cavities as in the work of Lundeen and Pipkin (1975) and Safinya et al (1980). The first problem is to obtain atoms which are in the $2S_{1/2}$ state. This is a metastable state with a lifetime $\sim \frac{1}{8}$s. The technique is therefore to take a beam of protons, between 21 keV and 100 keV, and to form a beam of hydrogen atoms by charge exchange collisions in hydrogen gas. Atoms which are excited into the $n = 2$ state decay to the ground state within less than a few tenths of a microsecond, with the exception of those atoms which are in the $2S_{1/2}$ state. The number of atoms in this state may be deduced by quenching it by the application of an applied electrostatic field. This causes mixing of the $2S_{1/2}$ and $2P_{1/2}$ states and rapid decay to the $1S_{1/2}$ state and the accompanying Lyman-α radiation may be detected with a photomultiplier.

Table 5.4   Experimental determinations of the Lamb shift interval
$$\mathcal{L} = n\,^2S_{1/2} - n\,^2P_{1/2}$$

Atom	n	Lamb shift		Author
H	2	1 057.77	(19)	Treibwasser et al (1953)
H	2	1 057.90	(10)	Robiscoe and Shyn (1970)
H	2	1 057.893	(20)	Lundeen and Pipkin (1975)
H	2	1 057.862	(20)	Newton et al (1979)
H	2	1 057.845	(9)	Lundeen and Pipkin (1981)
H	2	1 057.858 3	(22)	Sokolov (1981)
D	2	1 059.00	(10)	Treibwasser et al (1953)
D	2	1 059.28	(6)	Cosens (1968) and Vorberger and Cosens (1971)
H	3	313.6	(57)	Kleinpoppen (1961)
H	3	314.81	(5)	Fabjan and Pipkin (1971
D	3	315.3	(80)	Wilcox and Lamb (1960)
H	4	133.18	(59)	Fabjan et al (1971)
H	4	132.58	$\left(^{+36}_{-45}\right)$	Brown and Pipkin (1971)
D	4	133	(10)	Wilcox and Lamb (1960)
H	5	64.6	(50)	Fabjan et al (1971)
^{4}He$^+$	2	14 040.2	(45)	Lipworth and Novick (1957)
^{4}He$^+$	2	14 046.2	(12)	Narasimham and Strombotne (1971)
^{4}He$^+$	2	14 040.2	(29)	Drake et al (1979)
^{4}He$^+$	3	4 183.17	(54)	Mader et al (1971)
^{4}He$^+$	4			Bollinger et al (1981)

Atoms in the $2\,^2S_{1/2}, F = 1$ state were quenched by the application of a radio frequency field of about 1100 MHz which left atoms in the beam in the $2\,^2S_{1/2},$ $F = 0$ state. These atoms then entered a second RF field which was followed by a DC electric field. The latter quenched the $2S_{1/2}$ state and the resulting Lyman-α radiation was detected with a photomultiplier. When the RF field caused a transition from the $2\,^2S_{1/2}$ to $2\,^2P_{1/2}, F = 1$ state, the signal detected by the photomultiplier decreased markedly. Although Andrews and Newton used a single RF cavity, and Lundeen and Pipkin a double RF cavity, the advantage of the narrower resonance curve obtained by the double cavity technique was offset to a large extent by the much weaker signal. A major difficulty with this type of experiment is that the theory of the lineshape must be very accurately known, for the resonance curves are relatively broad and an accurate measurement of the Lamb shift involves estimating the resonance centre to about a thousandth of the width at the half-height intensity.

The results obtained for the Lamb shift were 1057.862(20) MHz, by Newton *et al*, and 1057.893(20) MHz by Lundeen and Pipkin respectively. These may be compared with the theoretical values of Mohr (1975a, b) of 1057.864(14) MHz and of Erickson (1971) of 1057.916(10) MHz. Although these results have an experimental uncertainty of 20 ppm, they depend on the cube of the fine structure constant and hence yield α with 6 ppm precision.

Sokolov (1982) reported a measurement which claimed the highest accuracy to date. He measured the $(2\,^2S_{1/2}, F = 0) \rightarrow (2\,^2P_{1/2}, F = 1)$ transition in atomic hydrogen. In his apparatus, protons were accelerated to about 20 keV and after velocity selection and charge exchange of the protons with molecular hydrogen, the atomic hydrogen beam had an energy spread of 3.25 eV or 1.6 parts in 10^4 of the beam kinetic energy. A critical part of the experiment was the measurement of the velocity of the atoms. This was deduced by inducing the transition to the $2\,^2P_{1/2}$ state and looking at the emitted Lyman-α radiation from consequent decay to the ground state as a function of distance along the beam. The velocity was then obtained from the simple relation for the intensity I of

$$I = I_0 \exp(-x/l_0)$$

where

$$l_0 = v/\tau$$

and τ is the lifetime of the $2\,^2P_{1/2}$ state. This has been calculated by Yakovlev to be 1.596 185 ns (see Sokolov (1982), after Yakovlev). Measurements of the intensity were made at some 26 points for

$$0.03 \text{ cm} < l_0 < 1 \text{ cm}$$

After passing through the velocity analysis region, which was inoperable during the Lamb shift measurement, the beam was purified to remove the $2\,^2S_{1/2}, F = 1$ state by passing it through RF fields of frequency 1.147 and

1.087 GHz. Finally, the beam passed through the double separated RF cavity region, where the $(2\,^2S_{1/2}, F = 0) \rightarrow (2\,^2P_{1/2}, F = 1)$ transition could be induced, before hitting the detector.

Although some 200 runs were examined, the velocity of the atoms was only considered to be sufficiently constant in 34 of these for inclusion in the final analysis. The overall spread of these results was some 11 kHz or 12 ppm. The value reported by Sokolov (1982) (which superseded a preliminary one reported earlier) was

$$\Delta E = 909.9003(22)\ \text{MHz}$$

for which

$$\mathcal{L} = 1057.8583(22)\ \text{MHz}$$

In view of the complexity of the expression for the Lamb shift, it is apparent that this type of experiment provides as much a test of QED theory as a measurement of the fine structure constant. The separated oscillatory field technique has also been applied by Fabjan and Pipkin (1972, 1980) to measure the Lamb shifts in hydrogen in the $n = 3$ state. The author would particularly recommend these recent measurements to anyone who is interested in RF and microwave metrology, for they involve some very careful work which even a study of the publications does not fully bring out.

5.4.3 Lamb shifts from higher Z atoms

Measurements of Lamb shifts are being extended increasingly further to high Z hydrogenic atoms, particularly with the advent of sophisticated laser sources. Such measurements provide highly sensitive tests of QED since the Lamb shift scales with $(\alpha Z)^n$, $n \lesssim 4$. Measurements have been made up to hydrogenic argon by Gould and Marrus (1978) and by non-resonance techniques by Kugel et al (1975) who reported a measurement of the $Z = 9$, F^{8+} system. More recently the same group, Wood et al (1982) reported a measurement of the $2S_{1/2} \rightarrow 2P_{1/2}$ Lamb shift in hydrogenic chlorine (figure 5.6). They used a relativistic beam of chlorine in an experiment which was broadly similar to the classic Lamb–Retherford experiment which has already been discussed. A 190 MeV beam of Cl^{14+} ions was passed through a $\sim 60\ \mu\text{g cm}^{-2}$ carbon film which completely stripped the chlorine. The Cl^{17+} beam then passed through a further carbon foil 0.5 to 3 $\mu\text{g cm}^{-2}$ thick, where it formed hydrogenic Cl^{16+} by picking up an electron. The beam was then focused into an interaction region. The beam focus was also at the mode waist of the radiation from a CO_2 laser which crossed the beam at an angle θ (150°). When the CO_2 radiation was at frequency v_L it caused resonance transfer of the atoms at the frequency v_m to the $2P_{1/2}$ state, where v_m was the frequency of the radiation in the rest frame and

$$v_m = v_L(1 - \beta \cos \theta)/(1 - \beta^2)^{1/2}$$

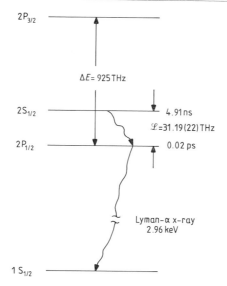

Figure 5.6 Partial energy level diagram of hydrogenic chlorine, Cl^{16+} (after Wood *et al* 1982).

where $\beta = v/c$. The consequent decay to the ground state was accompanied by the emission of the Lyman-α x-ray at 2.96 keV. The measurements were made at two beam crossing angles ($\theta = 150°$ and $\theta = 50°$) and at ion beam energies of 190 Mev and 150 Mev. The result of the measurements was

$$\mathscr{L} = 31.19(22) \text{ THz}$$

which was in good agreement with the value of Taylor *et al* (1969), equation (5.4.1), of 31.27 THz, or Mohr (1975a, b), but in poorer agreement with the calculation of Erickson (1977) of 31.93(13) THz.

Briand *et al* (1983) have reported a measurement of the 1S Lamb shift in hydrogen-like iron ($Z = 26$). The energy of the x-rays emitted in flight by the ions, after exciting them by traversing two carbon foils, was measured to 90 ppm with a flat crystal spectrometer. The spectrometer was calibrated using cobalt K-α x-rays. Their results were 3.4(6) eV and 4.1(7) eV respectively for the Lyman-α_1 and Lyman-α_2 lines, which may be compared with the theoretical value of 3.93 eV.

Gould and Marrus (1983) have improved their measurements on hydrogen-like argon whereby $4 \times 10^7 \text{ m s}^{-1}$ Ar^{13+} ions from the Lawrence–Berkeley laboratory Super-HILAC pass through a 400 μg cm^{-2} carbon foil target. Some emerge as stripped Ar^{18+} (62%), others as hydrogen-like Ar^{17+} (33%) and others as helium-like Ar^{16+} (5%) etc. Analysing magnets steered the stripped Ar^{17+} ions to a second target so that the hydrogen-like ions could be selected. The lifetime of the external electric field quenched $2\,^2S_{1/2}$ line, 3.2 keV (by mixing of the $2\,^2S_{1/2}$ and $2\,^2P_{1/2}$ states) was then measured as a function of the applied electric field. This lifetime depended on the Lamb shift between the two states. A measurement of the lifetime therefore enabled the Lamb shift to be calculated from the results. The latest result decreased the disagreement with theory, yielding 37.89(38) THz, within three standard deviations of both Mohr and Erickson's values, being lower than, but closest to, the former's calculated result.

5.5 The hydrogen hyperfine structure

The hyperfine splitting in atomic hydrogen, which is a measure of the interaction of the electron with the proton's magnetic field is an important link between the fields of high-energy physics and precision atomic physics. This is because this interaction is of short range and is sensitive to details of the proton structure and dynamics that are usually seen only in the high-energy electron–proton scattering experiments. Historically, the primary significance of the hyperfine structure has been as a probe of QED behaviour at short distances. The equation for the hyperfine splitting can be written in terms of the Fermi–Breit expression which has been corrected for vacuum polarisation and other QED effects, together with relativistic corrections, nuclear recoil and possible internal nuclear structure. Thus the expression is given by:

$$\Delta E_H = (16R_\infty/3)c\alpha^2(\mu_p/\mu_B)[m_p/(m_e + m_p)]^3$$
$$\times (1 + a_e + \tfrac{3}{2}\alpha^2 + \varepsilon_1 + \varepsilon_2 + \varepsilon_3 + \cdots). \qquad (5.5.1)$$

The radiative terms ε_1, $\varepsilon_2 \ldots$ are also expressible in terms of quantities involving the fine structure constant. The term $\tfrac{3}{2}\alpha^2$ is the Breit relativistic correction to the density of the electron wavefunction at the nucleus. The contributions to the terms including the QED term Q are listed in table 5.6. In the case of hydrogen there is also a term due to the possible polarisability of the proton, and internal proton structure. Here, theory and experiment suggest that it is close to zero within a few ppm. It is this term that limits the usefulness of measurements of the hyperfine splitting.

5.5.1 The hydrogen maser

The hydrogen hyperfine splitting may be measured experimentally to better than a part in 10^{12} with the aid of a hydrogen maser, but, as indicated above, uncertainties in the theoretical estimates of the proton polarisability limit the usefulness of the measurements as estimates of the fine structure constant to a few parts in a million. The hydrogen maser, however, is very useful as a very high precision frequency standard. The principle of the hydrogen maser is illustrated in figure 5.7.

The atomic hydrogen is generated by an RF discharge (~ 200 MHz) and

Table 5.5 The radiative terms in the expression for the ground state hyperfine splitting in hydrogen (which contribute ~ 103 ppm to $\Delta\nu_H$) and the nuclear size corrections (after Gidley and Rich 1981).

Term	Expression	Contribution to $\Delta\nu_H$ (parts per million)
ε_1	$\alpha^2(\ln 2 - 5/2)$	
ε_2	$-\dfrac{8\alpha^3}{3\pi}(\ln\alpha - \ln 4 + 281/480)$	-103
ε_3	$\dfrac{\alpha^3}{\pi}(18.4 \pm 5)$	
δ	nuclear size correction	-38.2
	recoil term	3.6
	nuclear polarisability	$2(3)$

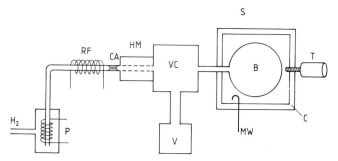

Figure 5.7 The general scheme of the active part of a hydrogen maser frequency standard. The hydrogen passes through the heated palladium leak P and the collimator CA, after being dissociated by the radio frequency RF. The beam is state selected by the hexapole magnet HM, and passes through the vacuum chamber VC into the PTFE coated bulb B. The bulb is shielded by the mumetal shield S and the microwaves MW coupled to the cavity C which is tuned to exact resonance by the tuning stub T.

atoms exit through a small orifice into the high-vacuum region. One of the $F = 0$ spin states is removed by passing the beam through the hexapole magnet and the beam enters the bulb in the microwave cavity through a small hole, sometimes a capillary array. The lifetime of the atomic hydrogen in the cavity is greatly increased by coating the surface of the quartz bulb with Teflon — similar to that used with kitchen utensils. The number of bounces made by the atoms depends on the ratio of the surface area of the bulb to the area of the orifice through which they enter and eventually exit. There is a small wall shift at the Teflon which is measured and corrected for by using bulbs of different area/aperture ratio. The shifts vary linearly with temperature above $23°$ C and they appear to become zero at about $100°$C. The shift for a 15 cm diameter bulb is about -23 mHz at $40°$C. If the flux of atoms entering the bulb is $\sim 10^{12}$ atoms s^{-1}, stimulated emission ($\sim 10^{-13}$–10^{-12} W), corresponding to the $(F = 1_1, m_F = 0) \to (F = 0_1, m_F = 0)$ transition frequency, takes place. The Q of the cavity must not be too great or it pulls the maser frequency. The first hydrogen maser was operated at Harvard by Goldenberg *et al* (1960) and many national standards laboratories now use them as very high-precision frequency standards whose short-term performance is better than caesium clocks.

Table 5.6 Contributions to the correction of the ground state hyperfine splitting in hydrogen.

Term ε	Value
$-\dfrac{\alpha}{\pi}\dfrac{3m_p m_e}{m_p^2 - m_e^2}\dfrac{2}{g_e}\dfrac{\mu_N}{\mu_p}\ln\dfrac{m_p}{m_e}$	$-0.000\,010\,2$
$-\dfrac{\alpha}{\pi}\dfrac{(\mu_p^2 - \mu_N^2)}{\mu_p/\mu_N}\dfrac{m_e}{m_p}16.5(0.6)$	$-0.000\,024\,0$ (9)
$(\alpha^2 \ln 1/\alpha)\dfrac{m_e m_p}{(m_e + m_p)^2}$	$0.000\,001\,1$
$\alpha^2 Q = \alpha^2\left(-2.5 + \ln 2 - \dfrac{\alpha}{\pi}(57.9 \pm 2)\right)$	$-0.000\,103\,4$ (2)
Unestimated recoil terms	$\pm 0.000\,000\,5$
Total	$-0.000\,136\,5(10)$

5.6 QED tests on the muon

The extent to which the muon may be represented as behaving similarly to a heavy electron may be tested in several ways, one of which we discuss in this section, namely the hyperfine splitting in muonium. Another way is the anomalous magnetic moment of the muon or mu-meson which is discussed in §5.9.

5.6.1 The hyperfine structure in muonium

The atom muonium comprises a nucleus, which is a positive muon, and an associated electron and provides perhaps the simplest system whereby the interaction between these nearly identical particles may be studied. As far as known, the system may be completely specified by quantum electrodynamics and the Bethe–Salpeter relativistic two-body equation. It has a hyperfine structure which is similar to that of hydrogen, but it avoids the problems of hadron structure which affect the comparison between theory and experiment. As a result, the precision measurement of the hyperfine structure in muonium provides a possible method of revealing differences between the muon and the electron. Hughes, at Yale University, has undertaken a systematic program of precision measurements of muonium for some twenty years and (with some eleven co-authors) reported (Casperson *et al* 1977), a measurement of the hyperfine splitting with an uncertainty of 0.52 kHz.

The hyperfine splitting of the ground state of muonium resembles that of hydrogen and the Breit–Rabi diagram (figure 5.8) shows the same type of Zeeman splitting pattern when a magnetic field is applied. In the experiment, the muons were produced by the parity violating decay of positive pions, $\pi^+ \rightarrow \mu^+ + \nu_\mu$. The parity violation was used to advantage, for it was used to provide polarised muons and also provided a method of monitoring the motion of their spins. The atomic muonium was formed by charge capture of an electron from a krypton gas target. The capture did not affect the spin direction. Thus the positive muon decays into a positron and two neutrinos and the positron is emitted preferentially in the direction of the muon spin. Spin

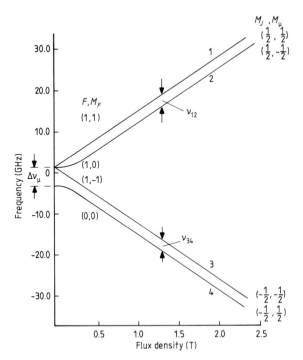

Figure 5.8 The ground state splitting of muonium in a magnetic field.

polarised muonium was formed inside the air-cored solenoid which produced a flux density of 1.35 T. The external magnetic field was in the opposite direction to the muon spin so that when the muons decayed the positrons were emitted preferentially in the backwards direction of the applied magnetic field. Thus $M_\mu = -\frac{1}{2}$ (figure 5.10) and the states $(\frac{1}{2}, -\frac{1}{2})$ and $(-\frac{1}{2}, -\frac{1}{2})$, labelled 2 and 3 respectively in the figure, were created in equal proportions. The muonium was then irradiated by two microwave fields which induced transitions at v_{12} and v_{34} between the states $2 \to 1$ and $3 \to 4$ respectively. This caused changes in the number of positrons emitted in the positive magnetic field direction. Of the two frequencies, the combination $(v_{12} + v_{34})$ was used to obtain the hyperfine splitting Δv_μ while $(v_{34} - v_{12})$, together with the proton spin precession frequency measured with an NMR water sample, was combined with the value of μ_p'/μ_B as an auxiliary constant in order to obtain the ratio of the muon mass to that of the electron. The results obtained by Casperson et al were

$$\Delta v_\mu = 4\,463\,302.35(52)\,\text{kHz}$$

and

$$\mu_\mu/\mu_p' = 3.183\,340\,3(44).$$

The latter was used to obtain

$$m_\mu/m_e = 206.768\,59(29).$$

5.6.2 The muonium hyperfine structure: theory

The theoretical calculations of the muonium hyperfine structure may be expressed in the form (to order α^3)

$$v_{th} = \tfrac{16}{3}\alpha^2 c R_\infty (\mu_\mu/\mu_B)(1 + m_e/m_\mu)^{-3}(1 + \tfrac{3}{2}\alpha^2 + a_e + \varepsilon_1 + \varepsilon_2 + \varepsilon_3 - \delta_\mu)$$

where

$$\varepsilon_1 = \alpha^2(\ln 2 - \tfrac{5}{2})$$
$$\varepsilon_2 = (-8\alpha^3/3\pi)\ln\alpha(\ln\alpha - \ln 4 + 281/480)$$
$$\varepsilon_3 = (\alpha^3/\pi)18.4(50)$$
$$a_e = (\alpha/2\pi) - 0.328\,48(\alpha/\pi)^2 + 1.184(7)(\alpha/\pi)^3$$

and are corrections which are also found in the hydrogen hyperfine structure. The δ_μ term is most relevant to muonium and may be written

$$\delta_\mu = \frac{m_e}{m_\mu}\left[\frac{3\alpha}{\pi}\left(1 - \frac{m_e^2}{m_\mu^2}\right)^{-1}\ln(m_\mu/m_e) + A\alpha^2\ln\alpha(1 + m_e/m_\mu)^{-2}\right.$$
$$\left. + B\alpha^2\ln(m_\mu/m_e) + \left(\frac{\alpha}{\pi}\right)^2\{C[\ln(m_\mu/m_e)]^2 + D\ln(m_\mu/m_e)\} + E\alpha^2\right].$$

Of these coefficients, A, B and C have been calculated to be 2, 0 and -2

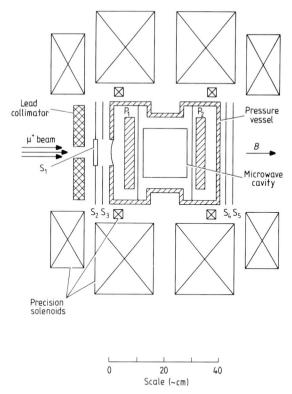

Figure 5.9 The experimental arrangement used by Casperson *et al* (1977) to measure the $v(\mu)$. P_1 and P_2 were proportional counters and S_1–S_5 were plastic scintillators. The signals were: (i) stopping muon: $S_1 + S_2$ and not P_2; (ii) forward decay positron: $S_4 + S_5 + P_2$ without P_1; (iii) backward positron $S_2 + S_3 + P_1$ without P_2.

respectively. Neglecting the D and E terms and using $\alpha^{-1} = 137.035\,987(27)$, the theoretical value becomes $4\,463\,297.9\,\text{kHz}$, which is essentially obtained using the Olsen and Williams low-field γ'_p result. This value is in good agreement with the experimental value of Casperson *et al* (1977) (figure 5.9) of

$$4\,463\,302.35(52)\,\text{kHz} \; (0.12\,\text{ppm})$$

and the subsequent improved precision of 3 parts in 10^8 of

$$4\,463\,302.88(16) \; (0.04\,\text{ppm})$$

by Hughes *et al* (1982) and Mariam *et al* (1982). Since the ratio of the muon to electron mass is also known with high precision from this experiment, the hyperfine splitting in muonium may be used to deduce a value for α^{-1}. Such a value was used in the 1973 evaluation of the physical constants. The object in giving the detailed expressions above was to demonstrate the complexities of the theoretical calculations to the experimental metrologist.

5.7 Positronium

The simplest possible 'atom' is that formed by the combination of an electron and a positron. This simple system does not have the complications which the involvement of baryons in a conventional atomic system bring to the theoretical understanding of the experimental situation. Consequently, both the spectroscopy and lifetime of positronium may be calculated with high precision. These calculations involve complex QED calculations and hence there is considerable interest in a comparison between theory and experiment. The lifetime of para-positronium is too short for an accurate experimental measurement, but ortho-positronium is suitable, for its lifetime is about $7\,\mu s$. We discuss the hyperfine splitting first.

5.7.1 The fine structure of positronium

The fine structure of positronium is of theoretical interest because it provides the most stringent test of the relativistic theory of bound states. This may also be expressed in terms of a power series, involving the fine structure constant, of the form

$$v_{th} = \alpha^2 c R_\infty [7/6 - (\alpha/\pi)(16/9 + \ln 2) + A\alpha^2 \ln \alpha + B\alpha^2 + ...].$$

The value of A was recalculated by both Lepage (1977) and by Bodwin and Yennie (1978) and while their calculations disagreed with earlier results both results for A were identical, being

$$A = -5/12.$$

The higher-order terms which contribute to the B term have only been partly calculated. These include effects from the analogue of the hydrogen hyperfine structure and two- and three-photon corrections. The first two of these dominate and contribute -18.10 MHz and $+13.3$ MHz respectively. The other contributions calculated so far are less than a MHz in magnitude. The theoretical value without the B term is

$$v_{th} = 203\,400.3(100)\,\text{MHz}$$

and this may be compared with the best experimental measurements by Egan *et al* (1977) and by Mills and Berman (1975) who obtained

$$v_{exp} = 203\,384.9(12)\,\text{MHz}$$

and

$$v_{exp} = 203\,387.0(16)\,\text{MHz}$$

respectively.

The methods were essentially to stop positrons which were emitted from a ^{22}Na source in a target gas of either nitrogen or SF_6. Positronium is formed with an efficiency of about 30%. The direct excitation of the ground state splitting of 203 GHz would require too much microwave power for present day sources. However, in the measurement, a magnetic field ~0.8T was applied over the region of positronium formation and, since the decay of the triplet $m = 0$ state was much faster than that of the $m = \pm 1$ states, the Zeeman transition was excited by applying an RF field, $f_{01} \sim 2\,\text{GHz}$ (figure 5.10). (Essentially the magnetic field mixes the singlet state in with the triplet state $m = 0$ so that it proceeds primarily by two-photon rather than three-photon emission.) The Zeeman resonance was detected as an increase in the two-photon coincidence counting rate of pairs of γ-ray scintillation detectors which were mounted at $180°$ to one another around the gas chamber. The frequency f_{01} is given by the Breit–Rabi type of expression, namely (writing Ps for positronium)

$$f_{01} = \tfrac{1}{2}\Delta v(\text{Ps})[(1 + x^2)^{1/2} - 1]$$

where

$$x = 2g'\mu_{\text{B}}B/h\Delta v(\text{Ps}).$$

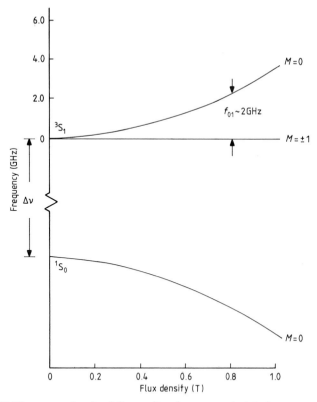

Figure 5.10 The energy levels of the positronium ground state in a magnetic field.

Hence we may obtain $\Delta\nu(\text{Ps})$ from the measured f_{01}. A further correction $\sim 9 \times 10^{-6}$ will be important for future measurements to allow for the effects of decay on the energy eigenvalues in a magnetic field (see Rich 1981). Since the positronium is formed in a gas it is necessary to extrapolate the measurements linearly back to zero pressure (the shift is ~ 7 MHz atm^{-1} in N_2). The experiments are not easy, especially since the natural linewidth of the transition is some 0.6% of the splitting.

It is apparent that the agreement between theory and experiment is at present quite reasonable, particularly since the components of the B term which have been calculated so far reduce the theoretical value to 203 395.45 MHz. However, a meaningful comparison between theory and experiment must await a more complete evaluation of the B term. Sapirstein (1983) has made improved calculations of the radiative corrections to the lepton lines in positronium and muonium hyperfine splitting, amounting to $-11.12(2)$ MHz and 2.64(7) kHz respectively. In a following paper, he made an improved estimate of the last term C in the correction to the hyperfine splitting of hydrogenic atoms. The correction ΔE to the Fermi splitting E_F is given, in terms of the fine structure constant α and the charge Z, by

$$\Delta E = [\alpha(Z\alpha)^2 E_F/\pi][-\tfrac{2}{3}\ln^2(Z\alpha)^{-2} + (37/72 + 4/15 - \tfrac{8}{3}\ln 2)\ln(Z\alpha)^{-2} + C]$$

where Sapirstein finds $C = 15.10(29)$. This is particularly useful in connection with the muonium hyperfine splitting, since the results of Mariam *et al* (1982) yield a value of

$$\alpha^{-1} = 137.035\,988(20)$$

in very good agreement with the other estimates of the fine structure constant. The present result for C leads to a reduction of the theoretical ground state hydrogen fine structure splitting by about 400 Hz and can be used to derive improved estimates of the proton electromagnetic radius and polarisability. At present the lack of knowledge of these contributes uncertainties of 0.3 and 3 parts per million to the ground state splitting respectively.

5.7.2 The decay rate of ortho-positronium

The lifetime of ortho-positronium may be expressed in the form of a power series involving the fine structure constant and is of the form

$$\lambda = \lambda_0(1 - 10.348(70)\alpha/\pi - \tfrac{1}{3}\alpha^2\ln\alpha^{-1})$$

where

$$\lambda_0 = (2/9)\alpha^6(mc^2/\hbar)(\pi^2 - 9).$$

The calculations of Caswell *et al* (1977) were further refined to include the higher-order term given above and lead to a predicted lifetime of 7.0379(12) μs^{-1}. This differs considerably from the earlier prediction of

Stroscio (1975) who obtained an expression of the form

$$\lambda = \lambda_0(1 + 1.186(45)\alpha/\pi) = 7.242(8)\,\mu s^{-1}.$$

It is interesting that the earlier determinations were in very good agreement with the last mentioned value, whereas the later measurements, Gidley *et al* (1978) and Gidley and Zitzewitz (1978), were in good agreement with the Caswell *et al* calculation. Evidently, we have had the interesting situation where both theory and experiment have led to changes in the results by similar amounts, $\sim 3\%$! One of the difficulties in the earlier work stemmed from the necessity to bring the positronium to rest before it decayed. This led to the need to extrapolate to the zero density condition. The earlier work by the Michigan group was probably affected by errors in their measurement of the density of the SiO_2 powder and also by afterpulses in their photomultipliers.

The experiment of Caswell *et al* made use of a ^{68}Ge positron source and the positrons were stopped in a mixture of 80% Freon-12 and 20% isobutane, the latter being chosen because it has a cross section for positron annihilation which is some 30 times greater than Freon-12 and hence quenches the free positron lifetime. The positronium decays into three γ-rays and these were detected with scintillators and associated photomultipliers. Measurements were obtained for gas densities between $1\,gl^{-1}$ and $10\,gl^{-1}$ (about 200 Torr to 1600 Torr) and the lifetimes extrapolated back to zero pressure or density. The lifetime at a density of about $10\,gl^{-1}$ was about $8\,\mu s^{-1}$ so that it is evident that accurate density measurements are required for this type of experiment. The overall value obtained by Caswell *et al* was $7.056(7)\,\mu s^{-1}$, which is in excellent agreement with the later theoretical values as also was the later measurement in SiO_2 by Gidley and Zitzewitz (1978). Further progress is likely to be made with all of this work in the near future.

5.8 The electron magnetic moment anomaly: $g - 2$ for the electron

The anomalous magnetic moment of the electron may also be written in terms of a series involving powers of the fine structure constant α. Thus

$$\mu_e/\mu_B - 1 = (g_e - 2)/2 = a_e$$

where QED predicts the value of a_e in terms of the power series in α/π as

$$a_e = A(\alpha/\pi) + B(\alpha/\pi)^2 + C(\alpha/\pi)^3 + D(\alpha/\pi)^4 + \cdots$$

$$+ \text{weak interaction corrections}$$
$$+ \text{muonic corrections}$$
$$+ \text{hadronic corrections}$$

where

$A = 0.50$	$C = 1.179 \pm 0.10$
$B = 0.328\,48$	$D = \text{unknown.}$

The coefficient C above is based on the weighted average of Levine and Roskies (1976) and Kinoshita and Roskies (1977). The hadronic correction is 1.4 ± 0.2 ppb while the muonic vacuum polarisation contributes 2.5 ppb. Although most weak interaction theories predict effects of 1 ppb or less, there are others which predict a larger contribution. Numerical contributions from QED are at present limited by an uncertainty of 0.12 ppm in the fine structure constant which in turn depends on the NBS weak-field measurement of the gyromagnetic ratio of the proton. It is to be hoped that the precision of this work will be improved by a further factor of five in the near future so that a value of α to 25 ppb will give the C coefficient to 0.2%. Since $(\alpha/\pi)^4$ is 30 ppb of a_e, the measurement of $g_e - 2$ to this precision also gives information about the D term. (Note: 1 ppb is an uncertainty of 10^{-9})

5.8.1 Experimental measurements of $g_e - 2$

The above sets a daunting target for $g_e - 2$ measurements in the next decade, but there are at least two groups that expect to meet this challenge. However, it is necessary first to discuss the present methods. The measurements at the University of Michigan stem from a series of measurements of $g_e - 2$ with ever increasing precision. They are based on the fact that the $g_e - 2$ term is relatively small. In consequence, the spin precession frequency of the electron is only slightly greater than the cyclotron frequency. The experiment therefore comprises trapping electrons in a magnetic field and measuring the amount by which the direction of polarisation of the electrons changes after the electrons have been trapped for a known time. There have now been two distinct types of measurement which are characterised by the energy of the electrons. This is

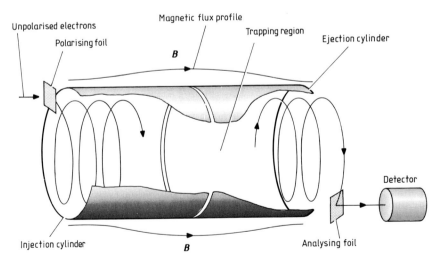

Figure 5.11 The experimental arrangement used in the Michigan $g_e - 2$ measurements by Rich *et al* (1973).

an important characterisation, for as will be discussed in § 5.11.1, the fact that the relativistic corrections are different has allowed the measurements to be regarded as being tests of special relativity.

5.8.2 The high energy measurement

This experiment has been performed at the University of Michigan and evolved from the early work of Wilkinson and Crane (1963). The experiment to be discussed is that of Wesley and Rich (1971) who succeeded in measuring the g-factor anomaly with a precision of 3 ppm.

The entire evacuated ($\sim 10^{-5}$ Nm^{-2}) apparatus (figure 5.11) was located inside a flux density 0.1 T which was produced by a large room temperature NMR stabilised air-cored solenoid. Polarised electrons were produced by Mott scattering of ~ 20 mA, 100 ns pulses of 100 keV, unpolarised electrons from a 10^{-7} m thick gold foil target. Some ($\sim 10^{-6}$) of those partially polarised electrons scattered elastically in a direction which was nearly perpendicular to the magnetic field. The trajectory of selected electrons was defined by an annulus to be in the form of a 2 mm wide beam on a 10 mm radius helix having a 1 mm pitch so that they drifted into a trapping region. The trap was a magnetic one and was formed by shaping the magnetic field so that it was some 60 ppm greater at the ends of the trap than it was at the centre. This region was enclosed by two coaxial metal cylinders. As the 100 ns pulse of electrons drifted through the gap between the cylinders, a 100 ns positive voltage pulse was applied to the injection cylinder. This caused the electrons to lose sufficient axial velocity to become permanently trapped by the magnetic field. While the electrons were in the magnetic field their polarisation P and their velocity v precessed at the spin precession and cyclotron frequencies ω_s and ω_c respectively, and the relative precession of P with respect to v occurred at the difference between the two frequencies. If v is perpendicular to the magnetic field B, the difference frequency has the simple form

$$\omega_D = a_e \omega_c = a_e \omega_0 / \gamma$$

where $\omega_0 = eB/m_e c, \gamma = (1 - v^2/c^2)^{-1/2}$ and $\omega_c = \omega_0/\gamma$. In practice, the motion was rather more complicated, and corrections were made to take account of the axial motion and the variation of the magnetic field, but the above equations illustrate the general principle.

The ejection was effected by a positive 2 μs pulse to the ejector cylinder and the process was designed so that the arrival direction of electrons incident on the analysing foil was independent of the trapping time T. Since there was asymmetry in the Mott scattering, the fraction of the electrons scattered into the detector varies as $P \cdot v$. Consequently, if the number of electrons ejected from the trap is independent of T, the counting rate at the detector will be given by

$$R(t) = R_0 [1 + \delta \cos(\omega_D T + \phi)]$$

where R_0 is the rate for unpolarised electrons and δ is the Mott asymmetry factor (which was typically 0.02) and ϕ was a phase constant. The difference frequency ω_D was determined by sampling the value of $R(T)$ at different times and fitting a periodic function by the method of least squares. If N oscillations of $R(T)$ are observed between times T_1 and T_2, then

$$\omega_D = 2\pi N/T(T_2 - T_1)$$

where $N \sim 2 \times 10^4$, $T \sim 2\text{–}5\,\text{ms}$. An average of 0.5 electrons of the ~ 5000 trapped electrons was detected per operating cycle so that a long observation time was required to achieve the required statistical accuracy. The zero energy cyclotron frequency was not measured directly. Instead it was inferred from the combination of the auxiliary constant (μ_p'/μ_B) with the magnetic field as determined by the spin precession frequency γ_p' of protons in a cylindrical water sample (whose relaxation time was reduced by adding a 0.2 molar solution of copper sulphate). Thus the frequency was given by

$$\omega_0 = (\mu_p'/\mu_B)^{-1}\gamma_p(H_2O).$$

The probe shape was identical to that used by Klein (1968) for his measurement of μ_p'/μ_B. A second NMR probe was used to evaluate the magnetic field over the trajectory of the trapped electrons in order to correct for the magnetic flux variations over the trajectory.

The result of the Wesley and Rich measurement was

$$a_e = \tfrac{1}{2}(g_e - 2) = 1\,159\,657.7(35) \times 10^{-9}.$$

5.8.3 The low energy measurement of $g - 2$: experiments on single electrons

One of the most remarkable developments in recent years has been the development of Penning traps which allow single electrons to be trapped for many hours (indeed, days) notably by the Dehmelt group at the University of Washington (Van Dyck et al 1976, 1977, 1978). Not only have single electrons been trapped, but it has been possible to make measurements on them with high precision. One such experiment has been the measurement of $g_e - 2$ for the electron.

In contrast with the Michigan experiment the energy of the electrons in the trap has been very low ($\sim$ meV), since the electron temperature was $\sim 4\text{K}$. Following the notation of the Washington group, the trapping potential distribution was of the form

$$\phi = A(r^2 - 2z^2) + B(r^4 - 24r^2z^2 + 8z^4) + \cdots$$

where the higher order terms were a result of the truncation of the hyperboloid electrodes. As a result of some initial experiments it was found that the z^4 term led to a hundredfold broadening of the resonances and so extra electrodes were added to reduce the effect of this term. The equations of motion for the electron in the trap may be written:

$$\ddot{x} - \omega_z^2 x/2 + \omega_c \dot{y} = 0$$
$$\ddot{y} - \omega_z^2 y/2 - \omega_c \dot{x} = 0$$
$$\ddot{z} + \omega_z^2 z = 0.$$

These lead to the solution:

$$x + iy = r_c e^{i\omega_c' t} + r_m e^{i\omega_m t}$$

Taking $v_c' = \omega_c'/2\pi = v_c - \delta_e$, $2\delta_e v_c' = v_z^2$, then, for a perfect trap, the magnetron frequency $v_m = \omega_m/2\pi = \delta_e$. Although small misalignments of the electrodes are inevitable, so that $v_m - \delta_e > 0$, the relation $2\delta_e v_c = v_z^2$ is essentially invariant. The system is in a very low energy state so that quantum mechanics must be applied and the system has equidistant energy levels

$$E_{m,n,k,q} = h[mv_s + (n + \tfrac{1}{2})v_c' + (k + \tfrac{1}{2})v_z - (q + \tfrac{1}{2})v_m]$$

where $m = -\tfrac{1}{2}$ and $n, k, q = 0, 1, 2, 3....$

The amplitude of the oscillations is quite small; the radius of the cyclotron orbit, determined by the temperature, is about 60 nm and that of the magnetron orbit about 50 μm to 3.35 mm (usually at the lower end of the range). Since the system may be regarded as comprising a single electron bound to the Earth via the magnet and trap electrodes, the Washington group have termed this pseudo-atom 'geonium'. The cyclotron and magnetron orbit radii are given by

$$r_c = (2n + 1)^{1/2} r_0 \qquad\qquad r_m = (2q + 1)^{1/2} r_0$$

with

$$r_0 \simeq (\lambda_c \lambda_c)^{1/2} \sim 20 \text{ nm}$$

where λ_c and λ_c are the cyclotron and Compton wavelengths of the electron respectively. In the apparatus a small inhomogeneity was added to the magnetic field, which was otherwise very homogeneous. This was applied by a nickel ring, and more recently by inducing a persistent current in a superconducting ring, and thereby provided coupling between the frequencies so that they could be detected. This bottle field led to a shift in the axial frequency by

$$\delta v_z = [m + n + \tfrac{1}{2} + (v_m/v_c)q]\delta$$

where

$$\delta \simeq \mu_B \beta/2\pi^2 m v_z \simeq 1 \text{ Hz.}$$

For large n and q this becomes

$$\delta v_z (r_c/r_0)^2 + (v_m/v_c)(r_m/r_0)^2 \delta/2.$$

The bottle field had the form

$$\phi_z = \beta(z^2 - r^2/2) \qquad \phi_y = \beta_{zy} \qquad \phi_x = \beta zx$$

with

$$\beta = 120 \text{ Tm}^{-2}.$$

The interaction of the bound electron with the cap electrodes shows essentially the properties of a resonant LC circuit which is resonant at v_z. Wineland and Dehmelt (1975) made use of this, and the attenuated output from a frequency synthesiser was connected to one electrode so that this signal was coupled through the LC circuit to the other electrode. This electrode was connected to a low noise gallium arsenide field effect transistor (GAT-1) operating at 4 K in a liquid helium bath, where it had a low noise temperature. This low-noise stage was followed by more conventional amplifiers and a phase sensitive detector.

A small probe electrode introduced the magnetron frequency and the beat between the free and driven axial motions could be detected quite sensitively. Beats could be seen for driven frequencies which were only 10 mHz from the free signal, which enable the magnetron frequency to be determined to $\sim 1\text{mHz}$. The effects of adding energy to the cyclotron frequency could similarly be detected, for when the electron energy was excited to higher levels it changed the axial motion of the electron. The cyclotron frequency v_c' was about 51 GHz and was generated as a harmonic of an 8.6GHz signal by means of a microwave diode.

The second important function of the magnetic bottle was to provide a mechanism for flipping the electron spin. Thus, in the reference frame of the electron, the magnetic field is seen as rotating at the cyclotron frequency v_c' and this frequency is modulated by the axial frequency thereby yielding sidebands at $v_c' \pm v_a$ (where the spin precession frequency is $v_s = v_c' + v_a'$). Each time that the spin flipped from $m = -\frac{1}{2}$ to $m = +\frac{1}{2}$, the noise pattern due to the thermally excited cyclotron resonances changed by 1 Hz. This could be monitored by looking at the induced signal as the frequency v_a' was varied.

It is interesting (such is the sensitivity of this experiment) that, despite the electrons having thermal energies at $\lesssim 4$ K, it was necessary to apply a relativistic correction $\sim -10^{-9}$. Electrons were introduced into the trap by producing them either by emission from a heated tungsten filament or by field emission from a tungsten point, both of which were on the axis of the device. Some electrons were able to enter the trap through axial holes and by applying suitable voltages to the electrodes the electrons could be gradually cooled inside the trap and the voltages substantially reduced.

When one reads the papers which have been published by the group at the University of Washington, their experiments tend to appear deceptively simple. It is apparent that since experiments were being performed on single electrons, even with all the frequencies properly adjusted, the signal to noise ratio was quite small. Thus many patient hours of experimentation, well into the night, were required before success was obtained. Initially, of course, it was possible to work with a number of electrons in the trap, which gave larger signals.

The interest of this experiment is that not only have experiments on single electrons been demonstrated, but they have led to the very accurate measurement of $g_e - 2$. The g-value of the electron is now the most accurately measured of all the fundamental constants. The result obtained by the Washington group was

$$a_e = \tfrac{1}{2}(g_e - 2) = \mu_e/\mu_B - 1 = 115\,965\,220\,0(40) \times 10^{-12}.$$

The work has been extended to positrons for a measurement of $g - 2$ for the positron and also to measure the ratio of the proton mass to that of the electron. Schwinberg et al (1981) have reported a measurement of the positron g-factor and a measurement of $g_{e^-} - g_{e^+}$. They used a double Penning trap. The first of these stored the positrons so that a pulse could be injected into the second and the g-factor of a single trapped positron measured.

The measurements based on four runs yielded

$$a(e^+) = 1\,159\,652\,222(50) \times 10^{-12}.$$

This may be compared with the theoretical value, based on

$$a_e = \tfrac{1}{2}\alpha/\pi - 0.328\,479\,0(\alpha/\pi)^2 + C_3(\alpha/\pi)^3 + C_4(\alpha/\pi)^4$$

where $C_3 = 1.176\,5(13)$ (Levine and Roskies 1982) and $C_4 = -0.8(2.5)$ (Kinoshita 1981), to give:

$$a(e^+)_{th} = 1\,159\,652\,460(\tfrac{127}{75}) \times 10^{-12}$$

(127: uncertainty in α; 75: uncertainty in theory). Thus theory and experiment agree at the two-sigma level. Comparison with the results for g_{e^-} gives

$$g_{e^+}/g_{e^-} = 1 + (22 \pm 64) \times 10^{-12}.$$

The use of Penning-type traps to perform experiments on single ions is likely to extend in a number of areas including measurements of fundamental constants. Thus measurements in conjunction with dye lasers on trapped He^+ ions would be of particular interest. This is a hydrogen-like atom whose theory is very well understood, and the Zeeman effects from the magnetic field and small Stark shifts from the electric field may not provide an insurmountable obstacle to measurements of Lamb shifts, the Rydberg constant etc. However, other methods are setting a daunting target.

5.9 The muon magnetic moment

As with the electron, the muon magnetic moment, in the form g-2, is calculable from quantum electrodynamics in a form expressible in terms of a power series involving the fine structure constant, i.e.

$$g_\mu - 2 = a_\mu = \tfrac{1}{2}\alpha/\pi + 0.765\,728\,23(\alpha/\pi)^2 + 24.45(6)(\alpha/\pi)^3 + 135(63)(\alpha/\pi)^4 \cdots .$$

Using the value of $\alpha^{-1} = 137.035\,987(29)$, this gives

$$g_\mu - 2 = 1\,165\,852.0(2.4) \times 10^{-9}.$$

From CPT conservation one expects that $g_{\mu^+} = g_{\mu^-}$. There are some additional correction terms to the above expression which are discussed after the experimental measurements.

5.9.1 The experimental value

The experimental determination of the muon magnetic moment was made possible through a series of measurements of ever higher precision at CERN. The inherent experimental problem was to contain the muons for sufficient time to enable the difference frequency to be measured to the required accuracy. Although the magnetic field in the CERN measurements contained the muons in two directions, a method of trapping them in the magnetic field direction was required. The first measurements used a gradient magnet to trap the muons. In 1961 this method yielded a value which had a precision of 2%, and further measurements which were published in 1962 (Charpak et al 1962) reduced the uncertainty to 0.4% (Alvarger et al 1964, 1966). New sets of measurements were made between 1965 and 1968. The improvements involved using a storage ring, with weak magnetic focusing, and yielded a result which had an estimated uncertainty of 270 ppm. In the final experiment (Farley et al (1977), figure 5.12), the muons were obtained from the decay of a beam of pions. The muon beam was deflected into a 14 m diameter storage ring which had electrostatic focusing for ~ 0.1 ms.

As a result of the pion decay process the muons were polarised in the forward direction and this polarisation rotated in advance of the cyclotron frequency by an amount

$$\omega_{a_\mu} = \tfrac{1}{2}(g_\mu - 2)eB/m_\mu c = a_\mu eB/m_\mu c$$

which was independent of the muon momentum. An electrostatic field was applied in the vertical direction to trap the muons vertically, i.e. to overcome the velocity in the magnetic field direction. This trapping field inevitably had

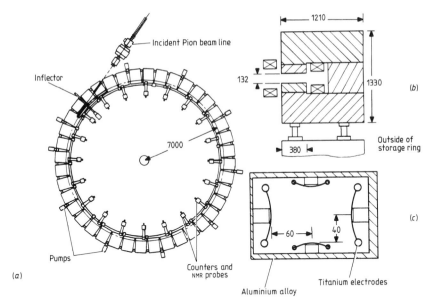

Figure 5.12 Scheme of the muon storage ring used by Farley *et al* (1973) in the CERN measurement of the muon $g_\mu - 2$ anomaly: (*a*) the storage ring; (*b*) cross section of the 'C'-magnets; and (*c*) cross section of the vacuum chamber and electric quadrupole. Dimensions shown are in millimetres.

components in the orbit plane (in order to satisfy $\nabla^2\phi = 0$, where ϕ was the electric potential) and these would normally affect the relationship between the polarisation and the velocity. However, a study of the equations of motion showed that for particles where there was a positive anomaly in the magnetic moment, i.e. where the polarisation turned faster than the velocity, the effect of the transverse electric field on the cyclotron frequency changed sign at a particular value of the relativistic term. This was at the so called magic γ which was given by

$$\gamma_{\text{magic}} = \sqrt{\frac{1}{a_\mu} - 1} = 29.3.$$

Fortunately, this value of γ corresponded to muons of a convenient momentum, namely 3.094 GeV/c. The actual beam energy had a small spread about this magic value, but the electric field effects on the non-optimum muons were of second order, amounting to less than a part per million. The decay electrons produced when the muon polarisation was in the forward direction of motion had the greatest energy, and so by restricting observations to decay electrons which had the highest energy, their decay rate was modulated in a manner which depended on the difference frequency $\omega_{0,\mu}$. The direction of the polarisation of the muons at the instant that they decayed was inferred by making measurements on the decay electrons. The value of the magnetic flux density was obtained by averaging measurements with an array of some 80 nuclear magnetic resonance probes uniformly arrayed over the muon orbit. Finally, the ratio of the proton spin precession frequency to the muon spin precession frequency was measured in a separate experiment by experiments with muons at rest in a magnetic field.

Measurements of $g_\mu - 2$ were made for both positive and negative muons and the final results were:

$$a_{\mu^- \text{exp}} = 1\,165\,936(12) \times 10^{-9}$$
$$a_{\mu^+ \text{exp}} = 1\,165\,910(12) \times 10^{-9}$$

which yield a mean of

$$a_{\mu \text{exp}} = 1\,165\,923\,(10) \times 10^{-9}$$

5.9.2 Corrections to the QED expression

Whereas the hadron contribution to the electron magnetic moment, through contributions to the photons vacuum polarisation was $1.6(2) \times 10^{-12}$, it is considerably larger for the muon, amounting to $a_\mu^{\text{hadron}} = 66.7(9.4) \times 10^{-9}$ and this, together with an expected weak interaction effect of $a_\mu^{\text{weak}} = 2 \times 10^{-9}$, and the τ-meson contribution $a_\mu(\tau\text{-meson}) \sim (1/45)(\alpha/\pi)^2 (m_\mu/m_\tau)^2 = 0.42 \times 10^{-9}$ should be added to the value of a_μ^{QED} given above in order to compare theory with experiment, to yield

$$a_{\mu \text{th}} = 1\,165\,921(10) \times 10^{-9}$$

which is in excellent agreement with the observed values.

5.9.3 Test of CPT theorem for the muon

The above values of a_μ may be used to test the validity of the CPT theorem for the muon. Thus we obtain

$$(g_{\mu^-} - g_{\mu^+})/g_\mu = 2.6(1.7) \times 10^{-8}$$

which consequently validates CPT invariance to high order.

Although the muon measurements enable the hadron contribution to be checked, and hence it is more than the validity of QED that is being tested, in order to check the weak interaction contribution it would be necessary to measure the g-factor anomaly to better than about a part in a million. This improvement is just beyond the reach of possible improvements to the present experimental technique, which would appear to be limited to 1–2 ppm. Moreover, it would be necessary to improve the accuracy of the calculations of the hadronic contribution. Further improvements appear unlikely for the present. While not wishing to restrict the pace of progress, it would be a pity if such a beautiful set of experiments was eclipsed too soon!

5.10 The current status of QED

In all of the high-precision test areas for QED which have been discussed there is satisfactory agreement between theory and experiment at the time of writing. The area continues to be an active one both experimentally and theoretically. One can but admire the work of the dedicated theoreticians who wrestle successfully with the complex computing required to evaluate getting on for a thousand Feynman diagrams. The very number of such diagrams involved suggests that some day the elegant Feynman diagram may go the way of the Copernican epicycles to be replaced by a theory whose beauty we can only guess at.

5.11 $g-2$ tests of special relativity

The general and special theories of relativity are central to modern physics and it is important to test them as precisely as possible. One normally associates relativistic effects with high velocities and high energies. Electrons, being light particles, attain high velocities at relatively low energies and, particularly in high-precision experiments, it is necessary to take relativistic effects into account. It was pointed out by Newman *et al* (1978) that the two measurements of $g-2$ for the electron, at the Universities of Michigan and Washington, provide between them one of the most sensitive tests of special relativity to date. The reason for this was that the electrons in the Washington measurement had an energy of about 5×10^{-4} eV ($\beta = 5 \times 10^{-5}$, $\gamma - 1 = 10^{-9}$) and that at Michigan the electron energy was 110 keV ($\beta = 0.57$, $\gamma - 1 = 0.2$). The claimed sensitivity of the method, i.e. the experimental uncertainty divided by the size of relativistic effect, was 3.5×10^{-9}.

The postulated breakdown of relativity was assumed by Newman *et al* to come from a breakdown of the energy–momentum relation at high energies. Consequently, the difference between the two measurements of $g_e - 2$, expressed as the difference ω_D between the spin precession frequency ω_s and the cyclotron frequency ω_c as a fraction of the cyclotron frequency, provides a test of relativity, thus:

$$\omega_D/\omega_C = (\omega_s - \omega_c)/\omega_c = g/2 - \gamma/\bar{\gamma}$$

where the cyclotron frequency is given by

$$\omega_C = eB/\bar{\gamma}m_e c,$$

the $\bar{\gamma}$ being introduced through the momentum–energy relation. The spin precession frequency is given by the sum of the term due to the interaction of the electron magnetic moment with the magnetic field and the second term which is due to the Thomas precession. This latter is a result of the kinematics of special relativity when applied to accelerated systems, that is $\omega_s =$

$geB/2m + (1 - \gamma)\omega_c$. The term γ is simply $\gamma = (1 - \beta^2)^{-1/2}$ with $\beta = c^{-1}dE/dp$. The term $\bar{\gamma} = (p/m)\,dp/dE$, where the electron energy and momentum are E and p respectively, and the rest mass is given by

$$\frac{1}{m} = \lim_{p \to 0}\left(\frac{1}{p}\frac{dE}{dp}\right).$$

The difference between the two experiments gave $1 - (\gamma/\bar{\gamma}) = (5.3 \pm 3.5) \times 10^{-9}$, which verified relativity to this precision at least for the particular mode of breakdown investigated by Newman *et al*. Since the accelerations were different in the two experiments, $1.3 \times 10^{-22}\,\mathrm{m\,s^{-2}}$ compared with $10^{-20}\,\mathrm{m\,s^{-2}}$, similar limits were placed on the effects of acceleration.

The publication of the above work stimulated a similar analysis of other $g - 2$ experiments. Cooper *et al* (1979) pointed out that a value of $g - 2$ for the electron had been obtained from an analysis of the polarisation of the high-energy longitudinally polarised beam of electrons produced in the Stanford Linear Accelerator Center (SLAC). After acceleration to a kinetic energy of $12\,\mathrm{GeV}$ the beam was magnetically deflected into the experimental area through an angle θ_c of $24.5°$ and the spin precessed relative to the momentum by an angle

$$\theta_a = a_e\theta_c.$$

In consequence, the longitudinal component of the polarisation of the beam was then given by

$$P(E) = P_0 \cos(\pi E/E_0 + \phi_0)$$

where P_0 was the magnitude of the initial polarisation vector $\boldsymbol{P}_0$ of the beam before the magnetic deflection, ϕ_0 was the projected angle of $\boldsymbol{P}_0$ with respect to the electron momentum in the plane of the bent trajectory, E was the electron energy and the parameter E_0 was defined as

$$E_0 = (180°/24.5°)m_0c^2/a_e \sim 3.2\,\mathrm{GeV}$$

where a_e was the g-factor anomaly and m_0 the electron mass. A best fit to their experimental measurements yielded the value

$$a_e = 11\,622(200) \times 10^{-7}.$$

Although this value was of lower precision than the Michigan value, it was obtained for electrons at $12\,\mathrm{GeV}$, or $\gamma = 2.5 \times 10^4$, compared with the electron energy of $100\,\mathrm{keV}$, or $\gamma = 1.2$, in the work of Newman *et al*. The Stanford experiment therefore provided an even more sensitive test of special relativity.

Combley *et al* (1979) also interpreted their measurements of the g-value of the muon as tests of relativity. They compared the g-value of their first experiment (Charpak *et al* 1965) with their latest result and also compared the lifetime of the muons in the storage ring with that obtained for muons at rest (Bailey *et al* 1977). Both sets of workers extended the phenomenological model of Newman *et al.* In particular, there are possibilities of different values of γ corresponding to different relativistic transformations for time, electromagnetic fields and mass and for the Thomas precession frequency, and the tests are of

$$\gamma_m : \gamma_t (= \gamma_t / \gamma_T).$$

The ratio of the difference between the precession and cyclotron frequencies was taken as

$$\omega_a (eB/m_0 c)^{-1} = \tfrac{1}{2} g - \gamma/\bar{\gamma} = a$$

and the different electron and muon velocities used to evaluate $\gamma/\bar{\gamma}$. The expression for $\gamma/\bar{\gamma}$ was set as a power series and the data used to evaluate the term C_1, and thus $\gamma/\bar{\gamma} = 1 + C_1(\gamma - 1) + \cdots$. The justification for the use of this series was based on the fact that in the low velocity limit, $\gamma \to 1$, it preserved the non-relativistic equivalence of γ and $\bar{\gamma}$. The values of C_1 deduced from the combination of the results at different values of γ are shown in table 5.7. It is seen that the results put quite tight limits on the value of the linear term in the equation above, but it should be noted that there have been suggestions by Redei (1967) that the leading term in the expression for the muon lifetime should be a quadratic one. However, the experiments may be regarded as testing special relativity within the constraints of the model which was adopted. As with all null experiments the results are open to reinterpretation.

Table 5.7 Summary of the use of the lepton $g-2$ experiments to test a postulated breakdown of special relativity. The figures in parentheses are the standard deviation uncertainty of the last digit quoted.

	Method	γ_{low}	γ_{high}	C_1
muon	1967 and 1977, CERN	12	29.2	$1.4(18) \times 10^{-8}$
electron	Washington and Michigan	1	1.2	$-2.6(18) \times 10^{-8}$
electron	Washington and Stanford	1	$(1.3 \pm 3.8) \times 10^4$	$-1.0(80) \times 10^{-10}$

5.11.1 Relativity effects in time measurement

The very high precision which is achieved in modern atomic clocks has reached the level where it is necessary to take several relativistic corrections into account. These corrections concern the rotation of the Earth, the altitude of the laboratory in which the standard is kept (since there is an effect due to the gravitational potential), and, if the clock is in motion relative to the Earth, for example in an aeroplane, then there is a further velocity effect. Carrying a portable clock eastwards around the Earth's equator will lead to a change of about -200 ns in returning to the original position. Such changes have been observed experimentally (Hafele and Keating 1972). For a typical jet aircraft, having a velocity of 270 m s^{-1} (604 mph), the time dilation effect $\frac{1}{2}(v^2/c^2)$ leads to a normalised correction of 4×10^{-13} (Ashby and Allan 1979). Most of the standards laboratories where there are high-precision atomic clocks are at altitudes of around 50 m and a typical altitude relativistic correction is between 0.3 and 0.8×10^{-14}. However, Neuchatel, Switzerland is at a geoid altitude of 487 m and the NBS Boulder laboratories are at 1634 m and the corresponding corrections are 5.3×10^{-14} and 17.8×10^{-14} respectively.

For a clock which is at rest on the surface of the Earth, radius a, angular velocity Ω, and at a latitude ϕ, the proper time Δt_s is related to Δt by

$$\Delta t = \Delta t_s \{1 - 2[U + (\Omega^2 a^2/2c^2)\cos^2 \phi]\}$$

where U is the gravitational potential and Δt is the elapsed coordinate time. Now

$$V = U + (\Omega^2 a^2/2c^2)\cos^2 \phi$$

is just the gravitational potential in a frame rotating with the Earth and since the gravitational acceleration g is given by

$$\boldsymbol{g} = -\Delta V$$

and we are interested in time differences relative to those on the geoid (the surface of a liquid Earth), the interval Δt is just

$$\Delta t \simeq \Delta t_s (1 - g(\phi)h/c^2)$$

where h is the altitude above the geoid. Consequently, a standard clock which is at rest only requires a correction for the red shift before being used to measure elapsed coordinate time at that point (the fractional red shift correction is 1.09×10^{-13} km^{-1}).

For a clock which is in motion, the elapsed coordinate time during transport may be calculated from

$$\Delta t = \int_{\text{path}} [1 - g(\phi)h/c^2 + \tfrac{1}{2}(V/c)^2 (\Omega a/c^2)V_E \cos \phi] \, dt_s$$

where V_E is the eastwards component of the velocity V. One problem inherent in this representation becomes apparent if we consider transport along a parallel of latitude, when $V = V_E$ and

$$V_E \cos \phi \, dt_s = L \cos \phi$$

where L is the distance travelled eastwards so that

$$\Delta t = \Delta t_s + (\Omega a L \cos \phi)/c^2.$$

This may be re-written as

$$\Delta t = \Delta t_s + 2\Omega A_E/c^2$$

where A_E is the area enclosed by the path projected on the equatorial plane (considered positive if it is traversed in the same sense as the rotation of the

Earth). This result implies that a clock carried eastwards completely around the globe shows an elapsed time which is less than a clock which has remained at 'rest' on the surface by

$$(2\pi\Omega a^2 \cos^2 \phi)/c^2 = 207.4 \cos^2 \phi \text{ ns.}$$

It is interesting that this correction does not depend on the speed with which the clock has been carried around the Earth and it can be appreciated that there are occasions when this step change of $207.4 \cos^2 \phi$ ns at a point can become an important consideration—almost a step uncertainty in the time. A further difficulty arises when transferring time via satellites, for it is necessary to remember that our time system is in a reference frame which is rotating with the Earth and hence is not in an inertial reference frame.

It is beyond the scope of this book to discuss these fascinating aspects of relativity in any detail, although other aspects are considered later. It is, however, interesting that the precision of modern atomic clocks is allowing the predictions of relativity to be verified with clocks as envisaged in the original *gedanken* experiments of Einstein and others. Of course, relativistic effects have long been an integral part of high-energy nuclear physics. So far all experimental tests of relativity have shown that the results are consistent with the theory. In the future we can expect more stringent tests as the precision with which we can measure time and other metrological quantities advances by a further decimal place.

6

Dimensionless constants whose roles have changed

6.1 Introduction

As we have discussed in chapter 1, the role of fundamental constants is an evolving one and in this chapter we consider three dimensionless constants whose roles have changed.

There are two natural units for magnetic moments; these are the Bohr magneton $\mu_B = \hbar e/2m_e$ and the nuclear magneton $\mu_N = \hbar e/2m_p$. These expressions yield the value that one would expect for a classically spinning particle of charge e, angular momentum $\frac{1}{2}\hbar$ and having one electron mass m_e or one proton mass m_p respectively.

In this chapter, we discuss the measurement of the proton magnetic moment μ_p in terms of the Bohr and nuclear magnetons, μ_p'/μ_B and μ_p'/μ_N respectively. The quotient of these is

$$(\mu_p'/\mu_B)/(\mu_p'/\mu_N) = m_p/m_e$$

and hence the two experiments lead naturally to the third constant which we discuss, namely the direct measurement of m_p/m_e.

6.2 The magnetic moment of the proton in terms of the Bohr magneton

The measurements of μ_p/μ_B may be divided into three distinct classes, depending on whether μ_p is measured for protons in atomic hydrogen or for protons in an oil or water sample, and also on whether μ_B is obtained from measurements on free electrons or on electrons which are bound in a hydrogen atom.

The correction from the bound-state electron g-factor $g_j(H)$ to the free electron g-factor g_e is given by the relativistic correction involving the fine structure constant

$$g_j(H) = g_e(1 - \alpha^2/3) \tag{6.2.1}$$

(for example Mott and Massey 1965). The proton g-factor in hydrogen $g_p(H)$ can also be corrected for the diamagnetic shielding using the theory of Lamb (1941) as:

$$g_p(H) = g_p(1 - \alpha^2/3) \qquad (6.2.2)$$

There are some further corrections in both cases (Grotch and Hegstrom 1971), but the g-factor corrections approximately cancel (the differences are of the order of parts in 10^8). We have

$$g_j(H)/g_p(H) \approx g_e/g_p$$

while from the definitions of the g-factors

$$g_e/g_p = \mu_e/\mu_p.$$

Now μ_e is obtained in terms of the g-factor anomaly of the electron, in Bohr magnetons, as

$$\mu_e/\mu_B = g_e/2$$

and, of course, the $g_e - 2$ experiments are now at the parts per million level so that μ_e/μ_B is known to one or two parts in 10^9 (see §5.8). It is apparent, therefore, that appropriate measurements on electrons and protons in atomic hydrogen are equivalent to measurements on free electrons and protons as long as the auxiliary constants mentioned above continue to be known with greater precision.

6.2.1 Measurements on free electrons

The ratio of the spin resonance frequency of protons v_s to the cyclotron frequency of electrons v_e measured in a similar magnetic flux density (B_s and B_e respectively) is given by

$$v_s/v_e = (2\mu_p/h)B_s/(e/\pi m_e)B_e$$
$$= (\mu_p/\mu_B)(B_s/B_e).$$

The experiments were commonly performed by a substitution method in essentially the same magnetic field and the ratio of B_s to B_e was readily evaluated experimentally since typically it differed by only a few parts per million from unity. For flux densities which could readily be achieved with electromagnets and permanent magnets the electron cyclotron frequencies were in the microwave region, i.e. for $v_s = 14$ MHz, $v_e \sim 9.2$ GHz. The width of the electron cyclotron resonance depends on the number of revolutions which the electrons execute in the microwave field. Taking $v_e \sim 10^{10}$ Hz, and a resolution of 10^{-6}, indicates that the electrons must be in the microwave field for $\sim 10^{-4}$ seconds. The velocity of a 0.01 eV electron is $\sim 5 \times 10^4$ m s^{-1} and it is clear therefore that for high resolution the electron energy must be as low as possible. Some of the measurements involved the confinement of the electrons and so space charge shifts in the cyclotron frequency became quite troublesome and had to be corrected for. The shifts were typically about 50 ppm.

Gardner and Purcell (1951) used a beam of electrons ($\sim 5 \times 10^{-10}$ A) which was emitted by a heated cathode and accelerated by a small voltage to pass through a hole into a microwave cavity, and excited by a further hole to hit a collector. The collected current decreased when the applied microwave frequency corresponded to the cyclotron frequency. The best resolution was obtained when the electron energy was less than about three volts. Franken and Liebes (1959) observed the power absorbed by the electrons while they were in the cavity. The electrons were emitted photoelectrically from the surface of a 5 mm diameter potassium-coated Pyrex bulb inside the cavity. They corrected for the space charge frequency shifts in the cyclotron frequency by measuring v_e as a function of $1/B^2$ and extrapolating linearly to $1/B^2 = 0$ (infinite flux density) over a range 0.075–0.17 T ($v_s \sim$ 2–4.5 GHz). An implicit assumption was that the space charge was not magnetic field dependent. Hardy and Purcell continued work at Harvard by measuring the power absorbed by an electron beam which was brought nearly to rest in the cavity by an appropriate decelerating potential. They worked at $v_s = 20$ MHz, $v_e = 13.3$ GHz and obtained resonances which were less than a part in 10^5 wide. The microwave power incident on their cavity was only 1–10 fW so that the energy gained by the electron was less than 0.05 eV. At this energy, the relativistic correction could be neglected. Whereas the other workers measured the proton spin precession frequency in water or oil samples, molecular hydrogen was used for this work. Further measurements were made by Sanders et al (1963), Honejager and Klein (1962) and Nosal and Kholin (1970). The most precise measurement on free electrons was that of Klein (1968) who measured the microwave absorption of a rather larger sample of electrons and hence had a space charge shift of the cyclotron frequency of only a part in a million. The halfwidth of his resonance signal was only 5 ppm.

6.2.2 Measurements on bound electrons and protons

The ratios μ_p/μ_e or $g_j(H)/g_p(H_2O)$ etc have also been obtained from measurements of the Zeeman splitting of the hyperfine structure of the ground state of atomic hydrogen. Two distinct methods have been employed, one using an atomic beam and the other relying on the detection of the absorption of energy in a microwave cavity.

The levels of the $1^2S_{1/2}$ ground state of atomic hydrogen are shown in figure 6.1. The magnetic moment of the proton causes the energy difference in zero magnetic field (expressed in frequency units). In a magnetic field the levels split into four as shown in the figure. The levels may be described by the quantum number $F = I + J$, where I is the nuclear spin and J is the total electronic angular momentum and its associated magnetic quantum number

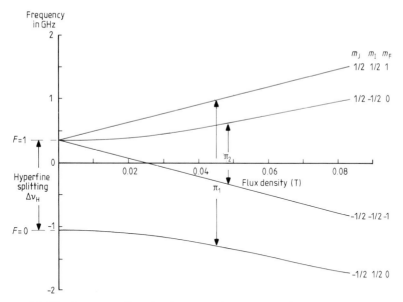

Figure 6.1 The Zeeman effect in the ground state hyperfine splitting of atomic hydrogen. The proton transitions v_a, v'_a are those between the upper and lower pairs.

M (for the electronic and nuclear momenta are coupled). At higher fields, I and J are decoupled and the levels are described by the magnetic quantum numbers M_I and M_J and

$$M_F = M_I + M_J.$$

The Breit–Rabi (1931) expression gives the energy levels at any field strength as

$$W_{F,M_F} = -h\Delta v_{\text{Hfs}} - 2\mu_p M_F B \pm h\Delta v_{\text{Hfs}}(1 + 2M_F x + x^2)^{1/2}$$

where

$$x = 2(\mu_J + \mu_p)B/h\Delta v_{\text{Hfs}}$$

and the positive sign is taken when $F = 1$ and the negative sign when $F = 0$. The frequency Δv_{Hfs} is the ground state hyperfine frequency (hydrogen maser). In the above expression, both μ_p and μ_J (the total electronic magnetic moment) are taken as positive. In the hydrogen ground state, μ_J is due to electron spin only. The two allowed microwave transitions are those labelled π_1 and π_2 in the figure.

Koenig *et al* (1952) measured the π_2 transition in a flux density of 0.15 T, at which flux density the frequency π_2 was about 3.655 GHz. They used an atomic beam method to measure π_2 and the flux density was obtained using an oil sample to obtain the proton frequency v_s. Beringer and Heald (1954) passed a Wood's discharge tube through a 9 GHz cavity which was between the poles of an electromagnet. The magnetic field was adjusted so that the frequencies π_1 and π_2, in turn, were at 9 GHz. The ratio of each of these frequencies to those of protons in molecular hydrogen was measured.

Lambe (1959, 1969) (with Dicke) measured $g_J(H)/g_p(H_2O)$ by a substitution method. The proton and electron spherical samples were each hollowed out of a cylindrical PTFE rod which could be moved so that one or other occupied the same portion of the magnetic field inside the microwave cavity (an RF coil was wound on the portion of the cylinder containing the water sample). An O-ring seal, which was shielded from the atomic hydrogen, maintained the vacuum tightness of the system. The π_1 and π_2 transitions were each measured with respect to the proton spin precession frequency in water. The microwave resonance was at about 9.2 GHz and was about 2 kHz wide. This reduced line-width was achieved by using molecular hydrogen as a buffer gas to increase the lifetime of the atomic hydrogen in the cavity. Both water vapour and oxygen (from dissociation of water in the discharge producing the atomic hydrogen) were present, and the latter played an important part in the thermal relaxation process.

Lambe's work was considerably refined in the work of Phillips *et al* (1977) who used fused silica spheres instead of PTFE. In addition they were able to interchange the two spheres so that the sphere used for the atomic hydrogen in the first series of measurements became the one used for the water sample in the second (figure 6.2). This exchange was quite important, since their two sets of results differed by 5–17 parts in 10^8, and their final results were the average of these. The difference underlines the importance of eliminating the effects due to the susceptibilities of the materials used to contain the hydrogen or water sample in high precision work. The Lambe and the Phillips *et al* results differ by 1.52 parts in 10^7, which is larger than one would like for such a key auxiliary constant. Both results are supported at a lower precision by the work of Klein.

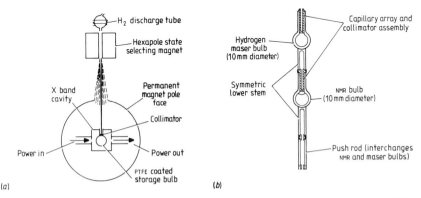

Figure 6.2 (*a*) Schematic diagram of the measurement of μ_p'/μ_B by Phillips *et al* (1974); (*b*) The method used by Phillips *et al* (1974) to interchange the water and atomic hydrogen samples.

There are analogous hyperfine splittings in the ground state of deuterium and Geiger *et al* (1972) obtained a value for $g_J(D)/g_p(\text{oil})$ while Walther *et al* (1972) measured the ratio $g_J(H)/g_J(D)$, which provided a check on the theory of Grotch and Hegstrom (1971) and others, to correct from $g_J(H)$ to g_e and $g_p(H)/g_p$. Thus

$$g_J(H)/g_e = 1 - \tfrac{1}{3}\alpha^2 + \tfrac{1}{4}\alpha^3 + \tfrac{1}{2}\alpha^2 m_e/m_p$$
$$= 1 - 17.705(m) \times 10^{-6}$$

and

$$g_p(H)/g_p = 1 - \tfrac{1}{3}\alpha - \tfrac{1}{3}\alpha^2(m_e/m_p)[(1 + 2\mu_p/\mu_N)/(\mu_p/\mu_N)]$$
$$= 1 - 17.733(n) \times 10^{-6}.$$

The estimated uncertainties m and n are thought to be of the order of a few parts in 10^9 through neglect of terms of order $(Z\alpha)^4 \simeq 3 \times 10^9$ in the Grotch and Hegstrom theory. Combining these equations and including the theoretical calculations of Grotch and Hegstrom, Faustov and of Close and Osborn and neglecting terms of the order of $\alpha^3 m_e/m_p$, the conversion to g_e/g_p then uses the relationship

$$g_e/g_p = \frac{g_J(H)}{g_p(H)}\left[1 - \frac{\alpha^2}{2}\left(\frac{\alpha}{\pi} - \frac{m_e}{3m_p}(1 - \mu_N/\mu_p)\right)\right]$$
$$= \frac{g_J(H)}{g_p(H)}(1 - 2.782 \times 10^{-8}).$$

The ratio μ_e/μ_B is known with high accuracy from the $g_e - 2$ measurements and so this may be combined with equations (6.2.1), (6.2.2) etc to yield the ratio μ_e/μ_p.

The results discussed up to this point were all measured with respect to protons which were in a water or oil sample, but Myint *et al* (1966) and Winkler *et al* (1972) have measured the ratio $g_J(H)/g_p(H)$ in the same magnetic field by measuring the ratio of the π_2 frequency v_a to the frequency v_b (see figure 6.1) at a flux density of 0.35 T. Their atomic hydrogen was produced in an RF discharge and hexapole state-selection magnets were used to form a beam of (m_J, m_I), $(-\tfrac{1}{2}, \tfrac{1}{2})$ atoms which entered the PTFE coated fused-silica spherical cavity by passing through a capillary array. The frequency v_b was observed by the quenching effect on the π_2 transition when the appropriate radio frequency was applied (640 MHz). They used pulsed techniques to excite and observe the transitions. The difficulty with this method lies in the requirement for a very precise measurement of v_b, since it essentially comprises the proton spin precession frequency v_s together with $\tfrac{1}{2}\Delta v_{\text{Hfs}}$. Consequently, in order to obtain v_s to parts per billion, the sum frequency must be measured to parts in 10^{11}. This means that it is essential to achieve an adequate lifetime for the atomic hydrogen in the microwave cavity. Myint *et al* and Winkler *et al*, whose experimentally observed electron spin resonances were about 100 Hz wide, obtained

(Myint *et al*) $g_J(H)/g_p(H) = 658.210\,49(20)$

and

(Winkler *et al*) $g_J(H)/g_p(H) = 658.210\,706\,3(66).$

6.2.3 Deduction of the spherical water sample corrections

The measurements of $g_J(H)/g_p(H)$ and $g_J(H)/g_p(H_2O)$ may be used to obtain an experimental measurement of the spherical sample shielding correction for water. Thus the Lambe and the Myint *et al* results yield

$$g_p/g_p(H_2O) = 1 + (26.0 \pm 0.3) \times 10^{-6}$$

and the Winkler *et al* and Phillips *et al* measurements yield

$$g_p/g_p(H_2O) = 1 + (25.637 \pm 0.067) \times 10^{-6}$$
$$= 1 + \sigma.$$

The latter result was for a spherical water sample at a temperature of 34.7 °C. This measurement may be corrected to other temperatures by using the results of Hindman (1966) between -15 °C and 100 °C; the temperature of the water sample in the Lambe measurement is not known. Thus Phillips *et al* obtained from Hindman's results, at a temperature t °C:

$$[\mu_p'(t) - \mu_p'(0)]/\mu_p'(0) = [0.2(15) - 11.18(18)t - 0.020(9)t^2] \times 10^{-9}.$$

These may be combined with the measurements of Thomas (1951) who obtained

$$\sigma(H_2O) - \sigma(H_2) = -6(3) \times 10^{-7}$$

and Hardy (see Liebes and Franken 1959) who obtained $-6(1.5) \times 10^{-7}$ and of Gutowsky and McClure (1951) who obtained $-3(3) \times 10^{-7}$. The weighted mean of the above yields

$$\sigma(H_2O) - \sigma(H_2) = -5.5(1.3) \times 10^{-7}.$$

Combining this with the experimental value for $\sigma(H_2O)$ yields

$$\sigma(H_2) = 26.19(1.3) \times 10^{-6}.$$

This is in excellent agreement with the theoretical values of Newell (1950), using some of Ramsey's results, of

$$\sigma(H_2) = 26.6(3) \times 10^{-6}$$

or of Harrick *et al* (1953) of

$$\sigma(H_2) = 26.2(4) \times 10^{-6}.$$

Thus the diamagnetic shielding correction for a spherical water proton sample is verified experimentally by the above measurements to about 0.07 ppm.

Even with this high precision it is likely that the measurements will continue to be referred to free protons, or water sample values, as appropriate.

The existence of this sample shape and temperature-dependent correction is likely to become an increasing obstacle to improving the accuracy of measurements of the gyromagnetic ratio of the proton. This will inhibit its use either as a transfer constant in experiments involving magnetic fields, or for measuring magnetic fields with high accuracy. Difficulties are likely to arise particularly at cryogenic temperatures. It may well be that a better alternative to γ_p' will emerge, such as, for example, the high sensitivities and long relaxation times that the optical pumping or SQUID detection of the magnetisation of ^{3}He might provide.

6.3 The magnetic moment of the proton in terms of the nuclear magneton

The proton, having a charge e, spin $\frac{1}{2}\hbar$ and mass m_p, would be expected from the simple Dirac theory to have a magnetic moment equal to $\frac{1}{2}\hbar e/m_p$, and this is a convenient unit, termed the nuclear magneton, in which nuclear magnetic moments may be expressed. As is well known, when Frisch and Stern (1933) measured the magnetic moment of the proton in their classic series of measurements they found that the value was between two and three nuclear magnetons, and thus μ_p differs considerably from μ_N. The proton moment in terms of the nuclear magneton is a dimensionless quantity, and consequently one might expect that it would be possible to calculate the value theoretically, as is, for example, the electron magnetic moment in terms of the Bohr magneton (which is the corresponding value expected from the simple Dirac theory for the magnetic moment of the electron). At present, our knowledge of nuclear forces is still far short of making such a theoretical prediction and the value may only be measured experimentally. No doubt some day prediction of the proton magnetic moment to six or more decimal places will provide a spectacular test of some refined theory of nuclear forces. The proton magnetic moment is therefore of interest mainly for the part it plays in allowing measurements which have been made at one flux density to be compared with those in a different type of measurement which were made at another flux density. Thus, until recently, the ratio of the mass of the proton to the mass of the electron was limited by our knowledge of μ_p'/μ_N, for we have

$$(\mu_p'/\mu_N)/(\mu_p'/\mu_B) = \mu_B/\mu_N = m_p/m_e.$$

The post-war determinations of μ_p'/μ_N all made use of the suggestion of Alvarez and Bloch (1940) that μ_p'/μ_N could be simply measured by taking the quotient of the proton spin precession and cyclotron frequencies in the same magnetic flux density. Thus

$$\mu_p'/\mu_N = (f_s/f_c)(B_c/B_s)$$

where B_s and B_c were the respective flux densities at which the two frequencies were measured and $B_s \simeq B_c$. The above equation is obtained quite simply, for we have

$$f_s = (2\mu_p'/h)B_s \qquad f_c = (e/2\pi m_p)B_c \qquad \mu_N = he/2m_p.$$

In using μ_p' we have assumed that the spin precession frequency is conveniently measured for protons which are in a water sample, as in conventional nuclear magnetic resonance. This frequency can be corrected to that of a free proton if required, as was discussed in the previous sections. A major contribution was made by the work of Sommer et al (1951) who measured μ_p'/μ_N with a precision of about 20 ppm.

6.3.1   The proton spin precession and cyclotron frequencies

As was discussed in the section concerning the measurement of the gyromagnetic ratio of the proton (chapter 4) the proton spin precession frequency may be measured by one of the standard techniques of nuclear magnetic resonance (e.g. Andrew (1958)). The difficult part of the measurement lies in the measurement of the proton cyclotron frequency, which we now consider.

The spin precession frequency is some 2.79 times greater than the cyclotron frequency and, if the flux density is sufficiently uniform, the linewidth of the former signal may be less than a hertz. A precise measurement of the proton cyclotron frequency requires (i) a sufficiently large time of flight for the proton cyclotron resonance signal to be narrow and also (ii) a detailed knowledge of the lineshape in order to be able to split the resonance with adequate precision. Unfortunately, the proton cyclotron frequency is very likely to be shifted from $\omega_c = Be/m_p$ owing to the effects of non-uniform electrostatic and radio frequency fields. An electrostatic field E causes the centre of the cyclotron orbit to drift with a velocity $(E \times B)/|B|^2$, which is in a direction perpendicular to that of the applied magnetic flux B. Thus, if the potential variations in a practical device are of the order of a tenth of a volt, over distances which are of the order of a centimetre, then in a flux density of one tesla the drift velocity is of the order of $10 \, \mathrm{m s}^{-1}$. This drift velocity is essentially mass independent and in the past has limited the time during which the cyclotron frequency must be measured to about a millisecond. These drift fields are particularly troublesome when the computation of the cyclotron frequency involves defining the orbit precisely, since they can then shift the cyclotron frequency.

An electrostatic field gradient $\partial E/\partial z$ in the magnetic flux direction, together with a space charge density ρ, perturbs the cyclotron frequency from the value given above to one typically of the form:

$$f_c' - f_c = -[(\rho/E_0)(-\partial E/\partial z)]/B.$$

It was the failure to take this shift properly into account which affected the measurements of e/m for the electron and proton in the 1930s and the elimination or measurement of this shift has proved a major factor affecting the more recent measurements. (It should be pointed out that although we have introduced $\partial E/\partial z$ above, the shift is more fundamentally a result of the radial electrostatic fields, but, if we neglect the space charge, the Laplace equation $\nabla^2\phi = 0$, where ϕ is the electrostatic potential function leads one to the term which we have given above.) The methods of eliminating this shift have mostly made use of the fact that the shift in the cyclotron frequency does not depend on the mass of the ion at resonance. Consequently, it may be eliminated by bringing ions of different mass successively to resonance and making use of the very high precision with which ion mass ratios have been measured in mass spectrometers.

The methods which have been adopted to measure the cyclotron frequency involve adding either (i) a small, or a much larger amount of energy to ions having essentially the thermal energy at the time of generation, or (ii) by accelerating or decelerating ions having a much greater initial energy, or (iii) by leaving their energy unchanged. It is convenient, from the point of view of both the chronology and the physics involved, to begin with the low-energy methods.

6.3.2 Low energy methods: the omegatron

The low energy methods of measuring the proton cyclotron frequency are derived from that developed by Sommer et al (1951) at the NBS. Since their device allowed the ion cyclotron frequency to be measured, it was called an omegatron. The device proved to have a utility which extended well beyond the measurement of a fundamental constant. The reason for this was that it was a mass spectrometer having a small volume and so could be used to analyse the gases evolved in small high-vacuum systems. The omegraton contributed, for example, to (i) the improved reliability of the repeater amplifiers used in the transatlantic telephone cable, (ii) the improved reliability of electronic valves and cathode ray tubes, (iii) the early work on the analysis of the gases in the upper atmosphere, (iv) putting a limit on the continuous creation of matter and (v) led to the technique of ion cyclotron resonance which has been used to study the chemistry of ion gas reactions. Consequently, there were several hundred papers concerning the operation of the omegatron and many more which involved its use. The number of papers may be taken to be an indication that the simplicity of the device was deceptive.

6.3.2.1 The variable ion mass determinations. The operation of an omegatron may be understood from figure 6.3. The electrons from a heated tungsten filament traversed the analysis region and finally hit the ion collector. In the process, some of the residual gas was ionised and, when radio frequency voltage of the appropriate frequency was applied to the electrodes, the ions gained energy from the RF field and the orbit diameter increased until they either hit the ion collector or the other electrodes. Since the ions were created with essentially the thermal energy of their parent atom, together perhaps with

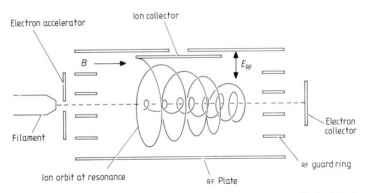

Figure 6.3 Schematic diagram of the omegatron mass spectrometer devised by Sommer *et al* (1954) to measure μ'_p/μ_N.

some dissociation energy, they had a velocity in the magnetic field direction and so an electrostatic field was applied to trap them in this direction. This potential distribution had to satisfy $\nabla^2\phi = 0$, which meant that there had to be components perpendicular to the direction of **B**. Thus there was an $E \times B$ term which represented drift of the orbit centre and a further term which shifted the cyclotron frequency. Here the experimenter had two choices: either to profile this field so that the ions remained trapped, or to allow them to escape. Since the ions were created continuously, indefinite trapping would only lead to the space charge fields building up until they overcame the trapping fields. This was undesirable since such conditions are difficult to control experimentally. In the Sommer *et al* omegatron the ion orbit guiding centre drifted along the diagonals of the rectangular box formed by the electrodes in the direction perpendicular to the magnetic field, with the drift velocity increasing as the ions moved away from the electric centre as illustrated in figure 6.4. In the omegatron used by Petley and Morris (1965, 1974) the box was made hyperbolic in shape and effectively infinitely long in one of the directions perpendicular to the magnetic field so that the trapping potential distribution was of the exact form

$$\phi = V_T[k_0 + k_1(x^2 - z^2)]$$

in contrast with the Sommer *et al* distribution which, to first order, was of the form

$$\phi = V_T[k_0 + k_1x^2 - k_2y^2 - (k_1 - k_2)z^2].$$

It should be noted that the mass independence of the orbit drift only applied when the electrostatic fields had the simple forms given above. Equally, it was important that the radio frequency electric field was uniform, otherwise mass dependent effects became important and the equations of motion became highly non-linear and could only be solved numerically. These effects not only

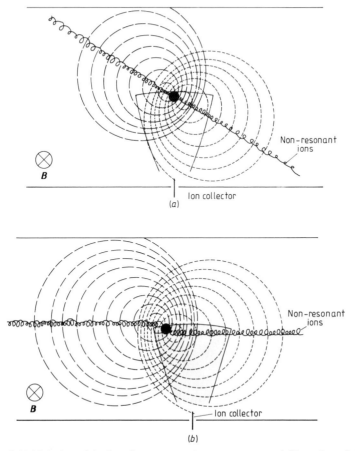

Figure 6.4 (a) Motion of the ions in a rectangular omegatron and (b) motion of the ions in a quadrupole omegatron.

led to harmonic acceleration effects, but they also led to shifts in the cyclotron frequency which depended on the ion mass, and in this way they affected the value deduced for μ'_p/μ_N.

In the Sommer *et al* measurement, the electrostatic field dependent frequency shifts were eliminated by taking the pairs of ions H^+ and H_2^+, H^+ and D_2^+, and H^+ and H_2O^+ with a magnet of flux density 0.47 T. Fystrom, using a similar design at about 1.4 T, worked with H^+, D^+ and HD_2^+ ions. One major difficulty with working with H^+ or D^+ ions comes from the considerable dissociation energy when H_2 is split into H^+ (~ 6 eV) compared with the thermal energy (~ 0.025 eV) of the other ions. The initial conditions had a major effect on the subsequent orbit drift and the elimination of the electrostatic field dependent frequency shifts depended on all of the ions concerned integrating the electric and magnetic fields in a similar manner. A further difficulty with the use of H$^+$ ions became important at the higher flux densities, for there was a relativity correction which depended on the inverse square of the ion mass. The filament which produced the electrons also produced a disturbing magnetic field. This shifted the cyclotron frequency by a few parts per million and so its effects were eliminated by reversing the filament current. Further small corrections were required to allow for the shielding of the interior of the omegatron by the materials used in its construction. Petley and Morris were able to make this correction by measuring the change in the spin precession frequency of a small NMR probe which could be inserted into their omegatron.

6.3.2.2 The variable magnetic field determination. The above methods required that the ions gained several keV of energy, the Boyne and Franken (1963) determination used similar techniques to the marginal oscillator method used in nuclear magnetic resonance to observe the ions. Their omegatron was of simple construction and the RF electrodes were bent into a 'u' shape, thereby producing a non-uniform RF field, but allowing the guard rings to be dispensed with. The H$_2^+$ ions were not accelerated very much by the RF field, ~ 1 eV, and remained in a tight bunch. Since trapping potentials were not applied the ion lifetime was rather short. The problem of the elimination of the electrostatic space charge shifts was solved by working at flux densities of between 0.8 T and 1.25 T and using the equation

$$f'_c = f_c(1 - \Gamma/B^2)$$

where Γ was the space charge term. The value of f_c was obtained as the intercept on a plot of f' as a function of $1/B^2$. The problem with their method was that it relied on the space charge remaining constant over a considerable range of flux densities and their uncertainty assignments made allowance for this.

6.3.3 The high-energy methods: deceleration

Bloch and Jeffries (1950) reported a measurement of μ'_p/μ_N which used a cyclotron in the decelerating mode. In the accelerating cyclotron (figure 6.5) the magnetic field must be non-uniform to counteract the defocusing effect of the radio frequency field which is applied between the 'dees' to accelerate the ions. Fortunately, in the decelerating mode the RF field acts in a direction which tends to focus the ions. This allows the use of a uniform magnetic field, which is almost essential for a high-precision determination. They used a flux density of about 0.53 T and the diameter of the dees was about 8.5 cm. The protons were generated outside the magnet and pre-accelerated to about 20 keV. Electrostatic fields were used to balance the Lorentz force on the protons until they entered the cyclotron. The RF voltage was essentially fixed by the dimensions of the entry slit, since the protons had to lose enough energy in the first revolution in order to clear it, and corresponded to a limiting

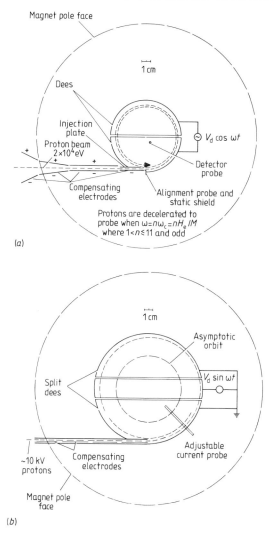

Figure 6.5 (*a*) The decelerating cyclotron used by Jeffries (1951) to measure μ_p'/μ_N (operating conditions: $B \sim 0.53$ T, $\omega = 8.1$–89 MHz, $V_d \sim 200$ V, $p = 10^{-6}$ mm Hg, probe current $\sim 10^{-12}$ A); (*b*) The modified decelerating cylotron of Sanders *et al* (1955) to measure μ_p'/μ_N (operating conditions: $B \sim 0.235$ T, $\omega = 8$th and 16th harmonics of 3.58 MHz, $V_d = 20$–30 V, $P = 10^{-6}$ mm Hg, probe current $\sim 10^{-14}$ A).

voltage V_c. This voltage entered into the expression for the cyclotron frequency. (It is in some ways similar to only observing the upper portion of the resonance in an omegatron where one would have $f_e' = f_c + \pi E_0 / R_c B$.)

When $V_d = V_c$, no ions can reach the collector and the value is obtained by measuring the collected ion current as a function of the dee voltage. There is a transit time effect which increases the resolution because the ions take a significant portion of an RF cycle to cross the gap between the two dees, and hence lose less energy per revolution as they approach the collector. The number of revolutions which the protons may make before reaching the ion collector is limited to about 200 revolutions as a result of defocusing and other effects. The resolution was increased considerably by making use of the fact that the protons left and re-entered the RF field at the dees so that it was possible for the RF field to be reversed many times while the protons were away from the gap between the dees. As a result, it was possible to use a radio frequency which was a harmonic of the ion cyclotron frequency and so the time of arrival of the ions at the dees became much more critical. Thus, Bloch and Jeffries obtained a resonance which was a few parts in a thousand wide by using about the ninth harmonic of the cyclotron frequency.

The proton magnetic moment was obtained from the equation for the measured (shifted) value of μ_p''/μ_N:

$$(\mu_p/\mu_N)'' = (\mu_p/\mu_N)\{1 - (f_r/nf_r)[(V_d/V_c)^2 - 1]^{1/2}\}$$

where f_r was the resonance width and nf_r the applied radio frequency. The measurement of this upper frequency was the dominant uncertainty (72 ppm) in their work, the other uncertainties amounting to about 15 ppm. Their work was later repeated by Trigger (1956) who made a more detailed analysis of the effects of the electric fields. However, his final uncertainty was still dominated by the difficulty of estimating the upper frequency edge of the resonance curves.

The decelerating method was improved by Sanders *et al* (1955, 1957 and 1963). They used a split-dee with the 14 cm diameter dees separated by a 3 cm wide central electrode. In this way the ions were decelerated by the RF field four times per cycle. After about 150 revolutions the ions reached an asymptotic orbit whereby the received no net deceleration per revolution. The RF voltages were only about 35 V (peak) compared with the ~ 200 V used by Blocj *et al*. Although the flux density was only about a half of that used in the other work, 0.235 T compared with 0.53 T, a fourfold increase in resolving power was achieved over that obtained by Trigger. Two thicknesses of injector plate were used and it was again necessary to deduce the voltages, 34.4 ± 0.7 V and 27.4 ± 0.7 V, needed to just clear these plates. The difficulty of estimating the upper frequency limit of the resonance was again a limiting factor in the final uncertainty of 26 ppm.

6.3.4 The high-energy measurements: acceleration

This method, by Mamyrin and Frantsuzov (1964) and Mamyrin *et al* (1972), was developed from the mass synchrometer of Smith (1971). Despite the fact that the ions made only two revolutions in a flux density of 0.13 T, the resolution was some four times better than that achieved by Sanders *et al*. They

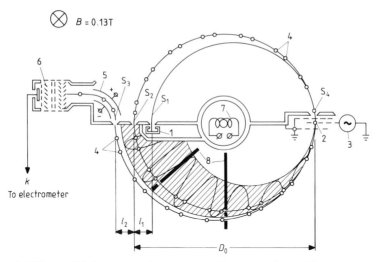

Figure 6.6 The modified mass synchrometer used by Mamyrin *et al* (1972) to measure μ'_p/μ_N.

accelerated ions to between 200 and 400 V sufficient to give them an initial radius of curvature of 9.8 cm in a flux density of 0.13 T. The ions passed through the slit S_1 (figure 6.6) where they were accelerated by an applied RF voltage of up to 1 kV amplitude. This spread out the ion energies and those which arrived at the appropriate phase passed through the slit S_2 and could again be accelerated by the RF voltage if the phase conditions were right. They then exited from the device and were detected by a shielded electron multiplier, gain $\sim 10^4$, which allowed currents of 10^{-15} A to be detected. Detection occurred whenever the applied radio frequency was a harmonic of the cyclotron frequency. About the 200th harmonic was used. The chosen ratio of l to D was 0.125 and so the velocity was increased by a considerable amount with each revolution. The velocity increment at S_4 did not very sinusoidally with time and, as can be seen from the figure, there were two pulses of ions per cycle which were able to traverse the slits. These did not receive exactly the same velocity increment and the velocity increment on the second revolution was less than that on the first (as a result of transit effects in the modulator), and so the distance l_2 is about 10% smaller than l_1. The position of the slit S_2 was adjusted until at a particular frequency f_{res} both pulses of ions were received simultaneously at the detector. The cyclotron frequency was then obtained from

$$f_{res} = f_c(n + k)$$

where n was the harmonic number (96) and k was a correction which resulted from the transit time effects. This was computed from a knowledge of the precise geometry of the modulator electrodes as being $-0.002\,671\,8(98)$. The shifts in the cyclotron frequency by the electrostatic field gradients and orbit drift were eliminated by using the successive ion pairs $^4\text{He}^+ - {}^{20}\text{Ne}^+$, $^4\text{He}^+ - {}^{20}\text{N}^{++}$ and $^4\text{He}^+ - {}^{40}\text{Ar}^+$. The magnetic field at the ion due to the filament current was eliminating by reversal. A reference NMR probe was located at the centre. Since the flux density at the orbit was about 7 ppm greater than that at the centre, a second probe was used to find the field difference between this and that over the ion orbit. The pairs of ions were produced by a double ion source which ensured that the potentials, and hence electrostatic fields, remained constant during the period of the measurement.

6.3.5 Time of flight measurements

The impetus for these measurements came from a discrepancy between the determination of certain proton–nuclear reaction energies at the Universities of Maryland and Zurich. The results obtained for protons from $\text{Al}(\rho, \gamma)^{28}\text{Si}$ and $\text{Li}(\rho, \gamma)^7\text{Be}$ reactions differed by more than the expected amount. It was suggested that this difference might be attributable to an error in the values assumed for μ'_p/μ_N. It was subsequently found that it was due to other effects. Essentially, the Maryland experiment required modulating the intensity of a beam of protons with a large alternating electric field before the beam entered a cavity containing an entry slit and an exit slit. The entry of the pulse of protons excited the cavity at its resonant frequency ($\sim 72\,\text{MHz}$, $Q \sim 4000$) and its exit caused a similar pulse. The conditions were adjusted until the two detected pulses were exactly out of phase, thereby producing a sharp null on the signal picked up by a detector coil which was located near the exit slit. The condition for this null was

$$L/v_p = (2N - 1)/f_{\text{res}}$$

where $N = 1, 2, 3$, L was the electrical length of the cavity (1.25 m) and v_p the velocity of the protons. It is interesting that this method is being explored at the PTB (Kose 1982) as a method of measuring high voltages.

The Zurich apparatus measured a quantity which was closely related to the beam velocity, namely the magnetic rigidity

$$B\rho'_p = m_p v_p/e[1 - (v_p/c)^2]^{1/2}$$

where ρ'_p was the radius of curvature in the magnetic flux B and e and m_p were the proton charge and mass respectively. The flux density was measured by conventional NMR techniques in terms of the spin precession frequency f_s. Rearranging the equation it became

$$\mu'_p/\mu_N = 2\pi f_s \rho'_p[1 - (v_p/c)^2]^{1/2}/v_p.$$

Hence by measuring v_p in Maryland and the remainder of the parameters in Zurich, Marion and Winkler (1967) deduced a value for μ'_p/μ_N. There was clearly a disadvantage in the two halves of the experiment being separated by the Atlantic ocean and so the Maryland apparatus was shipped to Zurich. Staub *et al* (1974) subsequently obtained a value for μ_p'/μ_N in good agreement with the other measurements.

There is an important moral here, for the motivation for their measurement of μ_p'/μ_N was an inadequate knowledge of μ_p'/μ_N at that time. This prevented the two groups from discovering the true cause of the discrepancy between their measurements.

6.3.6 *Summary of results and future developments*

The collected results for the μ_p'/μ_N determinations are given in table 6.1. Broadly speaking, the results are in good agreement. This was not the case a few years ago, principally because the uncertainty quoted by Sommer *et al* was treated as though it was several standard deviations instead of one standard deviation. This confusion arose because the authors were not specific enough about what they intended their uncertainty to represent. In that period there was not the rigid adherence to the standard error that there is today. Their work was discussed by Fystrom *et al* (1970). It is interesting that according to Hipple (1970) their paper was held to concern an instrument rather than a measurement of a physical constant and there was some difficulty in getting it accepted for publication in *Physical Review*. In consequence the paper was shortened, but at any rate the work was a classic piece of measurement which took nearly twenty years to surpass.

Table 6.1 Summary of measurements of μ_μ'/μ_v.

Publication date and author	Method	Value		Uncertainty (ppm)
1951 Sommer *et al* (Revised 1971 Fystrom *et al*)	Omegatron	2.792 711	(60)	21
1963 Sanders and Turberfield	Inverse cyclotron	2.792 701	(73)	26
1961 Boyne and Franken	Cyclotron	2.792 832	(55)	20
1970 Fystrom	Omegatron	2.792 783	(16)	5.7
1972 Luxon and Rich	Trapped ion	2.792 786	(17)	6.0
1974 Gubler *et al*	Velocity gauge	2.792 777	(20)	7.2
1972 Mamyrin *et al* (*revised* 1983)	Mass spectrometer	2.792 773 8	(12)	0.43
1974 Petley and Morris	Omegatron	2.792 774 8	(23)	0.82

Both μ_p/μ_B and μ_p/μ_N are really the old charge to mass ratio measurements for the electron and proton respectively in modern guise. We have discussed the μ_p/μ_N determinations at some length since these represented the best method hitherto for deriving the value of e/m_p or the proton cyclotron frequency Be/m_p. This sets the scene for the elegant work of the groups at the Universities of Washington and Mainz which we discuss next.

6.4 The proton to electron mass ratio

The masses of electrons and protons are far too small for them to be measured directly, either separately or as a ratio. The usual method has been to 'weigh' them in a magnetic field. Thus the cyclotron frequency is simply

$$f_c = Be/2\pi m.$$

Evidently, if the cyclotron frequencies are measured in flux densities, B_p and B_e, then one may obtain the ratio of the masses from the ratio of the frequencies as

$$m_p/m_e = (B_p/f_p)(f_e/B_e)$$
$$= (f_e/f_p)(B_p/B_e).$$

There is no problem, as we have seen, in measuring the ratio f_e/f_p, but the ratio of B_p/B_e must be obtained somehow as well. It has been usual to measure the proton cyclotron frequency in one apparatus and the electron cyclotron frequency in another. Ideally they would be measured in the same flux density so that to a very good approximation $B_p = B_e$. In practice, the problems of measuring the cyclotron frequencies in the earlier work were such that experimenters preferred to measure one or the other and to measure the flux density by measuring the proton spin precession frequencies $(f'_s)_p$ and $(f'_s)_e$. Consequently the experiments became determinations of μ'_p/μ_N and μ'_p/μ_B so that

$$m_p/m_e = (\mu'_p/\mu_N)/(\mu'_p/\mu_B)$$

which underlines the importance of the μ'_p/μ_N experiments which we have just discussed.

Despite the fact that μ'_p/μ_B is sufficiently well measured to be regarded as an auxiliary constant, this method of obtaining m_p/m_e is not as satisfactory as measuring the ratio of the two frequencies in the same apparatus, more or less at the same time, and thereby avoiding the need to substitute one set of apparatus for another. This ideal has been pursued by Gartner and Klempt (1978) and Graff et al (1980) and the latter have achieved a precision comparable with the indirect measurements. The group at the University of Washington (Van Dyck et al 1980) have also developed their ion trap technique to measure m_p/m_e. Petley and Morris in their μ'_p/μ_N measurement

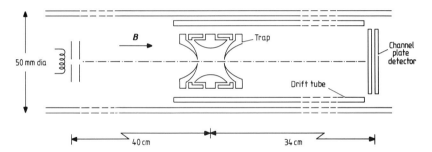

Figure 6.7 The quadrupole trap used by Graff *et al* (1980) to measure m_p/m_e.

were obliged to use a two-dimensional quadrupole ion trap because their method of ion detection required measuring an ion current. Because the sensitivity of the electrometer measuring the ion current was limited, there were a comparatively large number of ions present in their omegatron at any one time. Consequently there was only one option if the ions were to be created continuously and that was to ensure that the space charge did not build up to the point where it was a major factor in determining the ion lifetime or the ion trajectory. The types of ion trap employed by Gartner and Klempt, by Graff *et al* (figure 6.7) and also by Van Dyck *et al* (1981) and others for electrons were three-dimensional quadrupole traps. These relied on the space charge being very small for successful operation, so small in fact that one or two ions or electrons could be trapped for long periods and yet their cyclotron frequencies could be detected. In the Graff *et al* experiment the flux density (in the z-direction) was 5.8 T and the trapping potential distribution of the form

$$V(x,y,z) = V_0(x^2 + y^2 - 2z^2)/R_0^2.$$

In this configuration the ions behave like a three-dimensional oscillator, and in the relativistic limit they have the following frequencies

$$f_z = (2eV_0/mR_0^2)^{1/2}/2\pi$$
$$f_1 = f_c/2 - f_0$$
$$f_2 = f_c/2 + f_0$$

where

$$f_0 = (f_c^2/4 + f_z^2/2)^{1/2}$$

and the cyclotron frequency, including relativistic effects, is:

$$f_c = eB/2\pi\gamma m$$

where

$$\gamma = (1 - v^2/c^2)^{-1/2}.$$

For a trapping potential V_T of 1 volt the frequency f_z was about 16 MHz

(367 kHz), f_1 was 760 Hz (760 Hz), and f_2 was 163 GHz (89 MHz) for electrons (protons). The ion trap was constructed from OFHCL copper, and the ions were formed by electrons emitted from a heated tungsten wire source and accelerated to ~ 100 eV. The source was on the quadrupole axis and about 0.4 m away from it. These electrons entered and exited the quadrupole trap through ~ 1 mm diameter axial holes and produced both ions and secondary electrons inside the trap. The electrons rapidly lost energy by synchroton radiation in the strong magnetic field, and cooled to a thermal temperature of about 30 meV within a second. The apparatus was bakeable to 600 K. This temperature could be sustained for at least two days so that a system pressure of about 10^{-7} Pa (10^{-9} Torr) could be achieved. The quadrupole trap was mounted inside a drift tube and the whole assembly could be cooled to liquid helium temperatures by means of a flow cryostat. This drift tube was some 0.37 m long and a channel-plate detector was mounted at the end. The purpose of this detector was to detect the electrons and protons after they had been ejected from the quadrupole trap. The magnetic flux was produced by a superconducting magnet having room temperature shim coils to adjust the homogeneity. The field was stable to between 0.02 ppm and 1 ppm per day depending on whether the flux density was 5.28 T or 5.81 T respectively (these were the two values at which results were obtained).

There was a choice of three methods to obtain the cyclotron frequency:

(i)   measuring f_2 at different trapping potentials and extrapolation to zero trapping voltage
(ii)   measuring both f_1 and f_2, since their sum is f_c
(iii) inducing a direct transition at the frequency $f_1 + f_2$ (such a transition is possible at low electric field strength).

Method (iii) was chosen for the measurement of the cyclotron frequency since this was the least sensitive to misalignment of the electrodes, to space charge effects or to unknown surface potentials.

A sequence of pulses formed the cycle: electron injection, trapping, ion (electron) cooling, acceleration, and ejection. At ejection, the ions (or electrons) had been increased in transverse energy by the applied cyclotron accelerating field and had an axial energy of 100 meV (or 10 meV) respectively. The ions (or electrons) gained an additional energy-dependent acceleration from the inhomogeneous magnetic field as a result of this axial energy. The flux density varied rapidly along the axis of the magnet and dropped from about 6 T at a distance of 8 cm from the quadrupole to less than 1 T after 15 cm, and so it was quite small at the channel multiplier. Since the increased axial energy decreased the time of flight by $\sim 10\%$, the particle cyclotron frequency could be detected by studying the shape of the channel plate electron multiplier pulse as the applied microwave or radio frequency was varied. Thus at coincidence with the cyclotron frequency the electron time of flight decreased from about

3.7 μs to about 3.3 μs (depending on the microwave amplitude). The half-height width of the resonance was about 0.8 ppm. The electron cyclotron frequency was displaced from the required value by power broadening and also by loss of energy due to synchrotron radiation. It is interesting that although the electron energy was close to thermal energies, there were sufficiently large relativistic effects to affect the shape of the resonance and a correction of 0.05 ppm was applied for the electron relativistic mass increase corresponding to the mean energy of about 50 meV. Between three and 45 protons were trapped per pulse and the cyclotron frequency was studied as a function of as many as possible of the experimental parameters. Since H_2O^+ and N_2^+ ions were also trapped, these were ejected following the cooling cycle, but before the application of the cyclotron frequency, by applying a frequency f_z which added resonant energy to all ions heavier than protons. The assigned experimental uncertainty was 0.6 ppm, of which 0.4 ppm was due to uncertainties in the line shape. The result of the measurement was

$$m_p/m_e = 1836.1527(11) \qquad (0.6 \text{ ppm})$$

which is in good agreement with the sub-ppm indirect measurements (figure 6.8).

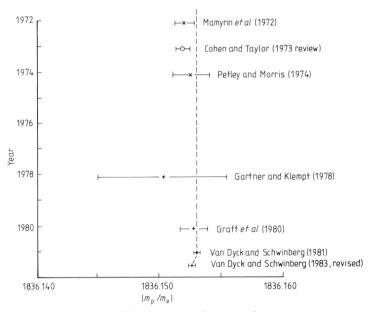

Figure 6.8 Recent determinations of m_p/m_e.

6.4.1 The Van Dyck and Schwinberg measurement

A preliminary measurement of m_p/m_e was reported by Van Dyck and Schwinberg (1981) using a compensated quadring Penning trap. Their

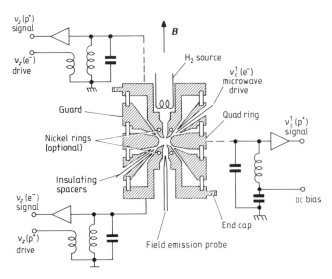

Figure 6.9 The quadrupole ion trap used by Van Dyck *et al* (1981) to measure m_p/m_e.

apparatus is illustrated in figure 6.9. It was modified version of a similar trap which was used to measure the electron $g - 2$ value by Van Dyck *et al* (1978). The flux density was about 5.05 T and was produced by a 0.3 m long superconducting magnet. The flux drifted by less than 2 parts in 10^9 per hour and was constant to better than 3 ppm over a 1 cm spherical volume.

The trapped charged particles oscillated axially in the electric potential well with a frequency f_z while rotating in a radial plane at the observed cyclotron frequency f'_c. A further effect of the crossed fields was the production of a slow, unbounded (but metastable) rotation of the centre of the cyclotron orbit at the magnetron frequency f_m. The cyclotron frequency f_c was obtained from the equations of motion as

$$f_c = f'_c + f_m$$

where $f_m = f_z^2/2f'_c$, B_0 was the applied magnetic field and $f_c = eB_0/2\pi\gamma m$ as before. Their quadring Penning trap was characterised by $R_0 = 1.59$ mm and $z_0 = 1.12$ mm respectively. The exciting and detecting circuits were connected to the electrodes as shown. Two small nickel rings served to produce a magnetic trap of the form $B_z(z^2 - r^2/2)$ which effectively coupled the magnetic moment of the charge to the axial resonance, thereby producing a shift in the axial frequency of 25 parts in 10^9 per quantum level which was used for detecting the electron cyclotron resonance condition ($v_{c,e^-} \sim 140$ GHz) Although the well depth was not sufficient to trap single electrons, measurements were performed with as few as 10 electrons and 40 protons in the trap (deduced from the observed proton linewidth of 0.2 Hz (in 76.4 MHz) compared with the 0.005 Hz width expected for single trapped protons).

The linewidth at present is limited by the observation time, which contributes to 80% of the width of the resonance. Since space charge shifts the frequency by about -1.2×10^{-10} per proton, it was necessary to extrapolate measurements made with various numbers of trapped protons back to zero.

The electron source was a tungsten point and the electric field emitted electrons entered the trap to produce both secondary electrons or H^+, H_2^+, H_3^+ and He^+ ions. The unwanted ions were prevented from being trapped by applying an axial drive frequency of the appropriate value.

The major uncertainty of their preliminary measurement was set by the different averaging of the magnetic field by the electrons and protons despite the small volume 10^{-7} cm^3 covered by their orbits. Their result was

$$m_p/m_e = 1836.153\,00(25).$$

6.4.2 The measurement of Wineland et al (1983)

Wineland *et al* (1983) have reported in interesting new ion trap measurement of m_p/m_e which required a tunable dye laser. In their experiment, some twenty $^9Be^+$ ions were stored in an ion trap in flux densities of 0.673 T, 0.764 T and 1.134 T. The ions were laser cooled and pumped to the $(M_I, M_J) = (-\frac{3}{2}, -\frac{1}{2})$ ground state by tuning the laser to the $2s^2S_{1/2}(-\frac{3}{2}, -\frac{1}{2}) \rightarrow 2p^2P_{3/2}(-\frac{3}{2}, -\frac{3}{2})$ transition at 313 nm. An optical double resonance technique was used to flip the field-dependent $(-\frac{3}{2}, -\frac{1}{2}) \rightarrow (-\frac{1}{2}, -\frac{1}{2})$ ground state hyperfine transition frequency v_s, and the cyclotron frequency ($v_c \sim 1.9$ MHz) of the $^9Be^+$ was measured almost simultaneously. The measured values of v_s(310 MHz, both at 1.134 T) and the values for the $^9Be^+$ hyperfine constant A and the nuclear-to-electron g-factor ratio g_I/g_J were then used to calculate $v_c(^9Be^+)$. This gives us the ratio

$$v_e(^9Be^+)/v_c(^9Be^+) = \tfrac{1}{2}g_J(^9Be^+)m(^9Be^+)/m_e.$$

Wineland *et al* interpreted their experiment in two different ways. The first was to take the Van Dyck *et al* value for m_p/m_e and use the equation to deduce a value of $g_J(^9Be^+)$ which could be compared with theoretical values obtained by Veseth and by Grotch. The other way was to use their theoretical values and to deduce a value for m_p/m_e. In each case the ratio of $m(^9Be^+)/m_p$ obtained from the atomic mass evaluation of Wapstra and Bos (1977) was used as an auxiliary constant. The value obtained for m_p/m_e in this way was

$$m_p/m_e = 1836.152\,38(62) \qquad (0.34\,\text{ppm}).$$

This value was in excellent agreement with the Van Dyck *et al* m_p/m_e measurement.

It is apparent that, in the spirit of this chapter, the role of this experiment is a dual one—either to test the theoretical value for $g_J(^9Be^+)$, or to regard it as a fascinating way of measuring m_p/m_e, but not of course both. It is clear though that this experiment, with the others, demonstrates that ion trapping has brought in a new era to the measurements of the fundamental constants that promises to open out into mass spectrometry as well. No doubt uncorrected electrostatic field-dependent shifts of the cyclotron frequency will be found to exist at some level, but for the present the experiments make the plausible assumption that such effects may be neglected, in marked contrast to the indirect determinations of m_p/m_e, where much higher ion densities were used.

6.5 The magnetic moment and other properties of the neutron

The properties of the neutron have been very carefully studied and subjected to measurements of increasing accuracy (see the review by Smith 1980). It is an important fundamental particle but measurements of its properties do not yield at present any information on the other fundamental constants. We discuss it here partly so as to mention in passing some of the very high precision measurements but also to show an interesting example of spin-off through the provision of a possible new method of measuring the gyromagnetic ratio of the proton. The neutron is electrically neutral so far as is known and an upper limit has been set on its charge by Gohler *et al* (1982) of about $-1.5(22) \times 10^{-20}$ of the charge on the electron. One of the difficulties of working with neutrons stems from their finite lifetime which has been measured as 925(14)s by Christensen *et al* (1972) and Byrne *et al* (1980) who detected the electrons and protons from the decay of a beam of electrons in flight. The study of the neutron lifetime provides a method of studying the strength of weak interactions, since its spontaneous decay is the result of interactions between its constituent elementary particles.

The measurement of the neutron electric dipole moment (EDM) is important because a finite result, however small, would imply a breakdown of both parity and time invariance in the weak interaction that is responsible for 3-decay. In a series of measurements, Ramsey and his collaborators established that the upper limit of the EDM was about 3×10^{-26} em (4.8×10^{-45}Cm) for which a neutron magnetic resonance spectrometer was used (Dress *et al* 1977).

In the latter measurement an electric field E was used and applied parallel and antiparallel to the magnetic flux B. For an electric dipole moment ed the change in the Larmor frequency when the direction of E is reversed was expected to be just $4eE \cdot d/h$ being the difference between $2(\mu_n B + eE \cdot d)/h$ and $2(\mu_n B - eE \cdot d)/h$.

The precision of the experiment was limited because neutrons passing through a field E with a velocity $v(\sim 150 \text{ m s}^{-1})$ would see a magnetic field that was proportional to $v \times E$ which, since it would reverse with E, would shift the Larmor frequency as though a finite EDM existed. Since one cannot ensure that neutrons traverse the apparatus exactly perpendicular to E the precision of the measurement is thereby limited. A group in Leningrad, Altarev *et al* (1980), have set a slightly lower limit of $4.0(75) \times 10^{-27}$ em ($6.4(12) \times 10^{-46}$Cm).

The advent of slow or 'cold' neutrons should enable an even lower limit to be set and a number of groups are working on the problem. Given a neutron storage time of 2000 s and a superconducting magnet with B stable to 10^{-14} T and an electric field of ~ 3 M Vm^{-1} the detection limit should reach $\sim 1.7 \times 10^{-30} em$ within the next decade, and this despite the fact that only about 14% of the incident neutrons would survive decay for the whole of the storage time. The above limit is of particular interest since theories predict dipole lengths ranging from 10^{-20} cm to 10^{-30} cm.

A further interesting measurement that is under way is that of the Compton wavelength of the neutron, which, given the ratio of the neutron mass to that of the electron or proton mass, would provide input data for an evaluation of the 'best values' if the measurement could be performed precisely enough. Weirauch (1978) has described a proposed experiment mounted by the PTB. The wavelength of a neutron of velocity v is given by the de Broglie equation

$$\lambda = (h/m_n)v.$$

The wavelength is to be determined by Bragg reflection of a modulated beam of neutrons by a single silicon crystal (the measurement of whose lattice spacing has already been discussed in §4.3). The velocity will be measured by a time of flight method, essentially by reflecting the neutrons back through the beam modulator. The measurement will use spin polarised neutrons and it will be the direction of the spins that is modulated, the spins being oriented perpendicular to a guiding magnetic field. The experiment is planned to achieve part per million accuracy.

6.5.1 The magnetic moment of the neutron

The magnetic moment of the neutron has been measured by Greene et al (1979). This measurement is of particular interest because, as we will discuss later, it led Greene (1980) to propose a new method of measuring the gyromagnetic ratio of the proton.

From the earliest observation of the proton and deuteron magnetic moments it has been inferred that the neutron magnetic moment could not be zero, for if the proton and neutron combined in a pure $^3S_1$ state to form the deuteron one would expect

$$\mu_d = \mu_n + \mu_p.$$

This was approximately in accord with experiment if one assumed that μ_n had a negative sign—the early measurements did not use rotating magnetic fields and hence could not detect the sign of μ_n (which was first established by Rogers and Staub in 1949). The discovery by Kellog, Rabi and Zacharias in 1939–40 that the neutron had an electric quadrupole moment indicated that the deuteron could not be a pure $^3S_1$ state, but must have some D-state admixture and hence the additivity of μ_p and μ_n ($\mu_d = \mu_p + \mu_n$) could not be exact. More recently the development of the quark models led to proposals by Beg, Lee and Pais and by Sakita, in 1961, that if the quarks have spins, and if also the internal symmetry group SU(3) of baryons is broken only by electromagnetism, then there must exist uniquely determined ratios among the members of the baryon octet. In particular, the ratio for the neutron and proton was predicted as

$$\mu_n/\mu_p \sim -\tfrac{2}{3}.$$

In fact, the uncertainty of this result exceeds the discrepancy between theory and experiment (see Greene (1978) for a simple derivation of this result).

The post-war measurements were conveniently of the ratio μ_n/μ_p since the magnetic flux was measured in terms of the proton spin precession frequency. Although the measurements by Cohen et al (1956) were probably adequate for the present level of theory, their measurement was limited by inhomogeneities in the magnetic field. The successes of the theory suggest that an improved measurement would be worthwhile, particularly since other related ratios, such as μ'_p/μ_B, $g_e/2$ and $g_\mu/2$ are known with thousandfold greater precision. The measurement which we now discuss, Greene et al (1979), was modified from that used by Dress et al (1977) to search for the electric dipole moment of the neutron. Essentially, the Larmor precession frequencies of neutrons and protons were measured in the same magnetic field. First, a beam of slow neutrons ($1.5 \times 10^5\,\mathrm{s}^{-1}$ intensity, velocity $\sim 180\,\mathrm{ms}^{-1}$), from the neutron source at the Institut Langevin in Grenoble, were polarised by reflection from a magnetised iron mirror. They then passed between two Ramsey coils 0.6 m apart which were between the poles of a permanent magnet (1.8 mT). After reflection by a further magnetic mirror, the neutrons were detected with a loaded ^{6}Li scintillator and photomultiplier.

The neutrons passed through the magnet along a 11 mm internal diameter glass tube and, after their Larmor frequency had been measured, water was flowed through the tube in place of the neutrons and the proton NMR frequency measured. The width of the central peak for the protons was only 0.8 Hz and this facilitated the alignment of the Ramsey coils with high precision. The field produced by the trim coils was also adjusted so that the magnetic field at the Ramsey coils was close to the average magnetic field between them. The apparatus could be rotated to allow phase differences and other effects to be removed by averaging. The neutron central peak was about 100 Hz wide and the centre was estimated to 0.01%. The essential experimental method is strongly reminiscent of that used in the caesium atomic clock.

The method of measuring the proton frequency is of interest elsewhere and so will now be explained in a little more detail. First, it was necessary to polarise the protons and this was done by first passing the water between the poles of a permanent magnet (0.2 T) where it passed through a 0.5 l chamber containing baffles to prolong the period in the magnet, the flow rate being about $0.8\,\mathrm{m\,s}^{-1}$. The water then flowed under gravity through the glass tube mentioned above and then a further distance to a second high homogeneity NMR magnet (0.4 T) in which it passed through a commercial NMR detector (figure 6.10). As the frequency applied to the Ramsey coils passed through the Larmor precession frequency the amplitude of the detected signal varied according to the characteristic pattern familiar with atomic clocks.

The method permits the averaging of a magnetic field over a distance set by the separation of the Ramsey coils and the detection sensitivity depends primarily on the strength of the polarising and detecting magnets. It thereby provides a sensitive magnetometer which may well be capable of wide application. This has been discussed by Pendlebury et al (1979).

One application that springs immediately to mind, in the context of the content of this book, is the measurement of the gyromagnetic ratio of the proton. This possibility has been discussed by Greene (1981) where he suggests that an overall experimental uncertainty of a few parts in 10^8 could well be achieved.

6.5.2 Possibilities for the measurement of γ'_p as a result of the neutron work

Greene's suggestion was that the Larmor angular frequency for Ramsey coils a distance L apart is simply

$$\omega_R = (\gamma'_p/L) \int_c^L B \, dx$$

where the water passes through a precision solenoid between the coils. For any

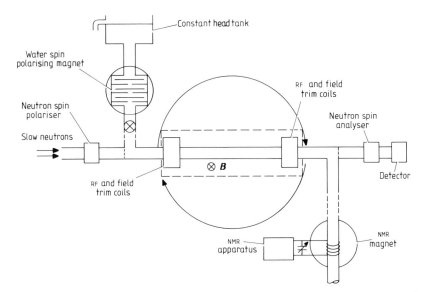

Figure 6.10 The apparatus used by Greene et al (1979) to measure the magnetic moment of the neutron.

Table 6.2 Measurements of the magnetic moment of the neutron. (The earlier measurements could not detect that μ was negative.)

Date	Authors	μ_n/μ_N		μ_n/μ_p		ppm
1940	Alvarez and Bloch (μ_n)	1.93	(2)			
1947	Arnold and Roberts (μ_n)	1.9103	(12)			
1949	Bloch et al (μ_n)			0.658 001	(30)	50
1956	Cohen et al			− 0.685 039	(17)	25
1979	Greene et al	− 1.913 041 84	(88)	− 0.684 979 35	(17)	0.25

closed path, Ampère's law can be written

$$\oint \boldsymbol{B} \cdot \mathrm{d}\boldsymbol{l} = \mu_0 N I$$

where N is the number of turns carrying a current I enclosed by the path. Since the field along the axis of a solenoid falls quite rapidly to zero the integral can be taken from $-\infty$ to $+\infty$ and closed at infinity. Hence combining the two equations

$$\gamma'_p = \omega_R L/\mu_0 N I.$$

Since the Ramsey coils have a finite extent it is difficult to measure L. However, by moving one or both of them through a distance to decrease the separation by ΔL the new frequency becomes

$$\gamma'_p = \omega'_R(L - \Delta L)/\mu_0 N I.$$

Hence by subtraction to eliminate L after rearranging:

$$\gamma'_p = \omega_R \omega'_R \Delta L/\mu_0 N I(\omega'_R - \omega_R).$$

In order to obtain his estimate of parts in 10^8 precision Greene assumed a solenoid 1 m long and 0.4 m diameter and the NMR water sample ($25 \pm 1\,°C$) flowing at 1.5 m s^{-1} through a 2 mm diameter axial tube and with ω'_R and ω_R $\sim 3 \times 10^5$ s^{-1} (~ 50 kHz) and that the change in distance L from 3 m to 2 m would be measured with a laser interferometer.

This method builds neatly on the earlier suggestion of this type of method by Williams and Olsen (1978). They however suggested using a SQUID magnetometer to perform the integration.

The above is a simplified scheme of the proposed experiment and there are several other necessary features. One is that there must be a small ambient magnetic field throughout the path between the polarising and detecting magnet in order to sustain the polarisation, and another is that there must be a shim field at both positions of the movable Ramsey coil so that the frequency width due to the transit time of a single coil $\Delta\omega_{RABI}$ is greater than the difference between the average frequency and that at the coil.

This section therefore illustrates quite well the theme of this chapter; the evolving role of the magnetic moment of the neutron to provide sensitive tests of theory, and also how one type of high-precision measurement leads to improvements in another. We can expect that the increasing access to beams of cold neutrons, of such low energy that they only rise to heights of a metre or less in the Earth's gravitational field before falling, will lead in time to further fascinating high-accuracy experiments.

6.6 Conclusion and possibilities for the future

The uncertainty reached for the m_p/m_e experiments discussed in § 6.4 is by no means the limit of the m_p/m_e methods and it is hoped to improve them further to better than a part in 10^8. One can only marvel at the ability and patience that will be required to trap and perform experiments on single electrons and protons. An ability to work through the night is an added requirement!

A further possibility of measuring m_p/m_e arises from a measurement of the ratios of the wavelengths or frequencies of corresponding lines in the spectrum of atomic hydrogen and deuterium. This ratio depends on the reduced masses $(1 + m_e/m_p)/(1 + m_e/m_d)$ and hence the measurement necessitates sacrificing a factor of nearly 2000 in precision. Nevertheless, the techniques of laser saturation and polarisation spectroscopy, pioneered by Schawlow and Hänsch at Stanford University for the measurement of the Rydberg constant, may well prove capable of further refinement to yield the required precision, especially since absolute wavelength measurements are not required. However, it may be that such measurements will serve more to check Lamb shift calculations and other corrections than to measure m_p/m_e —time will tell.

We have seen in this chapter examples of the changing roles of dimensionless ratios of the constants. Of these, μ_p/μ_B has become an auxiliary constant. On the other hand, μ_p/μ_N has just changed from being an important experimentally measured constant, because it limited the accuracy of m_p/m_e, to being a constant which is unlikely to be measured again. The dramatic improvement in the precision with which m_p/m_e may be measured directly suggests that this ratio will continue to serve as an auxiliary constant for the next evaluation of the constants at the greatly improved precision which will be required by the end of the decade.

7

The gravitational constants G and g

7.1 The universal gravitational constant G

Gravitational forces are among the weakest forces encountered in physics and yet their effects are, of course, very important on a terrestrial and astronomical scale. It is very difficult to perform gravitational measurements in the laboratory since the masses which can be assembled are necessarily miniscule compared with that of the Earth. Consequently, the force in any such measurement of G must be accurately perpendicular to that from g. The simplest method of achieving this is to suspended a mass from a very fine filament since this automatically hangs in the direction of the local vertical. Any horizontal motion of this system can then be attributed to the forces due to the masses involved in the experiment. The sensitivities of the measurements have been in the region of 10^{-10} g.

If one considers two spheres of radii r_1 and r_2 and densities ρ_1 and ρ_2 respectively, whose centres are separated by a distance d, the force F between them is

$$F = G16\pi^2 \rho_1 \rho_2 r_1^3 r_2^3/9d^2.$$

We must have $d \gtrsim r_1 + r_2$, and it is evident from this simple equation that if a large force is required, the densities should be as high as possible. It is not surprising, therefore, that most experimenters have tried to use dense materials such as lead, platinum, gold or iron in the construction of the masses, while tungsten is a more recent addition to the list. The measurements of G for the most part use a symmetric arrangement and measure the gravitational force as some form of torsional couple. Consider the system of figure 7.1, where the masses, angles and distances are as shown, and the torsional couple exerted by the suspension for a deflection θ is K_f. The couple exerted on the suspension is given by

$$K_f = \frac{\partial}{\partial \theta}[-2GmM/l(\phi - \theta)].$$

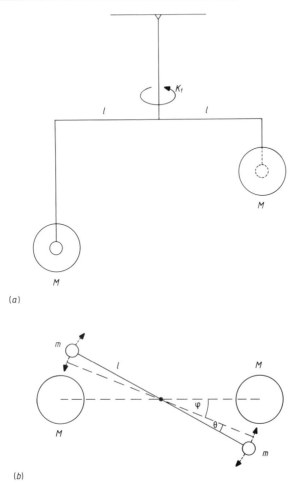

Figure 7.1 Schematic of the torsional pendulum method of measuring G. (a) Elevation and (b) plan view of the pendulum, illustrating symbols used in the text.

The experimental problem is to measure, or make use of, this torque in order to deduce G from the observations. One method is the torsion balance method of Cavendish (1798) which is well described in the literature. If ω is the period of free oscillation, θ_e the equilibrium angle of deflection of the beam and ϕ the angle between the equilibrium position of the beam and the line joining the centres of the fixed spheres, one obtains G from

$$\theta_e/G = mM/l^3\omega^2\phi^2.$$

If all quantities are scaled in the same ratio then mM/l^3 is constant so that θ_e/G is unchanged by the scaling. This property was noted by Cornu and exploited by Boys (1895) in his determination. He pointed out that if only the length of the torsion beam was reduced, while keeping the period and large masses the same, then the sensitivity would be improved. This method was limited by the cross attraction of the large masses on the remote one of the two suspended spheres, and also by the problems which are common to most of the determinations. These arise from the necessity to use a fine suspension, in order to increase the torsional deflection, while keeping the suspension strong enough to support the masses, which means that the suspension is close to its elastic limit. Boys developed the technique of making very fine fused-silica suspensions. He also reduced the geometrical problems by having the two attracting masses at different levels. His apparatus was operated in air, but Braun (1897) used a similar method with heavier masses (table 7.1) and

Table 7.1 The principle laboratory measurements of G.

Author	Country	Date	Method	$G(10^{-11} \text{ m}^3 \text{ s}^{-2} \text{ kg}^{-1})$	
Cavendish	UK	1798	Ts/Td	6.75	(5)
Reich	Germany	1838, 1852	Ts	6.64	(6)
Bailey	UK	1843	Ts/Td	6.63	(7)
Jolly	Germany	1873	Bb	6.447	(110)
Poynting	UK	1878, 1891	Bb	6.70	(4)
Boys	UK	1895	Ts	6.658	(7)
Eötvös	Hungary	1896	Td	6.657	(13)
Braun	Austria	1897	Ts/Td	6.649	(2)
Wilsing	Germany	1887, 1889	Bb	6.594	(150)
Richarz	Germany	1888, 1898	Bb	6.683	(11)
Burgess	France	1902	Ts	6.64	(?)
Heyl	USA	1930	Td	6.670	(5)
Zahradnicek	Czechoslovakia	1933	Td	6.66	(4)
Heyl and Chrzanowski	USA	1942	Td	6.673	(3)
Luther et al	USA	1971, 1978	Tc	6.6699	(14)
Facy and Pontikis	France	1972	Tr	6.6714	(6)
Renner	Hungary	1973	Td	6.670	(8)
Karagioz	USSR	1976	Td	6.668	(2)
Sagitov	USSR	1977	Td	6.6745	(3)
Luther and Towler	USA	1982	Td	6.6726	(5)
CODATA	—	1973	—	6.6720	(41)

Bb = beam balance, T = torsion balance, d = dynamic, c = compensated, r = resonance, s = static. The values in parentheses are the standard deviation uncertainties in the last digit quoted. At first sight there is little more than a tenfold improvement over one hundred years, but in many cases the uncertainties in the earlier values do not include adequate allowance for possible systematic effects.

evacuated his apparatus. Poynting (1891) worked straightforwardly by measuring the deflection of the arms of a bullion balance carrying a load of some 50 kg. His balance, incidently, is still in use today at the NPL for certain types of mass calibration.

The above methods of Cavendish and Poynting required the measurement of the equilibrium deflection position in the presence of the slow oscillations of the suspended masses on the torsional pendulum, but the other methods made use of the change in the period of torsional oscillations due to the gravitational force. Heyl (1930) developed the work of Braun at the NBS, later working with Chrzanowski (1942). It is now thought that this work, although a model of painstaking attention to detail, was none the less subject to a systematic error. This may have arisen because they used the oscillating method, and failed to take sufficiently into account the highly non-linear angular dependence of the couple on the beam when the gap between the stationary and moving masses was small. This determination also illustrates the difficulties associated with the use of fused-silica or fine tungsten wire suspensions. It suffered too from the disadvantages of working under conditions where the torsional couple from the suspension was large compared with the gravitational couple. Difficulties were also experienced as a result of the instabilities of the angular zero of the system.

7.1.1 The carrot and donkey method of Beams

There is an obvious advantage if the effects of the gravitational couple on the pendulum can be enhanced. A resonant method has been described by Pontikis (1972) in recent times, and earlier by Zahradnicek (1933). An alternative approach was developed by Rose et al (1969). The method of Beams allowed the couple on the suspended mass to be integrated up in an ingenious way by a type of carrot and donkey method. Essentially, the two large spherical tungsten masses were mounted on a table (figure 7.2) which could be rotated by a specially designed electric motor. Also mounted on the same table, inside an airtight chamber, was a horizontal cylinder of copper which was suspended by a quartz fibre. The gravitational attraction between the spheres and the cylinder would, of course, cause the cylinder to deflect slightly. However, this deflection was not allowed to occur, for the table was rotated so as to remove the deflection. In consequence the table gradually accelerated, taking about half an hour for the first complete rotation. In this way the gravitational forces could be integrated up, thereby improving the precision and reducing the noise level.

A theoretical analysis of the gravitational torque system (Rose et al 1969) gave

$$G = \frac{\alpha \pm \dot{\omega}}{A(1 + B + C + \cdots)}$$

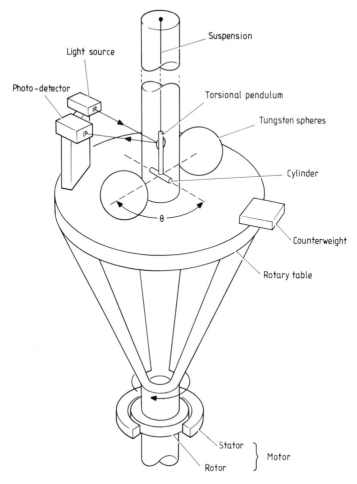

Figure 7.2 The levitated rotating table method devised by Beams and others at the University of Virginia

where

$$A = \frac{3M}{R^3} \frac{[\frac{1}{3} - \frac{1}{4}(2A/L)^2] \sin\theta}{[\frac{1}{3} + \frac{1}{4}(2a/L)^2 + (2/L)^2(I_s/m)]}$$

$$B = \frac{5}{6}(L/2R)^2[\frac{1}{5} - \frac{1}{2}(2a/L)^2 + \frac{1}{8}(2a/L)^4(7\cos^2\theta - 3)(\frac{1}{3} - \frac{1}{4}(2a/L)^2]$$

$$C = \frac{1}{48}(L/2R)^4 \frac{[\frac{1}{7} - \frac{3}{4}(2a/L)^2 + \frac{5}{8}(2a/L)^4 - \frac{5}{64}(2a/L)^6]}{[\frac{1}{3} - \frac{1}{4}(2a/L)^2]}$$

$$\times (1386\cos^4\theta - 1260\cos^2\theta - 210).$$

$\alpha(4.3299(22) \times 10^{-6} \text{ rad s}^{-2})$ and $\dot{\omega}(-0.3431(22) \times 10^{-6} \text{ rad s}^{-2})$ were the angular accelerations with the spheres on the table and removed from it, M

was the mass of one sphere (10.490 12(7) kg), R the distance from the axis of rotation to the centre of mass of the spheres (7.6178(5) cm), m the mass of the suspended cylinder (4.051 250 5(5) g), L its length (3.8105(4) cm), $2a$ its diameter (1.98 mm) and I the moment of inertia (0.044 510 3 g cm^2). The angle θ (0.7835(2) rad) was measured between the axis of the cylinder and the line joining the centres of the spheres. The terms in the denominator were simply the first three terms of the expansion of the gravitational potential of the cylinder system in terms of Legendre polynomials. Since the value of G depends on the cube of the distance of the centre of mass of the spheres from the axis of suspension of the cylindrical mass, one must make accurate measurements of the geometrical distance and also be certain that the spheres are very homogeneous: some effects (e.g. homogeneity) may be checked by rotating the spheres from one run to the next.

A simplified equation of motion for the torsional pendulum leads to

$$\ddot{\theta} = \tfrac{9}{2}(GmM \sin 2\theta / \lambda R^3)$$

where $\lambda = L/R$. It is evident that the acceleration was a maximum for $\theta = \pi/4$ and that both L/R and R should be small. The rotating table was very carefully constructed and aligned so that it had a runout (eccentricity) of less than 0.5 μm at a radius of 5 cm from the axis of rotation. The large spheres, of high density tungsten, were spherical to better than 0.12 μm. In most experiments, the rotation commenced when the angle θ was about 45° and the angular acceleration was between 4 and 5×10^{-6} rad s^{-2}, so that the table took about 30 min for the first revolution and was rotating at about 0.5 rpm after 2 h. The preliminary work of Rose et al (1969) was later improved and described by Lowry et al (1971) and (1975). Despite the considerable amount of effort and time spent in refining and uncovering the systematic effects the experiment as described did not lead to a final result. It is clear that the precision was better than that of the earlier determinations, but the accuracy still remained at around the same level.

7.1.2 The Luther and Towler measurement

Luther and Towler (1981), working at the NBS in Washington, have described a measurement of G using a modified form of the above apparatus. They used a more conventional method, i.e. they did not use the levitated table to integrate up the angular acceleration. Different forms of suspension were tried including carbon fibres of the type used in the construction of golf clubs. The gradual creep of the zero remained a major problem. A further source of error in the compensated rotating table method which was uncovered subsequently was that from motion of the point of suspension. If this motion was at 45° to the line joining the centres of the spheres it could add energy to the system and increase the torsional angle. This was a surprising result since motion along or perpendicular to the line joining the centres of the spheres gives a null effect in both cases. The lesson is that one should take nothing for granted in metrology!

The theory is essentially the same as that of Boys (1895). For a torsion balance with a moment of inertia I, angular frequency ω_f, which is supported by a fibre of torsion constant K_f, one has

$$\omega_f^2 = K_f/I$$

or

$$K_f = I\omega_f^2$$

where K_f is the second derivative of the potential energy with respect to θ, i.e. the torsion constant is the first derivative of the torque. If large masses are placed close to this torsion pendulum, the angular frequency is given by ω_{f+g} where

$$\omega_{f+g}^2 = (K_f + K_g)/I$$

or

$$K_f + K_g = I\omega_{f+g}^2$$

where K_g is the second derivative of the gravitational potential U evaluated at $\theta = 0$:

$$K_g = \left(\frac{d^2U}{d\theta^2}\right)_{\theta=0}$$

and

$$U = \int_V \frac{GM\rho\,dV}{R^2}$$

where U is the gravitational potential of the small mass system in the presence of the large mass system M and the integral is taken over the volume of the small masses, density ρ (Towler et al 1971). This only applies to an undamped oscillator, but their damping coefficient was very small since the system was evacuated. The small mass system (7 g) consisted of two tungsten discs 2.5472 mm thick and 7.1660 mm diameter mounted dumb-bell fashion on the ends of a 1.0347 mm diameter tungsten rod 28.5472 mm long (we illustrate the accuracy of the dimensional measurements). The torsion arrangement consisted of a top portion of 125 μm quartz fibre with an aluminium disc at the lower end which was magnetically damped. A spindle through the centre of this disc supported a second 12 μm diameter fibre which was 40 cm long. The torsional constant of the top fibre was 10^5 times greater than the lower one and hence this fibre served to damp out mechanical vibrations of the support, but not to damp the torsional oscillations. The period of the oscillations was about 6 min and the gravitational couple due to the large masses changed this by a few per cent.

The apparatus was mounted on a 5000 kg concrete block and was thermostated to ± 0.1 °C. The angular position was sensed by using a mirror and auto-collimator arrangement whereby light from a 15 μm slit was refocused, after a double reflection from the pendulum, onto a 1024 element photo-diode array. This allowed the motion over an angle of 2×10^{-2} rad to be analysed. The large 10.16 cm diameter spherical tungsten masses weighed about 10.5 kg and were separated by 14.059 454(3) cm. They could be removed and replaced to better than 0.05 μm RMS. Measurements were taken over a period of 50 to 75 h and data taken at 20 s intervals, the spheres being placed in position and removed at 6 to 12 h intervals.

The final value for G was $6.6726(5) \times 10^{-11}$ m^3 s^{-2} kg^{-1} and the four largest components of the uncertainty were those for (a) the overall length and thickness of the small mass system, (b) the uncertainty of the moment of inertia of the mirror used to sense the angular positon, and (c) (the largest contribution) was that due to the measurement of the 6 min torsional period. It is thought that the precision can be improved further by constructing a more simply measured small mass system, using a lighter mirror and using a thinner fibre to give a 20 min period. In this case the frequency change due to the large masses would be $\sim 20\%$.

Other measurements have also been reported. Thus Renner (1973) made a dynamic torsional pendulum measurement in which the period of oscillation was measured in the near and far positions in the usual way. His large masses comprised four identical hollow cylinders which were mounted on a table which could be rotated about the axis of the torsion balance.

Karagioz et al (1976) published a preliminary measurement by the dynamic method in which one large mass was shifted in the direction of the torsion beam. Although several materials were used for the large mass, only the results which were obtained with steel were included in the final result. Sagitov et al (1977) have reported a similar type of measurement with the large cylindrical mass being moved axially. Their pendulum torsional period was 2000 s to 2300 s. The deflection was recorded every 20 s and the angular resolution was 10^{-5} rad. This allowed the non-linear motions to be analysed in detail.

7.1.3 The resonant torsional pendulum of Facy and Pontikis

The method of Facy and Pontikis (1972) was a development of that used by Zahradnicek (1933). The suspended masses were 4.67 g spheres suspended by a 0.5 mm diameter tungsten wire about 1.3 m long. A mirror was fixed to the wire. This enabled the angular deflections of about 10^{-5} rad to be measured to 10^{-9} rad (a lamp focused a spot onto the mirror and the light was reflected onto a photodetector). The large masses weighed about 1.5 kg and were supported on a table which could be made to oscillate sinusoidally by means of a precision electric motor. The suspended spheres were inside an evacuated glass cylinder. The whole apparatus was located at the centre of a

1.9 m diameter Helmholtz coil arrangement which nulled out the static magnetic fields. A second torsional pendulum system was suspended about 1.5 m away from the first and the signal from the associated photocell was subtracted from that of the main photocell (in order to remove the effects of vibration of the supports, etc).

The period of the motion of the heavy masses was chosen so as to excite the torsional pendulum at its resonant frequency. The photocell signals were analysed by computer to give the best fit to an equation of the motion of the form

$$\frac{d^2\theta}{dt^2} + \Lambda\frac{d\theta}{dt} + \Omega^2\theta = \frac{GmM}{IR}C(\omega t - \theta).$$

Four different types of material were used for the heavy masses, namely silver, copper, bronze and lead. The values of G obtained for the four materials were not significantly different, the standard deviation uncertainties being about 800 ppm. The value quoted for the weighted mean of all of the results was $6.6714(6) \times 10^{-11}$ m^{3}s^{-2}kg^{-1} and was stated to be a preliminary result.

7.1.4 The scaled-up torsional pendulum

This determination (Cook 1968) is derived from those of Heyl and Braun, and the period of oscillations rather than the equilibrium deflection is to be measured. The experiment (which is still in progress), in collaboration with Professor Marusi, takes advantage of a natural site situated in the limestone tablelands which are just to the north of Trieste. The Grotte Gigante is a large cave which has a floor to ceiling height of about 100 m and the temperature is constant to within ~ 1 °C throughout the year, although the relative humidity is very high. It is used for other geophysical experiments and so has a well documented environmental history.

This G experiment can therefore make use of a much longer torsional pendulum than the other determinations and the planned experiment will use a tungsten or other wire which is some 80 m long and about 0.5 mm diameter. In order to avoid instabilities the restoring torque produced by this will be just greater than the maximum gravitational torque. The suspended masses are two 14 cm diameter spheres and are made from a stable copper alloy. They are supported on a 4 kg titanium beam and the moment of inertia will be obtained by making measurements at two separations of the spheres, 1.32 m and 0.66 m. The large bronze 500 kg masses each consisted of nine discs, 460 mm diameter and 27 mm thick, together with two small discs, 280 mm diameter and 50 mm thick at either end. The discs were all made from the same melt and were carefully machined to be flat to within a few μm.

The reason for the choice of this particular shape for the large masses may be appreciated if one expands the gravitational potential as a series involving zonal harmonics of the form

$$GM/r + (k_2/r^3)P_2(\cos\theta) + (k_4/r^5)P_4(\cos\theta) + \cdots$$

where M is the mass of the body and r the distance from the centre of mass to the point where the potential is evaluated. Close examination of the terms shows that some of them may be made to vanish by proper choice of the shape. Thus $k_2 = G(C - A)$, where C is the moment of inertia about the axis of symmetry and A the moment about any axis in the equatorial plane. The k_2 term becomes zero when $C = A$, which condition occurs when the mass is in the form of a uniform cylinder whose length is $2 \times 3^{1/2}$ times its radius. Similarly, the use of a cylinder with end caps allows the fourth and sixth zonal harmonics to be eliminated (the odd terms are zero from symmetry considerations). Consequently the potential due to the mass is close to that of a sphere, (GM/r) together with small correcting terms which are of the order of $1/r^9$ or less (see Cook 1968, 1970). The 500 kg masses were separated either by 2 m when they were in line with the suspended spheres, or by 1 m when they were perpendicular to this axis. The small departures of the discs from a perfect figure may be reduced by appropriate choices of their position during stacking. It should be noted that, although we concluded in §7.1 that the densities of the masses used in the G experiments should be as great as possible, the strength of the material is also an important consideration, for otherwise the masses will distort under the compressional forces due to their own weight.

The equations of motion for the torsional pendulum are significantly non-linear, of the form

$$I\ddot{\theta} + a\dot{\theta} + b_1\theta^2 + b_3\theta^3 + b_5\theta^5 + \cdots = \varepsilon$$

where ε is the net couple on the pendulum (using the notation of Cook 1970). The motion of the pendulum will be measured very accurately by using a laser fringe-counting Michelson interferometer system, for which purpose cube-corner reflectors will be mounted on the titanium beam of the torsional pendulum, and the results analysed in order to deduce the terms in the above expression. The period of the pendulum will be in the region of an hour and the amplitude of the deflection of the torsional pendulum between 1° and 3°. The apparatus will be under a reduced pressure of around 10 Pa and hence will be very lightly damped. Additional damping can be introduced as required by temporarily raising an oil bath until a needle suspended from the pendulum is immersed in it. The sophistication of modern computers should enable a good fit to be obtained between the actual and conjectured motion of the pendulum, ultimately leading to a more accurate measurement of G than hitherto.

The G measurements have been extensively reviewed by de Boer (1981) and the most precise determinations included in the 1984 evaluation of the best value of G by Cohen and Taylor. It is clear that the accuracy of the determinations has been improved by the application of modern data acquisition and measurement techniques, although many of the limitations of the early work survive in the more sophisticated measurements of today. The suspensions and vibration isolation are still a major limitation and we return to this in § 7.3 after a discussion of some of the tests of the nature of gravitation which are in progress.

7.2 Tests of the nature of gravitation

The gravitational force between an electron and a proton is only about 10^{-38} of the electrostatic force and it is the small size of the gravitational force which makes it so difficult to test experimentally in the laboratory. Any experiment must inevitably be affected by a large number of extraneous effects: magnetic, vibrational, convection currents and so on. The difficulty has not deterred experimenters from attempting some fundamental tests which we now discuss.

7.2.1 Tests of the inverse square law

One such test involves the test of the inverse square law to see whether the gravitational mass is always proportional to the inertial mass, or whether it varies from one substance to another. There has been considerable interest too for some time in finding a unified field theory for gravitational and other interactions.

One suggestion has been that there might be a particle having a finite mass (graviton, low mass axon, gluon exchange etc) which arises from some broken symmetry rule in the field theoretical description of gravitation. Thus the gravitational potential energy of two masses might not vary simply as $1/r$, where r is their separation, but instead might vary as

$$\phi_G = (G_0 Mm/r)[1 + \varepsilon \exp(-r/\lambda)]$$

(Wagoner (1970), Fujii (1971) and O'Hanlon (1972)), or alternatively as

$$\phi_G = (G_0 Mm/r)[1 + \alpha \ln(r/\lambda)]$$

(Long 1976). This would give an inverse square law of force relationship at large and small distances, but the constants of proportionality would be in the ratio of $1:(1 + \varepsilon)$. This is of course but one of many possible hypotheses.

Fujii (1971) favoured $\varepsilon = \frac{1}{3}$ and $\lambda \simeq 10$ to 1000 m, which would mean that values of G measured at laboratory distances between 10 and 30 cm, G^*, would be

$$G^* = \tfrac{4}{3} G_\infty$$

and also the mass of the Earth or other solar system bodies estimated from the measured values of $G_\infty M$ with the assumption $G_\infty = G$ would be $\frac{4}{3}$ of the presently accepted values. If $\lambda < 0.01$ m then $G_\infty = G^*$ and there would be no problem. Considerable interest was aroused by the suggestion of Long (1976) that the breakdown from the square law behaviour occurred at laboratory distances. It is interesting that he was able to obtain possible support for this by plotting earlier results as a function of the separation of the large and small masses. However, Stephenson (1967) has also found support for a possible secular variation of G from a similar analysis. Both may be due to systematic effects in the experiments. Any departure from an exact square law would, of course, have particularly important consequences for the laboratory measurements of G.

Long used an evacuated gravity gradient meter type of G apparatus (figure 7.3), which was similar to that used by Eötvös. The 50 μm tungsten torsion fibre was 48.6 cm long and the (50 g) attracted tantalum ball was hung 87 cm below the balance rod. An electrostatic 'puller' was located at the other end of the rod and the apparatus was used at constant deflection. This was sustained by applying an electrostatic voltage to counterbalance the gravitational force on the sphere. Two separate cylindrical rings were used. The near ring was of tantalum and weighed 1.225 kg. It was 1.78 cm thick with an outer radius of 4.55 cm and an inner radius of 2.75 cm. The sphere was arranged to be at the position of maximum gravitational force $(r/\sqrt{2})$ which was 2.6 cm from the face of the ring. The far mass was 57.58 kg of cast brass. It had a thickness of 7.6 cm, an outer radius of 27.1 cm, and an inner radius of 21.6 cm. The position of maximum force was 17.4 cm from the face of the ring. The forces on the 50 g sphere from the near and far rings were compared, any difference being attributable to a change in the value of G with distance. The experiment has been repeated by Spero *et al* (1980), with a greater sensitivity than was achieved in Long's experiment.

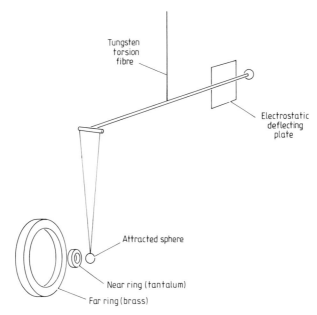

Figure 7.3 Apparatus used by Long (1976) to investigate the validity of the gravitational inverse square law at laboratory distances.

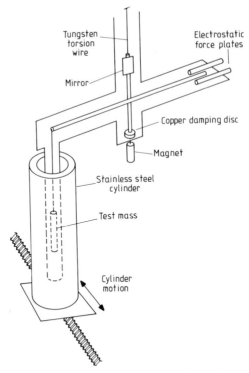

Figure 7.4 The apparatus used by Spero *et al* (1980) to investigate possible departures from an exact inverse square law of gravitation at short distances.

In their work, Spero *et al* used a modified form of torsion balance as illustrated in figure 7.4. The test mass was a 20 g high-purity copper cylinder, 4.4 cm long, which hung 83 cm below the 60 cm long torsion balance boom. The boom was suspended by a 32 cm long, 75 μm diameter tungsten wire. The whole apparatus was evacuated to ~ 10 μPa by an ion pump. A small force could be applied at the other end of the boom by means of a potential applied to the electrostatic force plates. The angular rotation of the boom was sensed by an optical lever system. The photocell signal was differentiated and fed back to the deflection system in order to damp critically any torsional oscillations from the system. The gravitational constant of the balance was obtained by using a 133 g copper ring of radius $r = 12.1$ cm at an on-axis distance of $r/\sqrt{2}$ below the suspended mass. Their method of testing the inverse square law was to insert the test mass inside a 60 cm long stainless steel cylinder. The cylinder weighed about 10.44 kg, its inside diameter was 6 cm and the wall thickness was 1 cm (the residual magnetic field was less than 5 nT). The cylinder was moved horizontally in a direction perpendicular to the axis of the boom and the change in the force on the suspended mass was investigated.

For an infinitely long cylinder and an inverse square law of force, the gravitational force due to the cylinder should vanish everywhere inside it. However, for a truncated cylinder, there are end effects. In the Spero *et al* measurement, the difference between the end effects of the near and far walls of the cylinder contributed a net force. This effect was at about the per cent level and so the magnitude of end-effect correction was only required to 0.1% for the inverse square law of force to be tested to a part in 10^5. Similarly the residual magnetic field was sufficiently small for it to be only necessary to know the motion of the cylinder to 0.1% in order to allow for eddy current effects to a comparable precision. The geometry and homogeneity of the cylinder had to be known to a part in 10^5. This was achieved by making measurements along different azimuthal axes and consequently the only correction required was that for the variations along the axial direction of the cylinder.

After applying several small corrections, Spero *et al* concluded that the experimental and theoretical gravitational forces on the suspended mass differed by $+2(14) \times 10^{-15}$ Nm. Hence, taking the form for the anomaly suggested by Long of

$$G(r) = G_0(1 + \varepsilon \ln r(\text{cm})) \qquad (7.2.1)$$

they found $\varepsilon \sim 1(7) \times 10^{-5}$ which may be compared with the value of $200(40) \times 10^{-5}$ obtained by Long for his data. Spero *et al*'s method was essentially a null test, whereas Long measured the force between cylindrical ring masses and a torsional balance at distances of 3 cm and 37 cm. Long (1980) suggested that there might be a gravitational analogue of the vacuum polarisation effect which produces departures from inverse square law behaviour in the electrostatic forces between charges and, if there was no polarising field in the region probed by Spero *et al*, the two results could still be compatible.

7.2.2 Tests using non-laboratory measurements of G

The earliest attempts to measure G were geophysical, for example, the change in g between sea level and on a plateau of known density or the deflection of the local vertical by a mountain could be used to compute G. It took some time before it was accepted that the laboratory measurements of Cavendish 1798 and Boys and others were more accurate than the large scale measurements. The next successful non-laboratory measurement was that of Airy (and others) in 1856 of the gravity profile within a coal mine in northern England. Gravity profiles have long been used to infer the geological structure of the surrounding material, but Stacy *et al* (1981) investigated the gravity profile both vertically over 950 m and along six tunnels which were available in both N–S and E–W directions in the metalliferous mine at Mount Isa, in Queensland, Australia, in an area which is at present remote from major excavations. The value obtained for G was

$$G = 6.71(13) \times 10^{-11} \text{ m}^3 \text{kg}^{-1} \text{s}^{-2}.$$

Although the uncertainty is, of course, considerably larger than that of laboratory experiments, the motivation for the work was to test for the variation of G with distance. This work put an upper limit for the coefficient in Long's expression (equation 7.2.1) of $\lesssim 0.003$ which is not quite precise enough to test Long's favoured value of $\varepsilon = 0.002$. More sensitive experiments are planned (Stacy 1981), and it is hoped to measure the change in g,

$$\Delta g = 4\pi G\rho h \sim 8.38 \times 10^{-6} \text{ m s}^{-2} \sim 10^{-6} g$$

as a fresh water hydroelectric reservoir is filled, and emptied, by mounting the apparatus on an electricity pylon at the centre as shown in figure 7.5. For 10 kg masses on the arms of a vacuum balance, one above high water and the other below low water, an additional mass of 8.5 mg would be required to restore the balance following a 10 m change in water level. It is hoped to use automatic interchange techniques to check for the stability of the length of the balance arms and to use capacitance sensing methods to detect the balance condition. The hoped for sensitivity on the 8.5 mg mass change is ~ 3 parts in 10^5. In principle it should also be possible to look at the variation of g with depth at sea. In this connection, a measurement by Drake and Delauze (1965) using a bathyscaphe in a deep trench in the Mediterranean points to the difficulties, and their measurements could be interpreted as $G_\infty > G^*$, for their measured vertical gradient was shallower than expected.

The results have significance for those theories which attempt to unify gravitational forces with the other forces. Such theories can be classified according to the mass scale of the particles which they introduce, or equivalently the length scale at which the phenomena occur. The mass scale can be expressed as

$$m_P(m_H/m_P)^n$$

where m_P is the Planck mass given by

$$m_P = (c\hbar/G)^{1/2} \sim 2.2 \times 10^{-8} \text{ kg (or } 1.2 \times 10^{19} \text{ GeV)}$$

and m_H is a typical hadron mass. These theories predict departures from

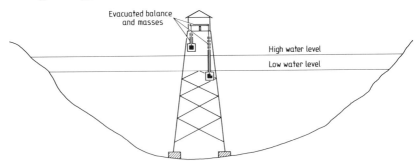

Figure 7.5 Scheme of the large-scale terrestrial measurement of G, proposed by Stacy, of detecting the local change in the acceleration due to gravity as an Australian reservoir is filled and emptied when used for hydroelectric power generation.

Newton's inverse square law at length scales

$$L \sim (m_P/m_H)\hbar/m_H c.$$

A value of $L(1 \text{ GeV}/m_H)^2$ km would correspond to the Compton wavelength of a particle of mass $\mu = (m_H/1 \text{ GeV})^2 \times 10^{-10}$ eV. The theories typically take n as 0 or 1, although some have taken $n = 2$ (which would have consequences at distance scales $\sim$ km). Gibbons and Whiting (1981) have analysed the data given here and also used lunar and planetary satellite data and concluded that on present evidence, m_H must be at least 1000 GeV (that is $L \sim 10^{-3}$ m and $\mu \gtrsim 2 \times 10^{-39}$ kg) for the theories to remain viable. Up to the time of writing the deviations from Newton's square law of gravitational force on larger length scales would appear to be less than 1%. This will continue to be an active area both theoretically and experimentally.

7.2.3 Recent Eötvös experiments

It is observed experimentally that acceleration in a homogeneous gravitational field does not depend on the material nature of the test mass. This property of gravitation has been tested to increasingly high precision and in all of the tests no difference has been observed at the level of accuracy of the experiments. This property has been incorporated into general relativity as a postulate. Consequently, the level, if any, at which the principle of equivalence is violated is essentially an indication of the extent to which the gravitational interaction is truly universal. One may anticipate, therefore, that increasingly sensitive tests of the equivalence principal will be devised and performed in the future. The principle of equivalence is also of fundamental importance in the G

Table 7.2 Limits on the equivalence of inertial mass m and gravitational mass M for materials A and B measured by the parameter $\eta(A, B) = 2[(M/m)_A - (M/m)_B]/[(M/m)_A + (M/m)_B]$.

Date	Author	Substances tested	Value for η
1686	Newton	Various	10^{-3}
1832	Bessel	Various	2×10^{-5}
1922	Eötvös	Various	4×10^{-9}
1923	Potter	Various	2×10^{-5}
1935	Renner	Various	2×10^{-9}
1964	Roll *et al*	Al, Au	3×10^{-11}
1972	Braginsky and Panov	Al, Pt	9×10^{-13}
1976	Koester	Neutrons	3×10^{-4}
1978	Worden	Nb-earth	10^{-4}
1981	Keiser and Faller	Cu, W	4×10^{-11}
Future	Space experiments	Various	10^{-15}–10^{-18}

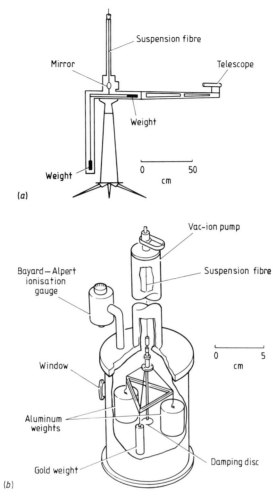

Figure 7.6 Apparatus used to measure the material independence of G: (a) the Eötvös torsional pendulum (also used for gravity gradient measurements) and (b) the apparatus used by Dicke (1960).

determinations although, as will be seen, it has been verified to more than adequate precision for the accuracy of the present G determinations.

Most of the experiments (table 7.2) have followed that of Eötvös *et al* (1922) (figure 7.6(a)). Notable among these are the measurements of Roll *et al* (1964) (figure 7.6(b)) and Braginsky and Panov (1972) (figure 7.7). These measurements used traditional types of Cavendish torsion balances in order to measure the difference in acceleration of two test masses. Essentially, if the principle of equivalence breaks down, an oscillating torsional pendulum which is falling along with the Earth in the gravitational field of the Sun will

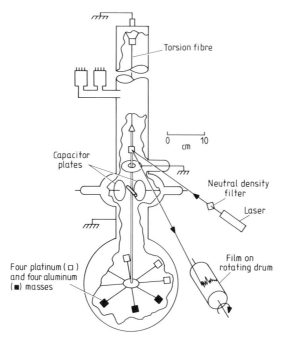

Torsion fibre

Capacitor
plates

Neutral density
filter

Laser

Four platinum (□)
and four aluminum
(■) masses

Film on
rotating drum

Figure 7.7 The Eötvös experiment of Braginsky and Panov (1972).

experience a mechanical moment which is proportional to the difference in the acceleration of the pendulum materials. Since the Earth is rotating, the magnitude of such a moment is given by

$$M = k_p I_0 r g_s(a, b)$$

where k_p is a constant whose value depends on the geometry of the pendulum, I_0 is the moment of inertia of the pendulum beam and r its radius and g_s (6.2 mm s^{-2}) is the acceleration towards the Sun. The dimensionless quantity

$$\eta(A, B) = (M/m)_A - (M/m)_B$$

is the difference between the effects on masses of materials A and B where m and M are the inertial and gravitational masses respectively.

As with many experiments, the sensitivity is ultimately limited by the effects of thermal fluctuations and, in the classical approximation (Braginsky and Manukin 1977) the minimum detectable acceleration a_{min} of a mechanical oscillator which has a relaxation time t_r which is longer than the measurement time t_m is given by

$$a_{min} = \xi(8kT/mt_r t_m)^{1/2}$$

where ξ is the coefficient which characterises the level of reliability with which the acceleration is detected. It is evident from this expression that for the maximum sensitivity the relaxation time of the oscillator should be as long as possible. In their experiment Braginsky et al (1972) achieved a relaxation time of $\sim 6 \times 10^7$ s and a period of oscillation for the pendulum of close to 24 h. The mass of their suspended masses was 4 g so that accelerations at the level of thermal fluctuations corresponded to $\sim 10^{-15}$ m s^{-2} which, with a value of 10 cm for the radius of the beam, meant that the corresponding moment which could just be detected was 4×10^{-19} N m. Their torsional pendulum was made from a 2.8 m long, 5 μm diameter tungsten wire. Considerable effort was devoted towards developing a vacuum thermal annealing process in order to remove any residual moments from the wire. Residual magnetic fields were reduced to the level of 3 μT by means of a Permalloy magnetic shield. It was found that the amplitude of the torsional oscillations changed by less than 3×10^{-3} in 72 h time interval, suggesting that $t_r \sim 6 \times 10^7$ s.

In this type of experiment the effects of local gradients in the gravitational field of the Earth can have important effects. These were greatly reduced by making the torsional pendulum in the form of an eight pointed star with masses of aluminium or platinum in opposite halves; the masses being identical to $\sim 4 \times 10^{-4}$. The oscillations of the pendulum were recorded by reflecting a low-intensity laser spot from a mirror on the pendulum onto a photographic film which was fixed to a rotating drum (which revolved once per week) and was about 5 m from the mirror. The angular sensitivity was such that a rotation of the pendulum of 10^{-7} rad produced a movement of the spot on the film of 5 μm. The amplitude of the oscillations of the pendulum was reduced to lie between 2×10^{-5} and 10^{-4} rad by mounting a small conducting ellipsoid on the fibre and applying appropriate electrostatic potentials to adjacent fixed-capacitor plates. The monotonic drift of the record spot corresponded to an angular drift of 4 μrad per day in the best records.

The value obtained for $\eta(A, B)$ from the Braginsky et al experiment was $-3(9) \times 10^{-13}$ at 95% confidence levels. The measurements of Eötvös et al and Renner, at a similar confidence level as re-analysed by Roll et al (1964), were 9×10^{-9} and 4.2×10^{-9} respectively and the Roll et al experiment yielded 3×10^{-11}. Thus, these and other similar measurements (at precisions comparable to the last quoted) all point to the verification of the equivalence principle to a high level of precision. It is possible, of course, as Roll et al point out, to take the view that the Einstein general relativity theory is so well founded that such experiments as the above are not worth undertaking and one's point of view concerning this will reveal whether one is an experimentalist or a theoretician! Certainly, to any metrologist it is a privilege to read about such careful attention to detail, and humbling to realise that in the original Eötvös experiment the gravitational effect of an overweight 100 kg metrologist at a distance of 6 m from the apparatus probably produced an effect which was some 80 times greater than the limiting sensitivity of the balance.

Although Eötvös *et al*, and Renner too, used Boys' modification of the Cavendish balance, their apparatus was particularly sensitive to gravitational field gradients. The Braginsky *et al* and Roll *et al* measurements were designed to eliminate such effects as far as possible. Although the Roll *et al* precision has possibly now been surpassed by the apparatus which was discussed above (see Everitt 1978), it is of interest to give a brief description of their apparatus. The torsional suspension was a 260 mm long 25 μm diameter quartz fibre having a torsion constant of 2.4×10^{-6} N m rad^{-1}. The masses suspended from the fibre were at the ends of an equilateral triangle of about 5 cm sides and the masses comprised two 30 g identical cylinders of pure aluminium (which were 3.2 cm long and 2.1 cm in diameter), together with a third cylinder of high-purity gold 3.2 cm long and 0.78 cm diameter. The build-up of electrostatic charge on the system was prevented by evaporating a thin layer of aluminium onto the suspension fibre (breaking strength about 200 g). The deflections of the system were monitored by a photomultiplier using a modulated light beam. The full width of the photomultiplier output at half maximum was 30 μm which, at the distance, used, corresponded to a deflection of 3×10^{-5} rad.

Keiser *et al* (1981) have proposed a test of the equivalence principle using an Eötvös type apparatus in which the torsional fibre is replaced by a flotation method. A 1.27 m diameter system will use approximately 490 kg test masses of copper and lead. Their system is to be floated in water which is at a temperature of 3.98 °C, where its density is a maximum, thereby reducing convection current effects. The apparatus can be turned and centred by applying AC potentials of ~ 1 kV to seven insulated balls in a centred hexagonal array. There are three further balls on the inside of the lid of the float chamber which are insulated from the lid. The inner ball serves as the centring electrode and the outside one serves to apply a torque. The 24 cylindrical masses comprise 12 of copper and 12 of lead. If there was a breakdown of the equivalence principle there would be a slight difference in the acceleration of the float towards the Sun, which would evidence itself as a 24 hour period variation in the position of the float.

A prototype apparatus on a 0.25 m diameter float using 3 kg test masses of copper and tungsten was tested by Keyser and Faller (1981) and yielded

$$\eta \lesssim 0.2(40) \times 10^{-11}.$$

Although down to frequencies of 10 cycles per day this system was limited by the Brownian motion of the float, at lower frequencies other effects dominated Brownian motion by a factor of ten. In fact, Brownian motion does not appear to have been the factor limiting the precision of any of the Eötvös experiments to date.

7.2.4 Extraterrestrial tests of the equivalence principle

It was pointed out by Dicke in 1961 that if the gravitational self-energy of a body varied with its position in the gravitational potential of another body, then the ratio of the gravitational to inertial mass of the body would differ slightly from unity. Thus, when the total mass of the body is a function of its position in a static gravitational field, then, in order for energy to be conserved, there must be an anomalous gravitational force acting on the body. In the Eötvös experiments this possibility is not ruled out since the gravitational self-energy contribution to their rather small mass would be too small. However, Nordtvedt (1968) suggested that this would result in a significant deviation of the semi-major axis of Jupiter's orbit from that predicted from Kepler's third law, given the orbital period, thereby allowing this aspect of the equivalence principle to be tested. In Nordtvedt's analysis, three possible experiments were suggested, one of which involved the monitoring of the Earth–Moon distance with high precision.

As a result of the Apollo and other lunar landings several arrays of cube-corner reflectors were mounted on the surface of the Moon. These have a sufficient area to allow some photons to be detected by a terrestrially mounted telescope when a pulsed laser is directed towards the cube-corner arrays. In typical experiments, the intensity is such that about one photon per pulse is detected by a photomultiplier. The time elapsed between the firing of the laser pulse and the arrival of the returning photon gives an accurate measure of the Earth–Moon distance. The precision of the analysis, ~ 10 cm, has been steadily improved over the years by the use of improved short pulse lasers and photodetection techniques. An interesting development for the detection apparatus has been the multiple moth-eye type of telescope. This comprises an array of small aperture lenses, typically as used for large amateur telescopes, all of these lenses being adjusted to bring light to a common focus. In this way, an aperture equivalent to that of a 100 inch telescope may be synthesised at much lower cost than is required for the large single mirror telescopes. The lunar ranging experiment is an ideal project for this type of telescope, since photon collection efficiency rather than image fidelity is the prime requirement. Williams *et al* (1976) have analysed the changes in the Earth–Moon distance taking all known perturbations of the orbit into account and interpreted the null result both in terms of a verification of the equivalence principle and of the constancy of G.

7.2.5 Other tests of the effects of physical parameters on G.

Aside from the time variations of G, the Eötvös experiments, tests of the equivalence principle and the inverse square law which are discussed in other sections, there has been a considerable amount of work with the aim of finding gravitational effects which are the analogue of those found in electromagnetism. The gravitational force is very weak anyway and consequently it is very difficult to establish a genuine dependence experimentally. Thus temperature dependent effects are plagued by the difficulties of removing convection currents.

The present situation appears to be that no claimed dependence on any physical parameter has stood up to the test of detailed scrutiny. The state of the art as reviewed by Gillies (1982) is summarised in table 7.3. It is apparent that in many cases this has not been a very active area of late. This reflects the need for a major advance in technique before the experiments can be repeated with greater precision, for in comparison with the advances in time and frequency measurement there has not really been much change during the last century in the sensitivity with which gravitational forces may be measured. Attempts to measure gravitational absorption or demonstrate a gravitational permeability have not been successful and, if the effects are indeed of the order of the gravitational permeability of free space predicted by Forward (1961), of $16\pi G/c^2 \sim 10^{-26} \, \mathrm{m \, kg^{-1}}$, this is not surprising.

If there was a dependence on the physical state of a mass, such as pressure or temperature, there might be major differences between the behaviour of the gravitational forces in a laboratory environment and that in stars. This applies to the temperature of the intervening medium as well. Spontaneous matter creation, which we discuss elsewhere, would have implications for G forces on an astrophysical scale since the masses of the attracting bodies would be changing with time. There have been suggestions too that G might depend on the state of quantisation of the attracting bodies and proposals have been made for G/h experiments (Nieto and Goldman (1980), Hawkins (1982), Page

Table 7.3 Some limits on the variation of G with various physical parameters (after Gillies 1982).

Date	Authors	Dependence of G on	Limit on $\dot{G}/G$
1905	Poynting and Phillips	Temperature dependence (common balance)	$< 10^{-10} \, {}^\circ\mathrm{C}^{-1}$
1909	Lloyd	Magnetisation	$\lesssim 5 \times 10^{-16} \, \mathrm{T}^{-1}$
1918	Zeeman	Coupling with radioactivity	$< 5 \times 10^{-8}$
1922	Shaw and Davey	Temperature dependence (Cavendish balance)	$< 10^{-6} \, {}^\circ\mathrm{C}^{-1}$
1922	Simons	Electrification	$\lesssim 1.2 \times 10^{-7} \, \mathrm{V}^{-1}$
1924	Heyl	Direction of force	$\lesssim 10^{-9}$
1962	Caputo	Gravitational permeability (or absorption)	$< 6 \times 10^{-16}$
1965	Curott	Time (laboratory)	$\lesssim 6.2 \times 10^{-7} \, \mathrm{yr}^{-1}$
1981	Van Flandern (review)	Time (astronomical)	$\lesssim -6(2) \times 10^{-11} \, \mathrm{yr}^{-1}$
1981	Will (review)	Physical and chemical states of bodies	$\lesssim 10^{-12}$

and Geilker (1981) and Page (1982)). This is an active and somewhat controversial area at present.

The whole area of null experiments such as these and others which we discuss in the next chapter provides a striking example of the strong interaction between high-precision measurements and theory. It also illustrates the experimentally verified precept (in high-energy nuclear physics, etc) that 'the wise theoretician consults a properly sceptical experimenter before rushing into print with an explanation of some claimed effect!' Of course, the converse applies just as much, for many high-precision experiments place heavy reliance on the theory being correct (for example, the Josephson effects, or the quantised Hall conductance, which have been discussed elsewhere); we all build on the work of one another —a point which we return to in chapter 9.

7.3 Current and future developments

A major limitation on both the measurement of G and the Eötvös experiments has been the presence of the strong force from the gravitational field of the Earth. The necessity to use very fine suspension fibres, in order to give a large deflection for a given gravitational couple, has conflicted with the necessity to make the fibre strong enough to support the suspended masses. It is of course very easy to exceed the breaking stress of the fibre while assembling the apparatus! The experiments must be designed to produce forces which are perpendicular to the Earth's gravitational field and local gravitational gradients are a further source of difficulty. Since the experiments are of long duration, care is required not to move apparatus, etc, in the vicinity. For example (de Boer 1981), a colleague driving to work and parking his 1 tonne car 91 m from the apparatus would cause a part in 10^5 change in the force on a 1 kg mass. If he then approached his desk at 42 m from the apparatus there would be a further part in 10^5 change. If the experimenter approached the apparatus and left a 0.5 g paper clip behind 72 cm from the mass, or if a paper clip at 30 cm was moved by 1.6 cm, the force would have changed by the same amount. Even at a distance of 9 m the (rounded up!) 100 kg experimenter could produce a similar change if he moved 1.5 m radially. Changes in the weather can effect the equipment as well as masses of air, etc, moving over it. It is apparent, therefore, that future experiments are likely to be performed at remote sites. One obvious solution is to perform the experiment in space by means of a satellite experiment. This produces further complications, not least being the addition of several noughts to the cost of the experiment, with the attendant need to convince many committees that the project is of sufficient interest to be worth funding! Any null result must accordingly be of a rather spectacular nature. It is therefore well worth considering whether any new methods might be available for Earthbound use.

7.3.1 Flotation method

One such attempt has been made by J E Faller at JILA in the USA. His thought was to remove the need for a fibre by floating the apparatus under conditions of neutral buoyancy and to introduce the required forces electrostatically. Initially, he tried using mercury as the float medium, but a persistent problem was the formation of a surface film. The mercury vapour also limited the voltages which could be applied to the electrodes. A further attempt was made using water, which was maintained at the temperature corresponding to maximum density, since this allowed convection currents to be greatly reduced. An extension of this technique down to cryogenic temperatures has also been proposed by Faller (1980), whereby the equivalence principle may be tested using four aluminium and four silver cylinders each of mass about 200 g which will be mounted on a float immersed in superfluid helium on a radius of about 13 cm. This might have an ultimate sensitivity in the region of 10^{-19} which, if achieved, would represent a seven orders of magnitude improvement on earlier work. One advantage of working at liquid helium temperatures is that it is possible to use superconducting materials in order to provide a high degree of shielding from outside electromagnetic interference.

A room temperature version of the equipment was designed with the aim of measuring G. The apparatus proved to have a high degree of sensitivity to gravitational field gradients and the apparatus responded to the presence of the experimenter and also the gradients produced by changes in the weather. It was therefore decided not to pursue this work further for the time being. Similarly, de Boer (1981) is developing the flotation method of measuring G at the PTB, using mercury as the displaced fluid; the surface tension forces also make his apparatus self-centring. The attractive force between fixed (40 mm diameter) and floating (50 mm diameter) fused silica discs, ($\rho \sim 2.5$ g cm^{-3}, mass ~ 1 kg) will be measured by applying electrostatic potentials ~ 10 V, to a butterfly capacitor arrangement (~ 32 pF rad^{-1}). The gap can be varied from 2 mm to 8 mm in 2 mm steps which may be measured with high precision in a laser interferometer. Contamination of the mercury surface is reduced by a layer of sulphuric acid.

7.3.2 Superspring

The great disadvantage of terrestrial measurements arises from the presence of sources of vibration, both man-made and of seismic origin. This can be overcome, as in the measurements of Sakuma and of Faller of g by suspending the critical part of the apparatus from a seismometer. Any seismometer of course must be suspended from some form of spring which has a response time. Faller has devised a spring system (figure 7.8) which is coupled to a feedback servo-system which enables any period to be simulated, thereby making a spring

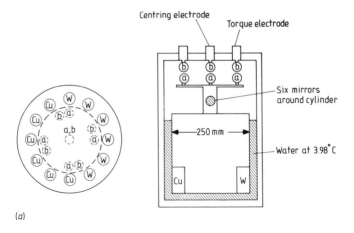

(a)

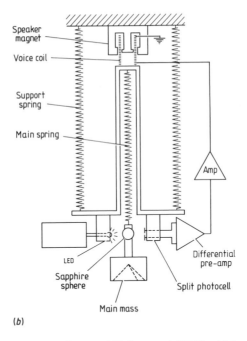

(b)

Figure 7.8(a) The Eötvös experiment of Faller *et al* (1979) which used a hydraulic suspension instead of the usual torsion fibre; (b) schematic diagram of the long period superspring devised by Rinker and Faller (1979) to give very good vibration isolation. A long period is achieved by means of electronic feedback to the loudspeaker solenoid—of course, this scales up the thermal expansion and creep as well—so very good control of the temperature etc is required.

electronically equivalent to a very long spring indeed. The fascinating aspect here is that one must take the change in g with distance into account and, if the spring has an effective length equal to half the radius of the Earth, the period becomes infinite. One difficulty is that the temperature must be very well controlled, for the expansion is also scaled up with the period. Thus a spring which is about a metre long has an equivalent length of many kilometres as far as the expansion is concerned. The achievement of a near infinite period is of metrological interest and requires the use of materials which have a very low temperature coefficient of elasticity.

The very high sensitivity to vertical motion was achieved by an adaptation of the sensing method used by Cook (1967) in his measurement of g. Faller attached an accurately ground sapphire sphere to the spring which functioned as an optical element so that it imaged a point source onto a split photodiode system. The amplified signal was then applied to move the point of suspension in the opposite direction to the sensed direction of motion of the sphere. The use of a spherical lens greatly reduced the sensitivity to the horizontal positions of the spring so that any unwinding of the spring produced minimal effect on the servo-system.

7.3.3 Satellite measurements

A number of satellite measurements have been proposed to test for gravity waves, to test the equivalence principle and for measurements of G. So far, these lie in the future and in this book we have concentrated on metrology which has been demonstrated to work, rather than on uncompleted experiments. One particularly interesting type of experiment concerns the use of a freely moving body which is shielded from extraneous outside effects by surrounding it by an outside body. This body is moved by appropriate servo-systems in such a way as to leave the inner body unaffected by outside forces such as gaseous drag, solar wind etc. If such bodies are operated in a low temperature environment, with the most sensitive measuring equipment, they may well represent the ultimate tests of physical theories in the future. However, they require many man-hours of effort and considerable expenditure, and it may be that the Earthbound metrologist will continue to survive by the exercise of ingenuity which can exploit the laws of physics, and yet accept the limitations provided by the terrestrial metrological environment.

7.4 The acceleration due to gravity

The acceleration due to gravity is not a fundamental constant in the same sense as the other constants which have been discussed in this book. It is, however, very valuable for comparing forces, and indeed it is very difficult to design an Earth-based force measurement which is not affected by gravity. The methods of measuring the gravitational acceleration have changed markedly since the advent of the laser and the earlier methods have been largely superseded. Traditional accelerometers are still used for relative measurements in geophysical surveying and form an important tool in the search for oil and other mineral resources.

Before the laser, the absolute measurements were at the part per million level, and it was necessary to set up a global net of stations where the acceleration due to gravity was well established and local gravity measurements with gravimeters were calibrated by transfer to these points. For many years an international gravity net was used which was derived from the absolute determination of Kuhnen and Furtwangler at Potsdam in 1906. It was made by the reversible pendulum method and their value $(9.812\,740\ \mathrm{m\,s^{-2}})$ was regarded as the International Standard. Evidence gradually accumulated that this value was too high by some 15 ppm.

Most users require relative values rather than absolute values and considerable confusion and inconvenience results if the values are changed too frequently. The worldwide net was last established in 1972 and values are denoted by IGSN72. Since that time there have been some improvements in the determinations which enable the net to be defined more precisely. Generally speaking there are two types of user. In metrology, values of g are required extensively for the force and pressure measurements and these are only just beginning to be required with precisions of less than a part per million. The other group of users, including geophysical surveyors, find it convenient to have calibrations approaching the part per billion level. The surveys for oil in the North Sea, for example, meant that the instruments might be calibrated on either side of the North Sea, thereby necessitating better gravity transfers from the United Kingdom to the rest of Europe.

Quite apart from the local variations due to mineral resources, there are variations of g with altitude and latitude and there are tidal gravity variations too, and all of these must be taken into account for accurate work. These effects must also be taken into account in reducing the absolute determinations to a communicable value. The outstanding contribution to absolute determinations came from the work of Sakuma at the BIPM which is discussed in the next section.

7.4.1 The basic principle of the free-fall methods

Although these methods are conceptually very simple they could not be used practically with high precision until high-speed timing techniques became available. There are two types of free-fall method, the simple free-fall and the symmetric free-fall.

(i) Simple free-fall method
The free-fall method is illustrated in figure 7.9(a). It is usual to begin the measurement after the object has fallen a little way. This is partly to avoid the

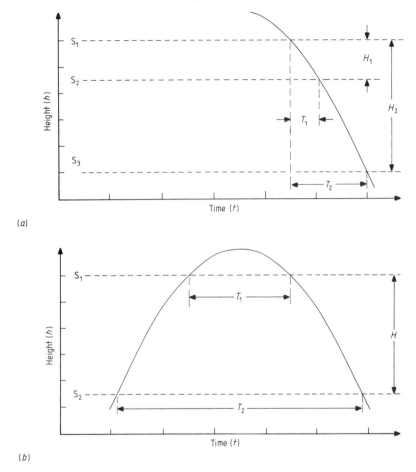

Figure 7.9 Schematic of recent methods used to measure the acceleration due to gravity: (*a*) free fall and (*b*) symmetrical motion.

inevitable vibration due to the release mechanism, but also to avoid difficulties in determining the position corresponding to zero initial velocity. As a result, the timings are made at the three points S_1, S_2 and S_3 shown in the figure, these times being T_1 to traverse H_1 and T_2 to traverse H_2 respectively. The acceleration due to gravity is then obtained from

$$g = \frac{2}{T_2 - T_1}\left(\frac{H_2}{T_2} - \frac{H_1}{T_1}\right).$$

(ii) Symmetric free-fall method

This method, which was proposed by Ch Volet (1946), is illustrated in figure 7.9(*b*). An object is projected vertically and in its upwards and downward path crosses the two horizontal positions S_1, S_2 at heights which are separated by the distance H. The two time intervals, T_1 between successive crossings of S_1, and T_2 between successive crossings of S_2, were measured and the value of the acceleration due to gravity g obtained from

$$g = 8H/(T_2^2 - T_1^2).$$

The two basic advantages of the method are first that the resistance due to motion through the air, or eddy currents from small magnetic fields, cancel out for they are equal and opposite in the upward and downward paths. The second advantage is that the timing pulses are similar in the upwards and downwards direction. This allows the timing errors which arise from the different shapes of the pulses at the two timing positions to be eliminated (the velocity is different). However, because the launching process involves accelerating the mass to its launching velocity, it is necessary to design the various mechanisms very carefully, both to ensure that the motion is accurately vertical, without any rotation or translation, and also to avoid vibrational shocks which might displace the timing positions.

7.4.3 Symmetrical fall absolute measurements of g

The first symmetrical up and down system, using quartz spheres 2.5 cm in diameter, was described by Cook (1967). His determination had an uncertainty of 1.3×10^{-6} m s^{-2}. The sphere was launched by a catapult and the sphere in flight imaged an illuminated slit onto a photomultiplier. There were two such slits which were separated by 1 m and this distance was measured with respect to a 0.2 m Fabry–Perot etalon. The passage of the spheres past the slits could be located to about 0.1 μm corresponding to an uncertainty of 10^{-6} m s^{-2} in gravity. The major uncertainty was that due to ground vibration, for the NPL site is on gravel and clay.

The method of Sakuma (1970) (figure 7.10(a)) is the most precise to date and is also based on a principle which is well known to small boys for it is based on the catapult. The method was to launch a cube-corner vertically by means of an elastic catapult and then to measure the time taken to traverse symmetric portions of the trajectory. Considerable precautions were required to ensure a vertical launch, with no sideways movements or rotations of the cube-corner. The centre of gravity of the system was adjusted to be at the optic centre of the cube-corner. Sakuma's system comprised a hollow cube-corner mirror system some 10 cm high and weighing about 0.43 kg. It was catapulted vertically and the cube formed part of a Michelson interferometer. Two mirrors, which were separated horizontally (by 0.8 m) by a fused silica spacer, formed two horizontal stations for the interferometer. As the cube-corner moved vertically, the coincidence of the path length with that set by the mirrors was indicated by the detection of the passage through the white-light fringe position. It was necessary to use a xenon flash tube as the white light source for the fringes, since the fringe frequency was in the region of 30 MHz and, in

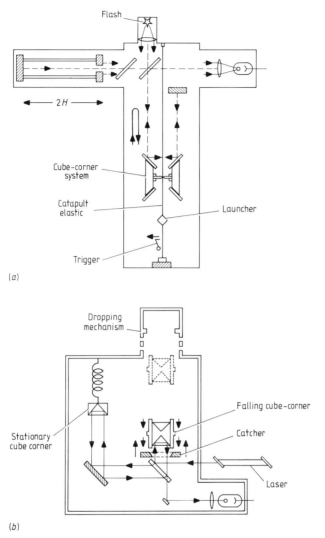

Flash

2H

Cube-corner
system

Catapult
elastic

Launcher

Trigger

(a)

Dropping
mechanism

Falling cube-corner

Stationary
cube corner

Catcher

Laser

(b)

Figure 7.10 (a) The catapult-launched, symmetric system used by Sakuma at the BIPM and (b) the falling cube-corner free-fall method of Faller. The latter technique has been refined in order to reduce the effects of drag by the residual gases by surrounding the cube-corner with an enclosure. This is servoed to fall synchronously with the cube-corner, but is spatially separated from it.

addition, the areas of the mirror surfaces were only about 0.2 cm², which reduced the available intensity. The halfwidth of the fringes corresponded to a distance of 0.06 μm, or about a part in 10^7 of the mirror separation. It was therefore only necessary to split these fringes to ∼1% to achieve precisions of parts in 10^9. The useful feature of the up and down method is that the effects of drag from the gas in the equipment are eliminated to first order, for in going up they aid gravity and in going down they are in the opposite direction. Despite this, Sakuma's apparatus was kept inside a well evacuated enclosure. The effects of microseisms were greatly reduced by mounting the sensitive parts of the apparatus on an antivibration mount. This mount was stabilised against vertical motion by means of a servo-system whose sensing element was a sensitive seismometer. Residual vibration effects were corrected for by deriving a signal from a long period seismometer which was mounted with the apparatus. The scatter of single measurements was 5×10^{-8} m s^{-2}. The 80 cm mirror separation was set by the need to determine their separation with the aid of a standard krypton-86 lamp and this was about the limit at which interference fringes were visible with this source.

7.4.4 Other free-fall methods

The acceleration due to gravity has been measured in free-fall with fairly conventional techniques by German (1970), Kitzunezaki *et al* (1969) and Tate (1966). Tate, working at the NBS, dropped a fused silica rod, 1 m long, having three slits which were separated by 30 cm. This method was of particular interest since the rod was falling freely inside an evacuated enclosure which was also falling almost freely in the atmosphere, being guided by two vertical rods. This reduced the effect of air resistance. Kitsunezaki *et al*, working at NRLM Kakioka, Japan, used a falling scale, which was 1 m long and was marked at 1 cm intervals, and the passage of the scale past a photoelectric microscope was timed accurately. Some vibrational problems were experienced, as the microscope was fixed to the walls of the vacuum chamber. The procedure employed by German at the PTB was to drop a fused silica rod, 2 m long, which was coated with a photographic emulsion. The rod fell freely inside a vacuum vessel and the emulsion recorded the image of an optical slit which was illuminated by 0.1 μs long light pulses at a frequency of 130 Hz. The uncertainties achieved by these experimenters were between 0.5 and 2.0 ppm and this probably represents the limit for this type of classical free-fall method.

The group at the Wesleyan University pioneered a method employing a laser, in which a cube-corner reflector was dropped. It was arranged to form part of a Michelson interferometer (figure 7.10(b)) and, as it fell, the cube produced interference fringes at a varying rate. In order to eliminate the effect of vibrations of the apparatus due to the release mechanism or from microseisms, the reference cube-corner was mounted on a seismometer which had a period of about 10 s. The apparatus was nearly 3 m tall and the cube was timed while falling a distance of about a metre. The first timing was begun after 0.1 s of free-fall and the two intervals T_1 and T_2 were usually 0.18 s and 0.36 s. The interference fringes were detected with a photomultiplier. This converted them into electrical signals which were counted with a frequency counter whose standard frequency was accurately known. The laser was a Lamb-dip stabilised type and was calibrated against a krypton-86 standard lamp. The apparatus was essentially portable and was taken to a number of sites, including several national standards laboratories throughout the world. This provided a new order of accuracy for tying in the sites where absolute measurements had been obtained in the past. The standard deviation of the results was from $4.5 \times 10^{-7} - 6 \times 10^{-7}\,\mathrm{ms}^{-2}$ depending on the site of the measurements.

It is necessary at these high accuracies to take into account a number of effects and make corrections for them, some of which may, at first sight, be unexpected. Thus there are transit time effects due to the finite velocity of light. This correction amounts to about 3 parts in 10^8 and is given by

$$\Delta g/g = -4g(T_1 + T_2)/3c + 2v_0/c.$$

Initial values of both g and the velocity v_0 may be calculated sufficiently accurately from the uncorrected results. A second correction arises as a consequence of the variation of the acceleration due to gravity with altitude.

The International Geophysical Association (1973) has agreed on the expression

$$g = 9.780\,131\,8(1 + 53.024 \times 10^{-4}\sin^2\phi - 5.9 \times 10^{-6}\sin^2 2\phi)\,\mathrm{m\,s}^{-2}$$

for the acceleration due to gravity at a latitude ϕ. This provides a reference value against which observations may be compared in order to identify local anomalies. At a location of elevation h (in metres) the value of g is given by

$$g_h = g_0 - 3.086 \times 10^{-6}h + 41.9 \times 10^{-8}\rho h\,\mathrm{m\,s}^{-2}$$

where ρ is the density of the Earth (in grams per cubic centimetre—in a laboratory environment, the acceleration due to gravity changes by about $3.6 \times 10^{-6}\,\mathrm{m\,s}^{-2}\,\mathrm{m}^{-1}$). In a falling method, the value of g_m is corrected (Faller 1970) by

$$g_0' = g_m(1 - \tfrac{1}{12}\gamma T_2^2(R^2 + R + 1) + \tfrac{1}{3}\gamma T_2\Delta t(R + 1) + \tfrac{1}{2}\gamma(\Delta t)^2$$

where T_2 is the longer time interval, $R = T_1/T_2$ and γ is the measured gravity gradient. This value is then corrected to give the value at the floor level. The above expression corrects the measured value g_m to the zero velocity position, Δt being the time interval before observations begin (~ 0.1 s). The effect of pressure drag due to the residual gas in the system may be calculated from

$$\Delta g = +0.10(5)P\,\mathrm{m\,s}^{-2}$$

where P is the pressure in torr ($\equiv 133.3\,\mathrm{Nm}^{-2}$). Since the pressure was less than 5×10^{-7} Torr in the Wesleyan results, this correction was usually in the region of $5 \times 10^{-8}\,\mathrm{m\,s}^{-2}$.

The above work has now been refined by Faller and the cube-corner falls inside an enclosure which is servoed by photo-detectors to fall synchronously with the freely falling body. This should greatly reduce the effect of air pressure.

In addition the precision of the timing of the interference fringes has been improved. It is hoped to achieve precisions of a few parts in 10^9 with this apparatus. This system has been specifically designed with portability in mind. A symmetrical type of absolute gravity meter has been described by Cerutti et al (1974) and by Alasia et al (1982).

This has also made extensive contributions towards improving the precision with which the acceleration due to gravity is known at specific stations throughout the world. There are an increasing number of sites at airports where g is known with high accuracy since these sites may be used to calibrate the conventional type of gravity meter. The precision of these absolute types of gravity meter is allowing the long term components in the tidal gravity variations due to the Moon and Sun to be investigated in much greater detail than hitherto. It may soon be possible to look at vertical movement of particular sites, for example the sinking of Naples, or the tilting of the east coast of England in the London region. Vertical motion is also believed to be a precursor to earthquakes in some parts of the world since folding of the Earth's crust due to accumulated strain would cause the site to lift or fall.

It remains to be seen whether the absolute type may be modified to show sufficient robustness and portability to become a major tool in geophysical surveying, although it may certainly be possible to have several cube-corners falling cyclically in order to obtain an essentially continuous measurement of g. Such a system would require better isolation of the reference cube-corner to remove the vibration generated by the act of dropping, catching and raising the other cube. This type of concept might be usable to allow gravity-free experiments to be undertaken in the laboratory. Such methods have been applied in the past in the making of lead shot. At present force measurements do not make very strong demands on the precision with which g is known, but it may be that developments connected with the realisation of the ampere, for example, will begin to make use of the precision which g measurements now allow to be achieved.

7.5 The gravitational acceleration of atomic particles

Although most research on gravitation has concerned assemblies of atomic particles there have also been some successful attempts at demonstrating the acceleration of atomic particles. The Mössbauer effect led to one such experiment. This effect uses the emission of γ-rays from ^{57}Fe atoms which are trapped in the crystal lattice, and hence have a very low recoil when the γ-ray is emitted, and this leads to a very narrow linewidth, $\sim 10^{-12}$ (10^{-8} eV in 10 keV). Sadly in some respects, the monochromaticity has not been applied to any great extent in high-precision metrological experiments. There is however one type of measurement which is of particular metrological interest and that is the use by Pound and Rebka (1959) and by Cranshaw and Schiffer (1960) to detect the gravitational red-shift of photons. (Incidentally, the latter work was criticised by Josephson (1961) on the grounds that the results would be nullified if the temperature difference between the source and detector was too great. At that time he was still an undergraduate!) In these experiments the photons were the Mössbauer γ-rays and the gravitational acceleration of the photons was evidenced by a change in their frequency as they fell. This meant that they were absorbed to a lesser extent by a second ^{57}Fe absorber. For a 30 m drop one would expect $\Delta v/v \sim 3 \times 10^{-15}$. The photons fell about 23 m and 12 m in the Harvard and Harwell experiments respectively. The change in the frequency of the photons was compensated by moving the absorber at an appropriate velocity. The experiments provided a test of the equivalence principle of relativity; theory and experiment agreed to $\sim 10\%$.

Witteborn and Fairbank (1977) have described an apparatus which they developed to observe the gravitational acceleration of electrons. Their apparatus was at 4.2 K which ensured that the metal enclosure was free from the patch effects which usually produce electric fields inside room temperature equipment. They produced a region which was about a metre long and was free from stray fields down to a level of about 10^{-11} V m^{-1}. This small field required particular attention to experimental detail. Thus the temperature drop along the surrounding copper tube had to be kept below 10^{-5} K, while the electrodes used to measure the electron drift velocity had to be shunted by very low resistances, $\sim 0.9\,\mu\Omega$, in order to keep the Johnson noise level sufficiently low. The experiment is also suitable for measuring the gravitational deceleration, by reducing their energy by a controlled amount, in order to produce a more monochromatic beam. The paper amply repays a careful study as there are many useful experimental details contained in the above reference. Very high-precision measurements of their gravitational acceleration are unlikely to be possible for some time, but they are likely to be used for precision measurements of the neutron lifetime (see Smith (1980) for further details). Measurements of the effects of gravitational fields on atomic particles are likely to be extended in the future. The effect of gravity on anti-particles is of particular interest. The gravitational deceleration of slow neutrons has been well studied and utilised (§6.5).

7.6 Gravity waves

Gravitational radiation is predicted from general relativity as being emitted from accelerated masses, in an analogous manner to the emission of electromagnetic radiation by an accelerated electric charge (the analogy is limited for there are important differences, as we do not have the gravitational equivalent of positive and negative charge). The velocity of the gravitational radiation is expected to be the same as the velocity of electromagnetic radiation in vacuum. The detection of such radiation, which is expected to be generated as a result of some large scale event in space, such as the collapse of a star, is difficult. (The principle is shown in figure 7.11.) This is because the gravitational analogue of a radio antenna cannot exist, for there is no common gravitational equivalent to positive and negative charge. A gravitational wave would produce a similar acceleration of all test objects at a given point, but because there is an associated curvature of space, it should be possible to detect a change in the separation of two separated masses. This has formed the basis of operation for systems which have been set up to observe gravity waves (Braginsky *et al* (1977), Drever (1977) and Everitt (1978)).

Interest in the subject was greatly stimulated by the work of Weber whose apparatus appeared to detect events at a rather higher rate than was expected. Later workers have failed to repeat his results and their efforts have been directed towards developing apparatus of greater sensitivity. The effects which are being sought are very small: for example, events such as stellar collapse in our own galaxy would change the length of a metre bar by only a part in 10^{17}, or by less than the radius of the hydrogen nucleus. Such events occur only about once per thirty years and the length change might well be a hundred times smaller than the above figure. The frequency of such events as a stellar collapse in the Virgo cluster of galaxies might be as high as ten per year, but the length change induced by these would be three orders of magnitude smaller than the above.

It will clearly be difficult to keep the apparatus working reliably at high sensitivity for a long period and so one would prefer to work with continuous sources of radiation, but these are even weaker. The best hope might be to

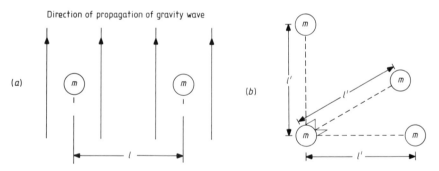

Figure 7.11 The principle of the operation of a gravity wave detector: (a) the distortion of the local space metric by the passage of a gravity wave changes the separation of two masses m by an amount δl which is proportional to their separation l. Since the strain $\delta l/l$ is smaller than 10^{-17}, it is advantageous to make l as great as possible. (b) Since, in general, the gravity wave will be inclined at an angle to the line joining the masses, it is better to use an orthogonal array of masses as shown. For laboratory detection of regular sources, one arm can be dispensed with, since the apparatus rotates with the Earth.

detect the radiation from binary sources. These produce length changes of the order of 10^{-20} to 10^{-21}, although their dominant frequency is less than a millihertz in most cases. Pulsars should typically produce length changes of the order of 5×10^{-26} to 10^{-27} at a more attractive frequency between 20 and 60 Hz. Clearly, the attempt to detect such signals is not for the faint-hearted metrologist.

The construction of gravity wave detectors has considerable metrological interest. In part, they demonstrate that our everyday measurements are a long way short of the quantum noise limited frontier, but they also enable experiments to be performed near to the quantum limit, and even stimulate interest in passing beyond what would normally be considered to be the natural quantum limits (Edelstein *et al* 1978). As an indication of the difficulties, the first generation instruments could detect an energy flux of $\sim 10^5 \, \mathrm{J \, m^{-2}}$ and frequently comprised an aluminium bar (which weighed between one and four tonnes) which was either a single cylinder or split into two at the centre (Aplin 1972). For certain aluminium alloys such bars can have in principle a quality factor Q of $\sim 10^6$. It is of interest that the amplitude of the thermal motion in the longitudinal mode of a one tonne bar is of the order of 10^{-16} m at room temperature.

The use of a high Q bar enables length changes much smaller than the thermal fluctuations to be discriminated and so the search for improved sensitivity leads naturally to the use of both low temperatures and high Q materials. Thus there are experiments at Stanford and Rome whereby masses of some six tons will be suspended at temperatures of some 50 mK and the motion detected with SQUID magnetometers or by changing the resonant frequency of a microwave cavity. An alternative approach at Moscow, by Braginsky and others, is to use a near-perfect crystal of sapphire. They have already obtained mechanical Q values in excess of 10^9 and values of 10^{13}–10^{17} should be possible for perfect crystals at low temperatures. Sensing the length change will not be easy and the bar will probably be coated with a superconducting film to form part of a microwave cavity. It is interesting that if a pulse of gravity wave energy of $10^{-3} \, \mathrm{J \, m^{-2}}$ were incident on a 1 kg sapphire crystal which was at 4 K and was initially in the lowest quantum level for longitudinal motion, it would not impart sufficient energy to the crystal to excite it to the next higher energy level. This problem is less acute if the bar is initially in a higher mechanical energy state, as it probably would be at 4 K. However, the detection of such a small energy change requires that the electromagnetic sensitivity of the microwave cavity is very high, for the change in the energy state of the cavity would be one or two quanta for the above gravity wave energy. If achieved it should enable pulses of millisecond duration to be detected from collapse events at distances beyond the Virgo cluster of galaxies (in contrast, the early detectors had sensitivities of about $10^5 \, \mathrm{J \, m^{-2}}$ integrated over the pulse duration).

An alternative approach to an integrated, relatively small-scale, apparatus is to attempt to use a separated mass system and here we can expect that in the longer term experiments will be set up in space. In a sense, the cube-corner arrays, which were set up on the Moon as part of the Apollo program in the USA and also by the USSR and France, already provide possibilities, but the present sensing accuracy is limited to a few centimetres (in 3.8×10^{10} cm) by the atmosphere, and also by gravitational and other effects. It should prove possible to set up interferometer systems in space which have a much higher sensitivity than this figure. For terrestrially based systems one may con-template masses of around 150 kg which are separated in vacuum by a distance of around 100 m. Drever's group at Glasgow is mounting such an experiment. Even at a pressure of as little as 10^{-8} Pa (10^{-10} Torr) the masses would recoil due to the Brownian motion of the residual gas. There would also be recoil effects from the reflection or absorption of the photons in the interferometer light beam. In such a system using multiple reflection (~ 100 times) of a 2 W light source (for example an argon ion laser) sensitivities might be better than 10^{-5} J m^{-2} to gravitational pulses lasting a few tenths of a second, being poorer by two orders of magnitude per decade frequency change (either up or down) (Drever 1978). The rate of such events might be several per minute. However, there will undoubtedly be more severe problems introduced by vibration and motion of the suspension. It is of interest that a prototype system of Moss et al (1971) achieved sensitivities of 10^{-15} m in a Michelson interferometer formed by mirrors which were mounted on suspended masses some 3 m apart. This sensitivity was $\sim 10^{-8}$ of an interference fringe. These are essentially null displacement precisions and, by way of comparison, the sensitivity of the interferometer in the Becker et al (1981) determination of the lattice parameter of silicon was such that a displacement of 120 optical fringes could be measured to a few millionths of a fringe, or about 10^{-12} m. It will be some time before distances of the order of 1 m are measured to this accuracy. However, it is apparent that gravitational wave detection is stimulating advances in the precision of null detection by interferometric and other techniques, which augurs well for the future of length metrology.

7.7 Time variation of g

There is is in a sense a gravity wave incident on the Earth from the gravitational effects of the Moon and Sun. These change with time and the reader will be familiar with the neap and spring tides of the sea. The surface of the of the sea remains an equipotential surface of the combined attraction and this would give an equilibrium tide of ~ 0.4 m amplitude. The lunar semi-diurnal change in the acceleration due to gravity has an amplitude of about 0.42×10^{-5} m s^{-2} at a latitude of $50°$ and the effects due to the Sun are a factor of about 0.42 times smaller. The lunar cycle undergoes a complete variation in a fortnight and there is also a monthly term due to variations of the distance of

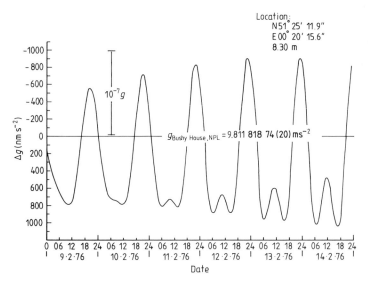

Figure 7.12 Illustrating how the acceleration due to gravity at a particular site varies with time as a result of the tidal attractions of the Moon and Sun (after Sakuma). For many purposes it is possible to compute these variations with adequate precision. For many years, tables of these variations were published as a supplement to *Geophysical Prospecting*, but the spread of sophisticated computers has rendered this unnecessary.

the Moon from the Earth. Similarly, the Sun produces diurnal and semi-diurnal tides which have a half-year period. It is evident from figure 7.12 that the acceleration due to gravity varies quite rapidly at certain times of day at the part in 10^7 level of precision. This variation will have to be taken into account in any future metrological experiments which involve making force and pressure measurements at this precision—one must, for example, choose the best time to make the force measurement in absolute determinations of current, voltage or *G*. Fortunately, corrections for these time variations have long been applied for in gravity surveys, and tables were published annually as a supplement to *Geophysical Prospecting* until the advent of microcomputers.

8

Some important null experiments

8.1 Introduction

We consider in this chapter some of the important null experiments which have a bearing on the fundamental constants and begin with a discussion of experiments related to the question of whether there is any frequency-dependent dispersion in the velocity of light. It is hoped that the way in which experiments in apparently different parts of physics come together will be of particular interest.

8.2 The mass of the photon

There are four types of experiment which may be used to investigate whether the photon has a finite rest mass μ (Goldhaber and Nieto 1971), for if it did:

(i) the velocity of light would vary with energy (that is, with frequency) so that there would be dispersion

(ii) the electrostatic field would derive from a potential of Yukawa form $(r^{-1}\exp(-\mu cr/\hbar))$ rather than the usual e/r Coulomb potential.

(iii) Ampere's law would be violated so that the corresponding Maxwell's equation would be

$$\nabla \times \boldsymbol{H} = \boldsymbol{J} - (\mu^2 c^4/\hbar^2)\boldsymbol{A}$$

where $\boldsymbol{A}$ is the magnetic vector potential: as a result, the magnetic field of a dipole, such as the Earth, would decrease with distance as $\exp(-\mu cr/\hbar)$ and

(iv) the dispersion relation between the frequency and the wave vector k (in frequency units) would become $\omega^2 = k^2 + \mu^2 c^4/\hbar^2$ in place of $\omega^2 = k^2$. The consequences of this would be, for example, changes in the arrival times of signals from a pulsar which depended on frequency. Since these pulses are well defined they provide accurate checks for the absence of any dispersion.

There are other less sensitive tests as well. Lowenthal (1973) has discussed the effect on the deflection of starlight by the solar mass $M_\odot$ and shown that the deflection $\theta_0 \sim 1.75$ seconds of arc would be modified to

$$\theta = \theta_0(1 + \mu^2 c^4/2\hbar^2 v^2)$$

where

$$\theta_0 = 4MG/Rc^2.$$

$R \sim$ solar radius and v the photon frequency. Hence, if the deflection may be measured to $\sim 10\%$, we have

$$\mu \lesssim (h/c^2)(20\%)^{1/2}$$

or

$$\mu \lesssim 1.5 \times 10^{-39} \text{ kg.}$$

As will become apparent, each of these effects gives upper bounds on the value of μ. The best precision comes at present from precise measurements of the shape of the Earth's magnetic field. This sets $\mu \lesssim 3 \times 10^{-15} \text{ eV}(\lesssim 4 \times 10^{-48} \text{ g})$ with the other methods giving upper limits between 10^{-10} eV and 10^{-14} eV.

It has been pointed out by Primack and Sher (1980) that most of these tests have applied to measurements which are at temperatures which are above or at the temperature of the cosmic black-body background temperature of $\sim 2.7 \text{ K}$; indeed many are at $\sim 300 \text{ K}$. They argued that since theoretical treatments generally considered that gauge symmetries were restored at high temperatures, particularly in the theories which unified the strong, weak and electromagnetic interactions (in extrapolating back to the 'big bang'), there remained the possibility that at sufficiently low temperatures the photons might acquire a non-zero mass through a breakdown of gauge symmetry at these temperatures. One way that this possibility might be investigated would be to study the change in the resonant frequency of a superconducting microwave cavity as the temperature was progressively reduced. The high quality factors or Q of such cavities would provide the required sensitivity, but it would be necessary to eliminate the possibilities of changes in the shape and leading dimensions of the cavity as the temperature was reduced. The sensitivities should be in the region of a few parts in 10^8 for temperatures as low as 0.05 K. Measurements at these temperatures should provide upper limits for the photon mass of the order of 10^{-9} eV. This type of prediction demonstrates very well how it will become useful to make very high-precision measurements at progressively lower temperatures in the years ahead. In the sections which follow we consider the measurements which have a bearing on the possible dispersion in the velocity of light. Clearly this might have implications for any definition of the metre which relied on a fixed value for the velocity of light.

8.2.1 Consequence of a non-zero photon rest mass

An ultimate limit on the mass of the photon may be set by the uncertainty principle (Goldhaber and Nieto 1971) as

$$\mu c^2 \Delta T \sim \hbar.$$

Here ΔT is the age of the universe! Taking ΔT to be $\sim 10^{10}$ yr leads to $\mu \lesssim 3.7 \times 10^{-66}$ g. Similar magnitudes have been set by de Broglie (1954) (who had a lifelong interest in the subject of the photon rest mass), from considerations of a spherical de Sitter cosmology, and also by Cap (1953).

If the photon has a rest mass it is necessary to modify Maxwell's equations which become

$$\nabla \cdot \boldsymbol{E} = [\rho - (\mu^2 c^4/\hbar^2)V]/\varepsilon_0$$
$$\nabla \times \boldsymbol{E} = -(\partial \boldsymbol{B}/\partial t)$$
$$\nabla \cdot \boldsymbol{B} = 0$$
$$\nabla \times \boldsymbol{H} = \boldsymbol{J} - (\mu^2 c^2/\hbar^2)\boldsymbol{A} + \varepsilon \partial \boldsymbol{E}/\partial t$$

where A and V are the space and time components of the four-vector potential A_μ. One important consequence of the above is the dispersion mentioned earlier. That is that light of wavelength λ will have a velocity differing from that at the short wavelength limit (at which the velocity is c), by

$$\Delta v/c = -\mu^2 c^4/2\hbar^2\omega^2$$

where $c = \lambda\omega/2\pi$ and the fourth- and higher-order terms have been omitted. A further consequence is that if the outer of two concentric conducting spheres is charged to a potential V, a potential difference ΔV exists between them which is given to first order by

$$\Delta V/V = (\mu^2 c^2/6\hbar^2)(a^2 - b^2)$$

where a and b are the radii of the outer and inner spheres respectively. The corresponding expression which would be obtained for a departure from an inverse square law of force, assumed to follow $r^{-(2+q)}$, would be (see Goldhaber and Nieto 1971)

$$q < (\Delta V/V)F(a,b)$$

where

$$F(a,b) = \tfrac{1}{2}\ln[(a+b)/(a-b)] - \tfrac{1}{2}\ln[4a^2/(a^2 - b^2)]$$

It is apparent therefore that tests of the inverse square law of electrostatics may also be regarded as being equivalent to tests for a photon rest mass.

8.2.2 Experimental tests of Coulomb's law

There have been a number of experiments (table 8.1) to verify that the attraction or repulsion between electrostatic charges follows the inverse square law assumed by Coulomb and Gauss. The verification by means of a null experiment date back to Cavendish (1773). Further verifications were made by Coulomb in 1785 and by Maxwell in 1873 who improved on the precision of the earlier work by some three orders of magnitude. It is usual to postulate the law of force as varying as $r^{-(2+q)}$, as indicated above, and to set an upper limit on the value of q as a result of the measurements. Thus Cavendish and Coulomb found $q < 2 \times 10^{-2}$ and $< 4 \times 10^{-2}$ respectively and Maxwell found $q < 4.9 \times 10^{-5}$. The now classical tests by Plimpton and Lawton (1936) set the value as $< 2 \times 10^{-9}$ while Cochran and Franken (1967) and Bartlett et al (1970) obtained a further improvement by setting limits of $< 9.2 \times 10^{-12}$ and $< 1.3 \times 10^{-13}$ respectively. The measurements of Williams et al (1971), which we now describe, improved on this limit by a further three orders of magnitude. Their experiment comprised a test of Coulomb's law at the relatively high frequency of 4 MHz. This test was based on the fact that an exact $1/r^2$ law of force leads to the absence of any electric field inside a charged hollow conductor.

The experiment comprised five concentric icosahedra (figure 8.1). A 10 kV peak to peak sinusoidal 4 MHz signal was applied between the two outer shells (4 and 5) which was established by feeding the output of a crystal-controlled transmitter into the resonant circuit formed by the shell capacitance and a water-cooled coil system. The presence of an electric field between the two inner icosahedra, which were also made from aluminium, was sensed by means of a battery-powered lock-in detector system. This amplified any signal developed across the coil which was connected across them. The reference signal for the lock-in amplifier was derived from the voltage between the outer shells and was fed to the inner icosahedron by first converting it to an optical signal and then sending it along a fibre-optic link. The phase of this

Table 8.1 Results of tests of Coulomb's inverse square law: limits on the value of q in $F = \varepsilon_0(e^2/r)(2 + q)$.

Date	Experiments	q
1773	Cavendish	$\lesssim 2 \times 10^{-2}$
1785	Coulomb	$\lesssim 4 \times 10^{-2}$
1873	Maxwell	$\lesssim 4.9 \times 10^{-5}$
1936	Plimpton and Lawton	$\lesssim 2 \times 10^{-9}$
1968	Cochran and Franken	$\lesssim 9.2 \times 10^{-12}$
1970	Bartlett et al	$\lesssim 1.3 \times 10^{-13}$
1971	Williams et al	$\lesssim 2.7(3.1) \times 10^{-16}$
	Geomagnetic data	$\lesssim 10^{-17}$

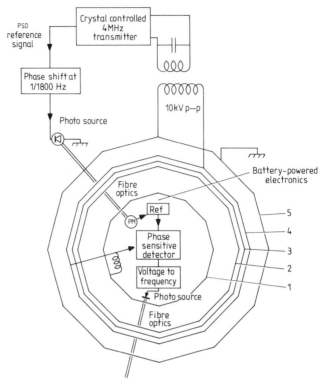

Figure 8.1 The apparatus used by Williams *et al* (1971) to verify the inverse square law of electrostatic force.

signal was changed linearly with time at a rate of about 360° per half hour. This signal passed through the shields by means of a pinhole which was much smaller than the cut-off size for 4 MHz. The phase sensitively detected signal was similarly converted into an optical signal and, on detection and amplification outside, was Fourier analysed for components at the phase shifting frequency.

Although the skin depth of the conducting shells greatly attenuated stray electric and magnetic fields, the soldered seams were not completely effective. Consequently the two copper shields (3 and 4) were added to provide extra shielding. It is important to the understanding of the experiment to recognise that all fields which are consistent with Maxwell's equations may be shielded out, but that any radially symmetric fields produced on the inner region as a result of departures from square law behaviour cannot be so shielded. It was found that closing shell number 3 completely reduced the detected signal by several orders of magnitude (to 2 nV). It was assumed that closing the outer copper shell number 4 would provide a similar reduction of the overall signal, so that the signal with both closed would be well below the Johnson noise level of 10^{-12} V. Observations over a three day period were analysed with respect to the phase shifted signal and the amplitude of the signal at this frequency was found to be $6.4(73) \times 10^{-13}$ V. From the amplitude of the applied voltage and the 1.5 m diameter of the outer shell and geometry of the smaller shells the above signal set an upper limit on the value of q of $2.7(31) \times 10^{-16}$.

Laboratory experiments of improved precision might allow a further worthwhile improvement. Thus it would be quite feasible to cool the inner icosahedrons and parts of the electronics to cryogenic temperatures, thereby reducing the Johnson noise considerably. The major difficulty at present would be that of providing some method of detecting and amplifying the signal in the absence of a cryogenic power source. Thus, although one might use a diode and galvanometer detector arrangement, it is almost certain that the flicker effect noise from the diode would be such that it would fall well short of the desired sensitivity. The above experiment of Williams et al can also be interpreted as setting a limit on the photon mass of $\lesssim 1.6 \times 10^{-47}$ g, or 10^{-14} eV.

8.2.3 The geomagnetic field

A further interesting consequence of any departure from inverse square law behaviour as a consequence of a finite photon mass would be found in the shape of the magnetic field from a magnetic dipole. In effect, an additional external magnetic field appears which is antiparallel to the direction of the dipole. The ratio of this 'external field' to the dipole field on the equator of a sphere centred on the dipole is given by

$$H_{ext}/H_{dipole} = \tfrac{2}{3}(\mu cR/\hbar)^2/[1 + \mu cR/\hbar + \tfrac{1}{3}(\mu cR/\hbar)^2].$$

Perhaps surprisingly therefore, geomagnetic data may be used to set a limit on the value of μ and this method has been used by Schrödinger (1943), Bass and Schrödinger (1945) and Goldhaber and Nieto (1968). In using such data it is, of course, necessary to make some allowance for known sources of external magnetic fields, some of which are parallel to the Earth's magnetic field. These include effects from the proton belt, the hot component of the current in the magnetosphere and currents in the geomagnetic tail. In addition there is an antiparallel component due to the compression of the geomagnetic field by the solar wind and there are further effects from the interplanetary field. Considering the overall data, Goldhaber and Nieto (1971) obtained an upper limit of 4×10^{-48} g for the photon mass.

8.2.4 Other limits

The study of interplanetary magnetic fields and the terrestrial field with satellites can also be used to set limits on the photon mass. Perhaps the smallest limit is that set by Williams and Park (1971) from considerations of the stability of the galactic magnetic field, for they argued that a filament of magnetic flux sustained by a partially ionised gas would decay exponentially

Table 8.2 Limits on the photon mass.

Date	Authors	Method	Photon mass (kg)
1971	Williams *et al*	Inverse square law	1.6×10^{-50}
1968	Goldhaber and Nieto	Geomagnetic field	4×10^{-51}
1971	Williams and Park	Intergalactic field	3×10^{-59}

as μ^2. This led them to conclude that $\mu \lesssim 3 \times 10^{-56}$ g. The limits on photon mass are summarised in table 8.2.

As with all null experiments, the reader should examine very carefully the assumptions which have been made concerning all of the data. Thus Goldhaber and Nieto (1971) mention that de Broglie made an (optimistic) error of 10^5 in the final part of his calculation, and they also suggest that a length estimate in another paper is wrong by a factor of 10^{10} and Williams and Park (1971) claimed that the Franken and Ampulski (1971) table top experiment to set an upper limit on μ was optimistic by a factor of 10^8! Primack and Sher (1980) have also pointed out that all of the limits apply to temperatures >2.7 K and that there might still be a breakdown of electromagnetic gauge symmetry and quantum chromodynamics at lower temperatures, with an associated graviton and mass for the photon. Evidently the subject is far from being closed. We move on to some related evidence.

8.3 Constancy of the velocity of electromagnetic radiation

Any adoption of a unit of length based on frequency and an agreed value for the velocity of light makes the implicit assumption that there is no dispersion in the values, i.e. c is not frequency dependent. An equivalent statement, as we have just seen, is that the rest mass of the photon must be zero. The question of possible frequency dependence of the velocity of light has been reviewed by Bay and White (1972). It has been usual to express the dispersion, or lack of dispersion, in terms of the expression.

$$p = (c/\Delta c)(\lambda_2/\lambda_1)$$

which relates measurements in different regions of the spectrum having different limits on the dispersion Δc in the velocity of light. Brown pointed out in 1969 that this was not a very satisfactory expression since it became infinite for $\lambda_1 = \lambda_2$. Bay and White further objected to this form on the grounds that it assumed a linear dependence of the velocity of light with frequency. They preferred a modified Cauchy expression of the form

$$n^2 = 1 + A/v^2 + Bv^2$$

or

$$n^2 = 1 + A'\lambda^2 + B'/\lambda^2$$

where n was defined by

$$c_{phase} = c_0/n$$

and c_0 was the velocity of light in the absence of dispersion. The actual experiments involve group velocities, but these may be readily calculated from the above. The assumption that time reversal symmetry holds means that there will be no odd powers in the expression and the higher terms may be neglected as long as one operates in regions remote from resonances. It should be noted that in vacuum the only possible dispersion formula which is compatible with special relativity would be one in which $B = B' = 0$ while A and A' may be non-zero. If the special theory of relativity is correct then causality would require that $A < 0$, for if A were to be greater than zero the group velocity would exceed c and the signal might appear to propagate backwards in time to some observers.

The above may be understood, following Bay and White (1972), by applying the relativistic theorem for the addition of velocities to the phase velocity c_{phase}, for this leads to the expression

$$n' = \frac{n + \beta}{1 + n\beta}$$

for the transformation rule for n, where $\beta = v/c$ and, for simplicity, the relative velocity between the two systems has been taken as being along the direction of light propagation. At the same time the frequency v transforms according to the relativistic Doppler formula as

$$\frac{v'}{v} = \frac{1 + n\beta}{(1 - \beta^2)^{1/2}}.$$

If β is eliminated from the above two equations we have

$$[n'(v')^2 - 1]v'^2 = [n(v)^2 - 1]^2.$$

It follows from this that, if $n(v)$ is to be the dispersion function which is characteristic of free space (vacuum), it should transform to itself and hence it follows that

$$n'(v') = n(v').$$

In consequence we must have that

$$(n^2 - 1)v^2 = A$$

where A is a relativistic invariant. This leads to the conclusion that the only possible dispersion formula for vacuum must be

$$n^2 = 1 + A/v^2$$

if the special theory of relativity is correct.

The above equation is quite general and should apply to many different types of wave. Thus it applies to the special case of electromagnetic waves where it is known that the only possible linear generalisation of Maxwell's equations which is relativistically invariant leads to this dispersion formula through the Proca equations. In the case of the quantised matter waves of de Broglie, or Klein and Gordon, the A term is simply $-(mc^2/h)^2$ where m is the mass of the wave. Because of the generality of the above equation it should also apply to gravity waves however they may be quantised. Tests of the dispersion in the values of the velocity of light are also a test of special relativity, to the extent that a limit may be placed on the magnitude of the B term.

Evidence for the dispersion, or rather the lack of it, in the velocity of light comes from several sources and these divide naturally into astronomical and terrestrial sources of evidence. Important astronomical evidence comes from a study of pulsars, for these emit a broadband spectrum of pulses which, if they leave the source simultaneously, should also arrive at the Earth simultaneously. It is of course also necessary to establish a reliable distance scale for the pulsars. Studies have been made of radio wave, visible and x-ray emissions from pulsars and the results analysed for possible dispersion in c. These studies are summarised in table 8.3. A study has also been made of the effects of the source velocity by studying the radiation from double stars (Brecher 1977). Studies of γ-ray velocities have been made by Alvarger et al (1966) and by Brown et al (1973). The former made a measurement of the velocity of 6 GeV γ-rays and the latter made use of the property of the Stanford linear accelerator whereby pulses of 7 GeV γ-rays were produced at the same time as pulses of synchrotron radiation produced by bending the electron beam. Both sets of radiation travelled about 1.31 km (4.3 μs) before being detected. The method was to generate the γ-rays from alternate accelerator pulses by inserting a copper target 0.38 mm thick (0.03 of a radiation length) instead of a beam-bending magnet. The timing and identity of the pulses with those derived from the accelerator electronics was established to 25 ps. The conclusion of Brown

Table 8.3 Limits on dispersion in the velocity of light (revision of Bay and White 1972).

Type of measurement	Frequency range (Hz)	$\Delta c/c$ visible	$\lvert A\rvert(\mathrm{Hz}^2)$	$\lvert B\rvert(\mathrm{Hz}^{-2})$
Pulsar emissions				
Radio wave	$1\text{--}4\times10^8$	$<10^{-10}$	$<10^6$	
Visible	$5\text{--}8\times10^{14}$	$<10^{-16}$	$<10^{14}$	$<10^{-45}$
X-ray	$4\text{--}24\times10^{17}$	$<10^{-14}$		$<10^{-50}$
γ-ray velocity	10^{24}	$<10^{-3}$		$<10^{-53}$
Photon mass limit	static field		<1	

et al was that

$$\frac{c(\text{GeV}) - c(\text{eV})}{c(\text{eV})} = \frac{\Delta c}{c} = \frac{\Delta t}{t} = \frac{(7.8 \pm 25) \times 10^{-12}}{4.3 \times 10^{-6}} = (1.8 \pm 6) \times 10^{-6}.$$

These results provide quite tight limits on possible dispersion in c as far as the B term is concerned. However, tighter limits come from experiments which establish an upper limit to the photon mass; thus Goldhaber and Nieto (1968) and Williams and Park (1971) established that $\mu \lesssim 4 \times 10^{-51}$ kg and $\mu \lesssim 10^{-51}$ kg respectively, which leads to the conclusion that the term A must be less than 1 Hz^2. Consequently, if Einstein's theory of relativity holds for all frequencies, then the velocity of light has been established as constant to 10^{-20} for frequencies greater than 10^{10} Hz. This is somewhat ahead of the requirements of today's technology and one may expect that even tighter limits will be established. For example, Hollweg (1974) studied the propagation of Alfven waves in interplanetary media from which he obtained $\mu \lesssim 1.3 \times 10^{-51}$ kg and a less certain upper limit of $\mu \lesssim 1.1 \times 10^{-52}$ kg.

8.3.1 Independence of the velocity of the source

The hypothesis that the velocity of light is independent of the velocity of the source has been the subject of a number of experiments. Astronomy is a natural branch of science in which to seek such effects for there are many examples of objects whose velocities are comparable with the velocity of light. However, measurements on visible radiation are complicated by the effects of the interstellar medium. The reason is that the light from a distant star is progressively scattered as it passes through the intervening matter and is replaced by light which is produced by the re-radiating dipoles of the medium. This radiation is propagated with the group velocity of the medium. Calculations show that for light in the visible region the characteristic length for the extinction of the original radiation is about two light years. This distance is less than that of the nearest star, and consequently astronomical evidence must be interpreted with care. The above argument does not apply to the same extent for x-rays, and Brecher (1977) made use of the 70 keV x-rays that were emitted from some pulsing sources in binary star systems. He examined three binary star systems and determined their distances, orbital periods and Doppler velocities. A small but finite variation in the value of k in the equation

$$c' = c + kv \tag{8.3.1}$$

where v is the source velocity, would produce variations in the eclipse times and affect the eccentricities of the orbits. Therefore, measurements of these quantities set upper limits for the value of k. The lowest limit for k obtained from these observations was $k \lesssim 4 \times 10^{-10}$.

A limit on the value of k may also be obtained from the measurement of the velocities of the γ-rays from decaying neutral pions. Pions with an energy of 6 GeV have been used to make estimates of k. Their velocities were very close to the velocity of light in the laboratory reference frame so that the equation (8.3.1) was tested at very high velocities indeed. These measurements have set an upper limit of $k \lesssim 10^{-4}$. These measurements were made by Alvarger *et al* (1964). One may expect that even lower limits will be set in the future. Since the constancy of the velocity of light, independent of the reference frame, is one of the corner stones of special and general relativity, any tests of relativity are also important (see §§5.2 and 8.4).

8.3.2 The isotropy of space and Michelson–Morley experiments

Present day measurements represent an evolution from those which were designed to check the hypothesis that light was moving at a fixed velocity in an all pervading aether with respect to which the Earth may be either stationary or in motion. If the Earth is in such motion then the velocity of light should show some directional dependence. The Michelson–Morley experiments (1881 to 1887), recently re-analysed by Handschy (1982), were originally devised to test this hypothesis.

The scheme of such experiments involves some form of Michelson interferometer (figure 8.2(a)) which is assumed to be moving in the direction SM_2 with uniform velocity with respect to the aether as in figure 8.2(b). In (a), the beams travel identical paths (say) so that $PM_1 = PM_2 = 2l$. In (b) the mirrors are at M_1 and M_2 when the light leaves P, but have moved to M_1' during the time that the light travels to the mirror and to M_2' when reflection occurs, respectively. In addition, P moves to P' while the light travels to M_1' and back. If the velocity of the apparatus with respect to the aether is v then it can be shown that the path difference between the two beams corresponds to lv^2/c^2. If the apparatus is now rotated through 90° there should be a displacement of the interference fringes. To date, such experiments have always led to a null result and, as Fitzgerald and Lorentz showed, a null result

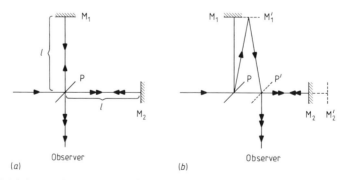

Figure 8.2 Scheme of the classical Michelson–Morley experiment: (a) stationary and (b) moving interferometer.

would be expected if the length of the arm which is in the direction of the velocity v contracted to a length $l/(1 - v^2/c^2)^{1/2}$.

In a sense, this contraction is included in Einstein's theory of relativity, but in that theory neither absolute rest nor absolute velocities have meaning and one cannot refer the velocity of light to an aether which is at absolute rest. Robertson (1949) showed that the null results of the Michelson–Morley experiments, later improved by Joos (1930), may be combined with the assumption of a null result for the Kennedy and Thorndyke (1932) and Ives and Stillwell (1941) measurements to lead unambiguously to the special theory of relativity. He showed that between two inertial frames which are moving in the x-direction the metric dimension transforms as

$$dt'^2 = dt^2 - c^{-2}(dx^2 + dy^2 + dz^2)$$

and

$$ds'^2 = (g_0 dt')^2 - c^{-2}(g_1 dx')^2 + g_2^2(dy'^2 + dz')^2.$$

The special theory of relativity corresponds to the case $g_0 = g_1 = g_2 = 1$. The Joos version of the Michelson–Morley experiment showed that $(g_2/g_1 - 1) < (0 \pm 3) \times 10^{-11}$.

Trimmer et al (1973) made a search for anisotropies of the form

$$1/c(\phi) = [1 + b_1 P_1(\cos \phi) + b_3 P_3(\cos \phi)]/c_0$$

where b_1 and b_3 are dimensionless parameters characterising the anisotropy, $c(\phi)$ was the vacuum phase velocity at an angle ϕ to the presumed preferred axis in space, c_0 the average speed of light and P_1 and P_3 were the Legendre polynomials

$$P_1(\cos \phi) = \cos \phi$$
$$P_3(\cos \phi) = (5 \cos^3 \phi - 3 \cos \phi)/2.$$

Any anisotropy in c would change Maxwell's equations and, in connection with the above experiment, Trimmer and Baierlein (1973) considered them to take the form

$$\nabla \times E = \rho/\varepsilon_0 \qquad \nabla \cdot B = 0 \qquad \nabla \times [E + (h_1/c_0)r \times B] = -(\partial B/\partial t)$$

and

$$\nabla \times [B + (h_2/c_0)r \times E] = \mu_0 J + (1/c_0^2)\partial E/\partial t$$

where r was the unit vector in the direction of the preferred axis. These equations give the desired variation with $\cos \phi$ with respect to a single preferred axis, satisfy parity conservation, and give a null result for the Cavendish inverse square law experiment, while preserving the linearity of E and B, as well as the symmetry between them.

Their scheme used a triangular path interferometer ($12 \times 12 \times 17$ cm) with counter-propagating beams. The long arm contained glass, so that any Fresnel drag due to aether drift would give asymmetry in the refractive index and thereby displace the fringes, yielding the P_1 term. The P_3 term is observable even in the absence of the glass. The position of the central white light fringe was observed at successive angular positions of the interferometer. The system was mounted in vacuum and the fringe positions were measured after each 60° rotation, once per minute for total observation times of the order of several days.

After computer analysis of the results, the b_1 term was found to be $(0.1 \pm 8.4) \times 10^{-11}$ and P_3 as $(2.3 \pm 1.5) \times 10^{-11}$. These were of course compatible with a null effect, but interestingly enough, the 'preferred axis' was found to lie within 6° of the axis of rotation of the Earth. Unlike a conventional Michelson–Morley interferometer, this type of interferometer is not sensitive to the even-order Legendre polynomial terms. Trimmer *et al* (1973) set a limit of $0.45(380)\,\mathrm{mm\,s^{-1}}$ on the velocity of the aether wind at the Earth's surface, or about one millionth of the Earth's orbital velocity. The above measurements represent about the limit achievable with the classical type of interferometer systems, but the advent of highly stable lasers has enabled the upper limit on the value of k to be lowered by several orders of magnitude.

The first of the new type of measurements was that of Javan and Townes and their colleagues at MIT (Jaseja *et al* 1964). They used rotating lasers which, although they had excellent intrinsic stability, unfortunately showed systematic effects as the apparatus was rotated. These were in the region of $275\,\mathrm{kHz}$ and limited their upper limit on the value of $g_2/g_1 - 1$ to about $\pm 2 \times 10^{-11}$.

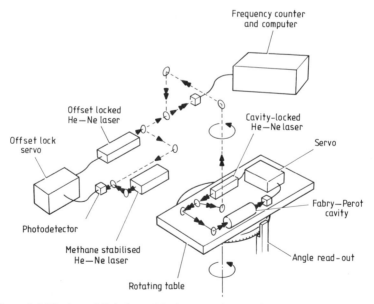

Figure 8.3 The laser Michelson–Morley experiment of Brillet and Hall (1979).

However, this value was already some 10^3 times smaller than that expected on the aether drift theory, taking the Earth's orbital velocity $\sim c \times 10^{-4}$ into account, and represented a small improvement over the precision of the Joos result. Their experiment has since been refined considerably by Brillet and Hall (1979) at JILA, which we now discuss.

In their experiment (figure 8.3) Brillet and Hall locked the laser to a passive Fabry–Perot etalon interferometer. Both laser and etalon were mounted on a solid granite slab, $0.95 \times 0.4 \times 0.12$ m, which could be rotated at a rate of about one revolution every 13 s. The etalon was about 0.305 m long and the radius of curvature of the 0.06 m diameter mirrors was 0.5 m. The mirrors were coated to give an interferometric efficiency of 25% which resulted in the interference fringes from the 3.39 μm helium–neon laser having a frequency width of about 25 MHz. Three yttrium–garnet isolators, each with a return loss of less than -26 dB, isolated the laser from the cavity in order to prevent frequency pulling. The laser frequency was about 10^{14} Hz and the frequency noise of the 200 μW cavity stabilised laser was about 20 Hz for a 1 s averaging time. The first harmonic locking technique was used to stabilize the laser and for this purpose the laser was frequency modulated by about 2.5 MHz peak-to-peak at a repetition frequency of 45 kHz. The etalon and laser were mounted inside an evacuated temperature stabilised cavity (± 0.2 °C) so that the etalon spacer, which was made of a low temperature coefficient fused silica material known as CERVIT led to a laser frequency drift of about -50 Hz s^{-1}.

As the table was rotated at one revolution per 13 s, there was some centrifugal stretching of the interferometer, corresponding to a frequency of about -10 kHz, and a gravitational stretching effect, because the axis of rotation was not quite vertical, which caused the laser frequency to vary at the rotation frequency by about 200 Hz peak-to-peak. The cavity-locked laser frequency was monitored as the table was rotated by reflecting the laser beam up the axis of rotation, and this laser signal was mixed on a photoelectric detector with that from another helium–neon laser. This laser was locked to be 120 MHz away from a methane stabilised helium–neon laser, which meant that the latter was completely isolated from any possible effects from the rotating cavity-locked laser. The beat frequency between the two lasers was taken in blocks of between 10 to 20 revolutions of the table in order to remove cross coupling of the systematic noise sources into the Fourier component of interest, that is at two cycles per revolution. The results were then analysed for any genuine sidereal effects over 12 h or 24 h and the duration of the experiment was about 245 days.

The results were analysed for aether drift effects with respect to the velocity of the Earth about the Sun. Their result was 0.13 ± 0.22 Hz which represented a fractional frequency shift of $(1.5 \pm 2.5) \times 10^{-15}$; this is also the limit on $\frac{1}{2}(g_2/g_1 - 1)$. Taking the Earth's velocity as $\sim 10^{-4}\,c$ leads to the conclusion that their result is some 5×10^{-7} smaller than the classical prediction, which represents an improvement by a factor of some 4000 over the previous lower limit. Brillet and Hall anticipate that this limit can be improved further, by at least an order of magnitude, by modifying the mechanical design; for example, by better stabilisation of the rotational velocity, and by better leveling of the table, and also by improving the vacuum and temperature stability.

In addition to the motion of the Earth around the Sun there has been an indication of a possible preferred direction in space as indicated by anisotropy in the 2.3 K black body background radiation. This anisotropy can be interpreted as arising from a Doppler shift from a $\sim 400\,\mathrm{km\,s^{-1}}$ motion of the Earth with respect to this radiation (see Smoot $et\ al$ 1977). This is a much higher velocity ($\sim 10^{-3}c$) than that of the Earth around the Sun ($\sim 10^{-4}c$) so that the sensitivity of the experiment, when analysed with respect to the direction of this velocity (11 h right ascension and $6°$ declination), is $\sim 3 \times 10^{-9}$.

This type of experiment tests the behaviour of rigid measuring rods and the effects differ from those which occur in atomic clocks when they are used to test relativity. It may well be that the experiment will be modified to test for gravitational effects in some way, for example by using a horizontal axis of rotation. The acceleration due to gravity varies with altitude, and a correction would be required to allow for this. However, the gravitational stretching of the apparatus would mask such effects at present.

8.4 Limits on the continuous creation of matter

An essential feature of some cosmological theories, namely those of Dirac (1938), Bondi and Gold (1948), Canuto and Lodenquai (1977) and Hoyle and Narlikar (1972), concerns the total mass of the matter in the universe and its constancy with time. If matter were to be continuously created it would have important cosmological consequences and if the rate were sufficiently high it would have to be taken into account in metrology as well. There have been some terrestrial experiments which put limits on matter creation and it is necessary to make certain assumptions in interpreting the results of these measurements.

In the Bondi and Gold model, a steady state is assumed whereby the mean density of matter in the universe remains constant while the universe itself expands at a rate determined by the red shift. In this model, the creation rate is proportional to the proper volume and is given by $3\rho/T$, where ρ is the mean density of matter in the universe and T, the Hubble time, is the reciprocal of the Hubble constant H. Taking $\rho = 2 \times 10^{-29}\,\mathrm{g\,cm^{-3}}$ and $1/T = 2.4 \times 10^{-18}\,\mathrm{s^{-1}}$ leads to a mass creation rate of $1.4 \times 10^{-46}\,\mathrm{g\,cm^{-3}\,s^{-1}}$ which is far too small to observe experimentally. Thus only $\sim 3\,\mathrm{mg}$ of matter would evolve in a volume equal to that of the Earth during 10^9 years! However, one may modify the theory to assume, perhaps more reasonably, that the rate of mass creation depends on the amount of matter within the volume, which in Bondi's model (1952) would be $7.2 \times 10^{-18}\,\mathrm{g\,s^{-1}}$ per gram. The difficulty with testing such a hypothesis experimentally lies in our lack of knowledge of the form that the newly created matter might take, but since hydrogen is the most abundant element it might be reasonable to assume that the matter appears as hydrogen. On this basis, Cohen and King (1969) searched for the creation of hydrogen in mercury. Mercury was chosen because it has a high density and can be vacuum distilled, thereby enabling any generated hydrogen to be detected. The experimental procedure adopted by Cohen and King was to use liquid nitrogen to freeze the mercury for period of 25–100 hours and then to vacuum distil it and measure the amount of hydrogen given off. Taking the Bondi figure, the distillation of one mole of mercury would release 9×10^8 hydrogen atoms for each second that the mercury was in the frozen state, and in a system volume of $0.6 \, \mathrm{dm}^3$ this would give a pressure increase of about $5 \times 10^{-9} \, \mathrm{Pa \, s^{-1}}$. It was of course necessary to have a vacuum system which was capable of being sealed off and maintaining a sufficiently low outgassing rate during the distillation period of about 30 minutes. The amount of hydrogen expected was very small and it was therefore necessary to keep the system volume as small as possible. Interestingly enough, a modified omegatron mass spectrometer (which we have encountered in chapter 6 in the measurement of μ'_p/μ_N) was used to detect the evolution of hydrogen from the mercury. Cohen and King were able to set an upper limit on the rate of matter creation of $\sim 4 \times 10^{-23} \, \mathrm{g \, g^{-1} \, s^{-1}}$ with their apparatus.

Other methods have been suggested for setting upper limits to matter creation. Thus Gittus (1976) suggested that matter created interstitially in rocks would diffuse to the dislocations. Towe (1975) disagreed with this since it would result in some 30% of the atoms in old terrestrial rocks occupying the interstitial positions. Gittus (1975) proposed that a sensitivity of about 3×10^{-18} atoms per atom per second could be achieved by determining the ratio of the shear modulus to the crystal viscosity in crystalline materials. In geophysics much could be explained by an expansion of the Earth of about 0.6 mm per year and Wesson (1975) has shown that such a rate would require a matter creation rate of around $7 \times 10^{-18} \, \mathrm{g \, g^{-1} \, s^{-1}}$. If one takes the Cohen and King upper limit and the mass of the Earth as 6×10^{24} kg then only about 200 kg of matter would be added to the Earth's mass per second. Over a period of 4.5×10^9 yr the accumulated mass, if the rate were sustained, would total some 3×10^{19} kg, or only a small fraction of the present mass of the Earth (Wesson 1978). An obvious objection to the Cohen and King experiment is that they postulated that hydrogen would be evolved and it would be rather more satisfactory to have an experimental method which was independent of the nature of the matter evolved.

The late Professor J W Beams had a considerable interest in performing experiments on levitated objects. Thus, in addition to his accelerated mass method of measuring G, discussed in § 7.1.1, he held a place in *The Guinness Book of Records* for some years for spinning a 1 mm levitated steel needle at some 6 million revolutions per second. It was appropriate, therefore, for his group to suggest (Ritter *et al* 1978) that it might be possible to detect the continuous creation of matter as affecting the angular momentum of a spinning levitated rotor. The principle of the experiment is to have a magnetically levitated cylinder inside a hollow cylinder which is mounted on a levitated table which is spinning at about 10 radians per second. Initially, the two cylinders would spin at the same rate, but the evolution of matter in the inner cylinder would lead to a slowing down of the inner cylinder (it is necessary to make certain assumptions about the angular momentum of the created matter of course). The interpretation of the results of the experiment when completed will obviously be open to debate. Meanwhile there is the interesting metrological problem of suspending a rotor to have an inertial decay time of 10^{11} yr (10^{18} s). In order to achieve decay times of this order of magnitude it will be necessary to largely eliminate such sources of energy loss as

   (i)   magnetic hysteresis in the support
  (ii)   interaction with external magnetic fields
 (iii)   viscous losses to the residual gases in the vacuum
 (iv)   change in cylinder radius due to thermal expansion
  (v)   change in the equilibrium position with respect to the outer cylinder due to thermal fluctuation effects.

The viscous losses and magnetic hysteresis effects will be largely eliminated by using a magnetic shielding feedback system involving a pulsed laser to sense the relative angular positions of the inner and outer cylinders so as to maintain their relative angular velocity as close to zero as possible. The inner cylinder will be constructed from a material having a very small expansion coefficient, $\sim 10^{-7}\,°C^{-1}$, so that by maintaining the temperature constant to 10^{-3} K the radial stability will be at the part in 10^{-10} level. The thermal angular fluctuation will be given by

$$\tfrac{1}{2}I(\delta\omega_n)^2 = -\tfrac{1}{2}kT$$

where I is the moment of inertia of the cylinder and kT the thermal energy. Consequently, if the duration of the experiment is τ, the random walk from the equilibrium angular position will be

$$\Delta\theta_n = \delta\omega_n\,\mathrm{d}t = (kT/I)^{1/2}\tau.$$

If the matter is created with the same angular velocity as the rotating cylinder then there would be no observable effect, but one can hypothesise, for example, that it will be created with an angular velocity appropriate to that of the rest of the universe. Ritter *et al* (1978) showed that a number of different hypotheses lead to an angular displacement of the inner cylinder relative to the outer, one of which is of the form

$$\delta\theta = \omega\tau^2 = -\gamma\omega(\dot{M}/M)^2$$

where γ is a constant of the order of unity. The key feature of this expression is that the angular position changes quadratically with time, whereas most of the expected experimental changes in the angle, due to dimensional drift, hysteresis and so on, are expected to change linearly with time. Ritter *et al* expect the sensitivity

$$\dot{M}/M = [(\delta\theta/\Delta\theta_n)(kT/I)^{1/2}]/\gamma\omega\tau^2$$

without feedback to be $\dot{M}/M \leq 10^{-8}$ with $I = 25 \text{ kg cm}^2$ and $\omega = 10 \text{ rad s}^{-1}$. This sensitivity can be improved by a further factor of $\bar{G}^{1/2}$, where $\bar{G}$ is the gain of the feedback system, and by cooling the apparatus. The sensitivities and decay times expected for the rotating system are similar to the limits set for mechanical oscillators from noise considerations (see Braginsky 1972). Feedback will be accomplished by using an appropriately directed beam of light to add or subtract angular momentum to or from the inner cylinder.

8.5 The isotropy of inertia

A certain class of gravitational theories implies that clocks of different electromagnetic nature will interact with local gravity differently and thus result in slightly different gravitational red-shifts. One way of investigating this which is being actively pursued (Cheung and Ritter 1981) is to rely on the constancy of the moment of inertia of a rotating body. This is being investigated at the University of Virginia by a co-rotating double system whereby a levitated rotating inner body is surrounded by an outer body which is servoed to rotate at the same angular velocity. In this way, long decay times, $\sim 10^{10}$ s have already been achieved, and it is expected that a scaling up of the apparatus might lead to a decay time of $\sim 10^{18}$ s (compared with the spin-down time of the Earth of $\sim 10^{17}$ s with short term variations $\sim 3 \times 10^{-8}$ per day).

The above work lies in the future, but one of the few microscopic rotor tests of the nature of gravity has been provided already by the Hughes (1960) and Drever (1961) tests of the anisotropy of space. These have important consequences for other parts of metrology for they test whether the gyromagnetic ratio of an atom depends on the orientation of the magnetic field with respect to the gravitational fields of the Earth, the Sun or the galaxy. These effects may be expressed in terms of the gravitational contribution to the spin precession frequency. The limits obtainable from present data have been discussed by Gallop and Petley (1983) who also suggested further sensitive experiments, particularly using ^{3}He precessing in a magnetic field trapped by a hollow superconducting niobium cylinder. The spin precession frequency could be detected with a SQUID magnetometer and the sensitivity should be in the region of $\lesssim 1 \text{ µHz}$.

It is important to place such limits on the gravitational influence on the gyromagnetic ratios of atomic particles since some experiments are performed with the flux vertical and others with the flux horizontal. The possibility of such effects is a particularly important consideration whenever the flux densities in an experiment are small. If there is such an effect, then the linearity of the dissemination of the magnetic flux scale could be affected on the one hand, while Mach's principle could be tested on the other. This type of null experiment indicates very strongly the interaction of modern precision metrology with fundamental physics.

8.6 The neutrality of matter

In our description of the physical world in terms of fundamental constants there is the implicit assumption that these constants are universal, that is all electrons have the same mass, charge and spin and so on. If this were not so there would be important consequences which would manifest themselves by changes in the observed phenomena. Thus we have seen that a non-zero mass for the photon would necessitate using a different form for Maxwell's equations. If the electronic charge was not the same everywhere there would be consequences. One possibility which has excited particular interest over the years is the question of whether the electron and proton have an exactly equal and opposite charge. Interest in this question followed the famous Millikan oil-drop experiments in the 1920s.

Einstein suggested in 1924 that a small charge difference between the electron and proton would account for the observed magnetic fields of the Earth and Sun. He estimated that this difference should be of the same order of magnitude as the gravitational attraction between the electron and proton. Thus he expected that the ratio of the proton to electron charge, λ, would be

$$\lambda = -1 + (3 \times 10^{-19})$$

i.e. the charge on the proton should be slightly greater than that on the electron. This hypothesis was investigated by Piccard and Kessler (1925). They took a 27 l cylinder which contained CO_2 at a pressure of 7 or 8 atm and allowed the gas to discharge into a large vacuum vessel. By measuring the change in the potential of the cylinder as the gas was discharged they were able to infer that

$$\lambda = -1 \pm (5 \times 10^{-21}).$$

Some 350 g of CO_2 was discharged, the sensitivity of their electrometer was 2.5 mV and the electrostatic capacity of the cylinder was ~ 50 pF. Their experiment therefore showed that a molecule of CO_2 did not have a charge greater than $2 \times 10^{-19} e$ (where $-e$ was the electronic charge) and was taken as refuting Einstein's hypothesis.

Interest in the subject was rearoused following the suggestion by Bondi and Littleton (1959) that if the electron and proton charge differed by a part in 10^{18} it would account for the expansion of the universe. Despite the above result there was a possible objection to the interpretation of the Piccard and Kessler experiment which stemmed from their choice of CO_2 because it contains equal numbers of neutrons and protons to within 0.1%. They had implicitly assumed in their work that the neutron behaves similarly to a hydrogen atom whereas their null result might also be accounted for if neutrons behaved oppositely to a hydrogen atom. Hillas and Cranshaw (1959) therefore set out to repeat the Piccard and Kessler experiment with gases which contained unequal numbers of neutrons and protons. They used argon (18 protons, 18 electrons and 22 neutrons) and nitrogen (14 protons, 14 neutrons and 14 electrons). They found that the charge per nitrogen atom did not exceed 12×10^{-20} e and on argon 8×10^{-20} e. From this, by treating the charges on nitrogen and argon as the sums of the charges on electrons, neutrons and protons, they deduced that the proton charge was $(1 + 10^{-20})e$ and that the neutron charge was less than 4×10^{-20} e.

The Hillas and Cranshaw experiment was broadly similar to that of Piccard and Kessler (figure 8.4). The gas cylinder was mounted inside an aluminium container which was insulated from a surrounding aluminium container and the potential difference between the two containers was monitored as the gas was discharged to the atmosphere. As it discharged the gas passed between plates which were connected to an isolated battery, and since the gas spent about 5 ms in this region, which was about 12 times the time taken for an ion to traverse the gap, it was considered that the gas leaving the cylinder was ion-free. In addition, after leaving the ion trap the gas did not pass over any more insulators before it left the system. The ion trap was demonstrated to be necessary since the gas contained an excess corresponding to about 3 ions per litre (at NTP) when the ion trap voltage was switched off. The sensitivity of their electrometer was about 1.5×10^{-14} C and they discharged 81 l of argon and 58 l of nitrogen at NTP for each experimental run. Although there was some criticism of their experiment by Bondi and Littleton (1960), which was refuted by Hillas

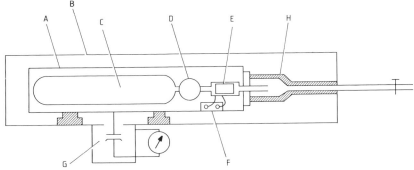

Figure 8.4 The apparatus used by Hillas and Cranshaw (1959) to verify the neutrality of matter. C, gas cylinder; D, reducing valve; E, ion trap; F, battery; G, electrometer; H, double-walled outlet tube containing thermal insulation; the hatched areas are polystyrene insulators.

and Cranshaw (1960), the work gained general acceptance by the scientific community.

In addition to the gas efflux method other methods have been used. Thus the electrostatic levitation method used on oil drops by Millikan in his measurement of the electronic charge was extended, through the levitation work of Beams and colleagues (Beams *et al* 1955) to the levitation of diamagnetic and ferromagnetic spheres and lately to the levitation of superconducting spheres. These experiments were all used to set limits on the neutrality of matter.

The limits in table 8.3 have been supplemented by those relating to recent quark searches (table 8.4). The Stanford group continues to be the only one to have observed possible fractional charges (and interestingly the Cabrera 'Valentine day' monopole as well), in their levitated superconducting niobium sphere apparatus and they have increased the mass of niobium studied to 1.1 mg (LaRue *et al* 1981). Marinelli and Morpurgo (1982) reviewed quark searches and extended their total mass of levitated steel to 3.7 mg. Hodges *et al* (1981) have used the Millikan method on 175 μg of mercury drops and the same group, Joyce *et al* (1983), used 51 μg of water from various sources, including the sea. Sea water is a favourite candidate for quark searches since quarks are expected to remain in solution. Ogorodnikov *et al* (1979) attempted a considerable amount of scaling up of their sample size, achieving a total of 134 kg of sea water and ocean sediments. They relied on the quarks coming off into heated vapour and then being collected on filter electrodes and being transported from thence to collector electrodes. The rate of desorption of charge from the heated collector in the presence of an applied electric field then sets limits on the quark concentration.

Dylla and King (1973) devised an acoustic method whereby microphones were used, together with phase sensitive detection, to test for acoustic vibration at the same frequency as an electric field was applied to SF_6 in the space between concentric spheres (40 cm and 2.5 cm diameter) and forming a resonant acoustic cavity, $Q \sim 1000$, at audio frequencies. Polarisation and

Table 8.4 Results of searches for particles of charge $\frac{1}{3}e$ in different samples of material (after Joyce *et al* 1983).

Authors and date	Sample material	Mass studied	Free quarks per nucleon
Ogorodnikov *et al* (1979)	H_2O	134 kg	$< 4.0 \times 10^{-28}$
Boyd *et al* (1979)	He	20 μg	$< 6.4 \times 10^{-16}$
LaRue *et al* (1981)	Nb	1.1 mg	2.1×10^{-20}
Hodges *et al* (1981)	Hg	175 μg	$< 2.8 \times 10^{-20}$
Marinelli and Morpurgo (1982)	Steel	3.7 mg	$< 1.3 \times 10^{-21}$
Liebowitz *et al* (1983)	Steel	720 μg	$< 6.9 \times 10^{-21}$
Joyce *et al* (1983)	H_2O	51 μg	$< 9.8 \times 10^{-20}$

Table 8.5 The experiments having a bearing on the electrical neutrality of matter (1) electrostatic, ferromagnetic or diamagnetic levitation (2) gas efflux method (3) acoustic cavity (4) beam method.

Method	Authors	Particle or molecule	Upper limit for ε	(Charge per nucleon)$/e$
(1)	Millikan (1935)	10 pg H_2O		$\pm 10^{-16}$
	Gallinaro and Morpurgo (1966)	1 µg C		10^{-17}
	Stover et al (1967)	1 µg Fe		$\pm 8 \times 10^{-19}$
	Rank (1968)	0.1 ng oil drop		10^{-18}
	Braginsky et al (1970)	0.2 mg Fe		10^{-16}
	Gallinaro et al (1977)	1 mg Fe		2×10^{-20}
	LaRue et al (1979)	0.7 mg Nb		10^{-17}
(2)	Piccard and Kessler (1925)	CO_2	$\pm 2 \times 10^{-19}$	$\pm 5 \times 10^{-21}$
	Hillas and Cranshaw (1960)	Ar	$4(4) \times 10^{-20}$	$1(1) \times 10^{-21}$
		N_2	$6(6) \times 10^{-20}$	$2(2) \times 10^{-21}$
	King (1960)	H_2	$1.8(5.4) \times 10^{-21}$	$+0.9(27) \times 10^{-21}$
(3)	Dylla and King (1973)	SF_6	2×10^{-19}	1×10^{-21}
(4)	Hughes (1957)	CsI	4×10^{-13}	2×10^{-15}
	Zorn et al (1963)	Cs	$1.3(5.6) \times 10^{-17}$	$1(4.2) \times 10^{-19}$
		K	$-3.8(11.8) \times 10^{-17}$	$-1(3) \times 10^{-18}$
		H_2	$\pm 2 \times 10^{-15}$	$\pm 1 \times 10^{-15}$
		D_2	$\pm 2.8 \times 10^{-15}$	$\pm 7 \times 10^{-16}$
	Fraser et al (1965)	Cs	$\pm 1.7 \times 10^{-18}$	$\pm 1.3 \times 10^{-20}$
		K	$\pm 1.3 \times 10^{-18}$	$\pm 3.3 \times 10^{-20}$
	Shapiro and Estulin (1957)	n	$< 6 \times 10^{-12}$	$< 6 \times 10^{-12}$
	Shull et al (1967)	n	$-1.9(3.7) \times 10^{-18}$	$-1.9(3.7) \times 10^{-18}$

other effects occur at the second harmonic frequency. Molecular beam methods, in which the molecules are deflected by an electric field, have also been used. Similar techniques have been applied for neutrons. It is interesting that the work also led to Stover *et al* (1967) setting a lower limit for the charge on 2 eV photons of $10^{-16}e$ per photon. Their limit was an order of magnitude lower than that set by Grodkins *et al* (1961) using 14.4 keV photons. The results of the experiments are summarised in table 8.5.

Overall, the results set a limit which is lower than that required by either Einstein or by Bondi and Littleton. However, since the strength of the electromagnetic interaction is $\sim 10^{39}$ times stronger than the gravitational interaction, any small asymmetry could have a profound effect. Further, although there might be charge equality, there might still be effects if the forces of attraction were as little as 10^{-36} times greater than the forces of repulsion. Gold pointed out that this possibility had been propounded by Hli (1959) to provide an electrical origin to account for gravitational forces and also, according to Poincaré had been visualised by Benjamin Franklin. As we have seen with the photon rest mass, any asymmetry or small inequality in the proton and electron charges and electromagnetic forces could have profound and far-reaching effects.

The story does not however end there, for the work has some bearing on another quest in physics, namely the search for a particle having a charge of $e/3$. Thus the above work can be invoked as setting limits on the number of quarks in the gases which were used in the experiments. In fact, according to the review by Jones (1977) they gave the tightest limits for the searches for the existence of quarks in stable matter. This trail in turn crosses the quest for the magnetic monopole, mass $\sim 10^{16}$ GeV, postulated by Dirac (1931), for if this exists, it can be shown quite simply that electric charge must be quantised. The Dirac monopole should have a 'magnetic' charge given by

$$g_D = h/4\pi e \simeq 3.3 \times 10^{-16} \text{ Wb}$$

and this would produce an ionisation which was some 4700 times greater than that produced by a singly charged particle. Cabrera (1982) caused considerable excitement by observing a signal from his SQUID detection system which was consistent with the passage of a monopole through the system. If genuine this would set an upper limit on the monopole flux of $6.1 \times 10^{-6} \text{ m}^{-2} \text{ sr}^{-1} \text{ s}^{-1}$. This was for monopoles of all velocities. Bartelt *et al* (1983a, b) have set a lower limit for monopoles of velocity between $10^{-3}c$ and $10^{-4}c$ of $<4.1 \times 10^{-9} \text{ m}^{-2} \text{ sr}^{-1} \text{ s}^{-1}$ (c is the velocity of light). Of course, one event, which, although the work was carefully performed, could be due to other causes, does not necessarily prove anything. We can safely write: 'to be continued' after much of the content of this chapter!

The group at Stanford University (Fairbank *et al* 1981) have reported results consistent with the presence of quarks on magnetically levitated niobium spheres and, if these results are confirmed by other groups and so far they have not (see Marinelli and Morpurgo (1980) and Stevens *et al* (1976)), they will have exciting implications for physics. Here too the trail crosses another much older one, for Hodges *et al* (1982) have modified a Millikan oil-drop type of apparatus and combined it with automatic data acquisition methods to look at the electric charge on mercury droplets. In all, 10 000 drops from 3.5 to 6.5 μm diameter were examined, in total 65 μg of refined mercury and 115 μg of native mercury, and no fractionally charged droplets were observed. Millikan's apparatus therefore lives on in different guises and even in its conventional form: as a study of Millikan's law of particle fall in gases by Kim and Fedele (1982) has revealed, there are still anomalies to be discovered which repay further study. Incidentally, Fairbank (1982) has re-examined Millikan's notebooks to settle the argument as to whether Millikan saw any fractional charges (quarks?) in his experiments and concluded that he probably did not.

8.7 The proton decay rate

A direct consequence of the grand unified gauge theories, which unify the strong, electromagnetic, weak and gravitational interactions, has been the prediction that the proton is unstable with a lifetime of the order of 10^{31} to 10^{33} years. The lifetime of the universe since the 'big bang' is only about 10^{10} years so such an instability is hard to detect. It is clearly of fundamental importance if the proton has a decay rate, for it might well lead to the situation where all matter was unstable. This is unlikely to be of any significance in our immediate metrology, but it does affect whether or not we regard the mass of the proton as being a fundamental constant.

The experimental bound set by Learned *et al* (1979) was $\sim 10^{30}$ yr. Experiments are being undertaken in several places with about 1000 tonnes of matter, which contains $\sim 5 \times 10^{32}$ protons and neutrons (Georgi (1981) and Bartelt *et al* (1983a)). These would yield about 50 events per year if the average proton lifetime was 10^{31} years. Clearly the measurements will be difficult and, if cosmic ray events are to be excluded, it is necessary either to work deep underground, or with the apparatus surrounded by a large body of water. Good discrimination against background radioactivity will also be necessary. There have been some reports of decay events at about the rate predicted by the above proton lifetime, but this tale is another one which is to be continued.

9

Experimental uncertainties and the evaluation of the 'best values'

Both direct and indirect measurements provide information about the values of physical constants. In a direct determination the value of the physical constant is measured experimentally and does not require knowledge of any other physical constants. Examples of this type of measurement would be determinations of the velocity of light, the electronic charge by Millikan's method, the fine structure constant, the universal gravitational constant and so on. In the indirect measurements the constants are measured in combination with other physical constants. Thus with the Josephson effects one measures the value of $2e/h$, while a measurement of the fine structure constant essentially yields a value for e^2/hc. Each measurement is subject to experimental errors, and consequently a problem arises when there are more experimental determinations than there are constants to be evaluated, for a unique set of values must be derived from these measurements which may be used by scientists throughout the world. It is necessary for the recommended values for e and h, with their associated uncertainties, to be consistent with the recommended value for $2e/h$ and so on. Of course, given a set of over-determined data on the constants, statistical methods may be used to derive the best values (§9.3). In practice we have additional information which cannot be quantified, but which equally cannot be ignored, and this provides an obstacle to treating the experimental data in a totally abstract fashion. Therefore, before considering how the best values are obtained from the data, we consider the uncertainties in the experimental measurements in a little more detail.

9.1 The uncertainty of the measurements

The result of each experiment makes some reliance on theory; for example, a spectroscopic measurement of a fundamental constant which makes no allowance for Lamb shifts will not be totally correct. On the other hand, the theory may be correct, but the experimental realisation not quite as envisaged.

All of the measurements discussed in this book are distilled from a considerable amount of experimental work. Much of this work does not contribute directly to the final result, but it does yield information which is taken into account when assessing the uncertainty, expressed as a standard deviation, that is ascribed to the finally quoted number. This uncertainty is traditionally divided into two groups, which are combined as the root sum of their squares. One group is often termed the 'random' uncertainty and the other the 'systematic' uncertainty.

9.1.1 The uncertainty estimated from the fluctuations in the experimental results

The random uncertainty is assumed to be backed by the gaussian or some other law of errors and the science of statistics (Topping 1962) and as such is accompanied by a vast body of literature.

In most experiments it is possible to take sufficient observations for the magnitude of the random uncertainty to be comparable with or less than the limits that it is possible to place on the systematic effects and it is not at all unusual to find that in practice the overall experimental uncertainty is comparable with the standard deviation of a single measurement. One reason for this is that one can only set limits on a possible dependence on an experimental parameter at the level of $\sigma_s/(n-1)^{1/2}$ where σ_s is the standard deviation of a single measurement and n_s is the number of observations contributing to the investigation of the particular systematic effect. In most of the experiments discussed in this book $(n-1)^{1/2}$ has been between three and ten so that, when the uncertainties ascribed to the various null hypotheses are recombined as the root sum squares, the final result tends to be comparable with the original σ_s. Hence the causes of the 'random' scatter should be considered very carefully and reduced where possible by improving the design of the experiment. It is this type of consideration that led Rutherford to pronounce (see for example MacKay 1977): 'If your experiment needs statistics, you ought to have done a better experiment'.

The random fluctuations in the result may come from the superposition of several different types of noise. Thus the noise frequency spectrum of a high quality source can be represented by the superposition of independent noise processes (see Audoin 1976) which belong to the types

$$S_y(f) = h_\alpha f^\alpha \qquad \text{and} \qquad \sigma_y^2(t) \propto t^\mu$$

in the frequency (f) and time (t) domains respectively, where

$$\alpha = -2, -1, 0, 1, 2 \qquad \mu = 1, 0, -1, -2, -2$$

and $S_y(f)$ is the one sided normalised spectral density of the fractional frequency fluctuations. These modes may be identified as shown in table 9.1, the dominant mode depending on the averaging time. Such fluctuations are

Table 9.1 The values of α and μ for different types of noise affecting frequency stability.

Type of noise	α	μ
Random walk frequency noise	-2	1
Flicker frequency noise	-1	0
White frequency noise	0	-1
Flicker phase noise	1	-2
White phase noise	2	-2
Uniform drift		2

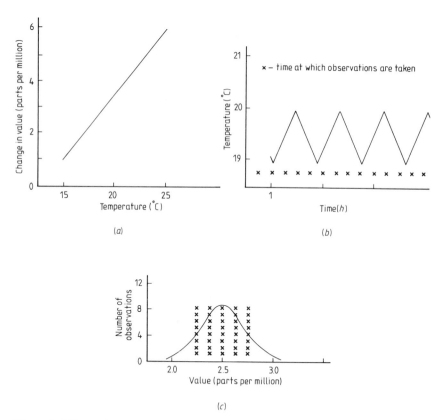

(a)

(b)

(c)

Figure 9.1 Illustrating how a systematic effect, if undiagnosed, may lead to a 'random' uncertainty: (a) the result actually depends on temperature (for example) so that when (b) the temperature varies in a cyclic fashion (e.g. because it is thermostated), this leads to (c) a rectangular distribution which would be difficult to separate out from the other 'random' effects if they were of the magnitude shown by the curved distribution. Moral: measure as many parameters as possible and search for systematic variations.

found both in lasers and in frequency standards and may well apply elsewhere, affecting our estimates of the mean and standard deviation. Blick (1980) has investigated the fluctuations in two-dimensional conductors of various shapes and Cabrera (1979) investigated the $1/f$ noise in squids down to $1\,\mu Hz$!

9.1.2 Uncertainties: random or systematic?

The uncertainties can be divided into a number of different groups. One of these comprises systematic effects which vary between one observation and the next, and in this way are able to contribute to the scatter of the observations.

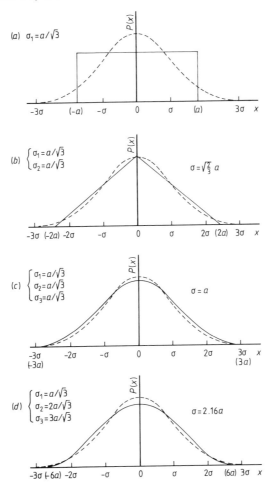

Figure 9.2 Illustrating the rapid convergence to a distribution of gaussian form (dashed lines) as different types of rectangular distribution are combined: (a), (b) and (c) show one, two and three rectangular distributions of the same width and (d) a combination of three rectangular distributions of width a, 2a, and 3a respectively.

To see how these may arise, consider the simple case where there is a systematic effect that depends fundamentally on temperature. The experimenter tries to reduce this type of effect by introducing temperature control so that the apparatus is changing in temperature cyclically (figure 9.1). (We assume that we are unable to measure this temperature.) Observations are taken at various times with respect to this cycling and, if enough observations are taken, they will build up some form of distribution. If the temperature is varying in a linear fashion between well defined limits then a rectangular distribution will be built up. This distribution will combine with the other randomly varying effects and the resulting distribution will have a shape depending on the relative magnitudes of the rectangular and random error distributions. However, as is readily apparent from figure 9.2, several rectangular distributions can combine quite rapidly to give a remarkably close resemblance to a gaussian distribution (except in the extreme wings of the distribution). In a typical metrological experiment a search is made for systematic effects, and it is frequently found that as the temperature regulation, the pressure regulation, vibration isolation or the electrical stability, and so on are improved, the scatter in the results decreases, until at length the scatter in the results does not appear to be reducing any further. At this stage, the final results are obtained, but there are still undiscovered effects. Table 9.2 shows how a sequence of eight measurements can test for a null effect of three variables.

Table 9.2 The sequence of combinations for a single factorial replication of three variables, where A, B, C are high values and a, b, c low values of the respective variables (the order is not important). It is evident that the results may be used to look for systematic effects by combining them in different ways.

abc	*abC*	*aBc*	*aBC*
(1)	(2)	(3)	(4)
Abc	*AbC*	*ABc*	*ABC*
(5)	(6)	(7)	(8)

9.2 Estimating the uncertainty

From the above considerations it appears probable that many, but not all, of the fluctuations which we call 'random' are almost certainly the result of systematic changes of some variables which we have not recorded. The difference between the random and many of the systematic uncertainties is related to the time spectrum of the fluctuations. Thus fluctuations that occur in times which are of the order of the time to take a measurement are the 'random' ones and those taking the whole duration of the experiment are 'systematic'. An example of the latter type might be the drift with time in the value of a standard which was measured only once during the experiment. Those effects fluctuating in between these times are partly random and partly systematic. From these simple considerations it is apparent that 'systematic' and 'random' uncertainties are not necessarily sufficiently different in character to justify keeping them totally separate and it is reasonable to combine them. However, they should always be quoted separately before they are combined to give a single measure of the uncertainty. An important reason for this is that the random uncertainty has a measure of its reliability, characterised by the number of degrees of freedom, whereas the systematic uncertainties at present do not.

9.2.1 The standard deviation

The normal error curve has the form

$$y = \frac{1}{\sigma(2\pi)^{1/2}} \exp[-(x - \bar{x})^2/2\sigma^2]$$

where σ is the standard deviation. The probability of an observation outside the range $\bar{x} - \sigma < x < \bar{x} + \sigma$ is 0.3173, i.e. 68.27% of the results lie within $\pm\sigma$ of the mean. Similarly, 95.43% lie within $\pm 2\sigma$ and the probability of an observation lying more than $\pm 3.09\,\sigma$ from the mean is 1 in 500. Since the experimental results provide only a small sample from such a distribution, the experimental estimate of the standard deviation is uncertain, and for N results may be written as $\sigma[1 \pm 1/(\sqrt{2N} - 1)]$ and the standard error of the mean as $(\sigma/\sqrt{2N})[1 \pm 1/\sqrt{2(N - 1)}]$ where $\sigma^2 = \Sigma^N(x - \bar{x})^2/(N - 1)$.

If a certain quantity is measured a large number of times N_i with each measurement X_i having a standard deviation uncertainty σ_i due to the random fluctuations, then the weighted mean $\bar{X}$ and the uncertainty of the mean σ_m are given by

$$\bar{X} = \sum_{i=1}^{N} (X_i/\sigma_i^2) \left(\sum_{i=1}^{N} (1/\sigma_i^2) \right)^{-1}$$

$$\frac{1}{\sigma_m^2} = \sum_{i=1}^{N} (1/\sigma_i^2).$$

For identical σ_i, σ_m reduces to

$$\sigma_m = \sigma_i/N^{1/2}$$

and this form is used in the analysis of the results of many experiments.

9.2.2 Other measures of the uncertainty

In measurements which conform to the normal error distribution one expects half of the observations to lie within $0.6745\,\sigma$ of the mean, and in earlier work this measure, known as the *probable error*, was often quoted instead of the standard deviation. However, in the method of least squares the standard deviation has a wider applicability. It is the appropriate measure of the width of distributions other than the normal distribution and this measure of the uncertainty is almost invariably used today.

Another measure which has been used is the average deviation σ_i given by

$$\bar{\sigma}_i = \sum_{i=1}^{N} |X_i - \bar{X}|/N.$$

This is related to the standard deviation by

$$\bar{\sigma}_i = \sigma_i\,(N/2)^{1/2}$$

for a normal distribution, but is different for other distributions, for example for the rectangular distributions of figure 9.2 it is $1.15\,\sigma$. Other experimenters have used the term *limit of error* to express the uncertainty in their results and it is not always clear what probability this represents. Taylor *et al* (1969) took the term to mean three times the probable error, or two standard deviations, when reducing such uncertainties to standard deviations. On the other hand, DuMond and Cohen interpreted the uncertainty quoted by Sommer *et al* (1957) as three standard deviations.

(An illustration of the usefulness of the standard deviation over other measures is given by a consideration of the combination of two rectangular distributions to form the triangular probability distribution (figure 9.2(b)). The standard deviation of the rectangular distribution of width $2a$ is $a/\sqrt{3}$ and this correctly predicts the standard deviation of the triangular distribution to be $a(2/3)^{1/2}$ (the root sum squares of the two component standard deviations). However, the probable error of the triangular distribution is $(2 - \sqrt{2})a = 0.586a$, and this *differs* from the probable error obtained from the root sum squares of the probable errors of two rectangular distributions, for the latter would be $a/\sqrt{2} = 0.707a$. It may be argued that the difference is small considering the uncertainties in the experimental estimates, but it helps to show why the probable error has fallen into disuse.)

9.2.4 The fallible metrologist: intellectual phase locking

The velocity of light measurements between 1935 and 1941 were all about $20\ \mathrm{km\,s^{-1}}$ too low compared with present day values (see table 9.3) and the tendency for experiments in a given epoch to agree with one another has been described by the delicate phrase 'intellectual phase locking' by Franken and others. Most metrologists are very conscious of the possible existence of such

Table 9.3(a) The four determinations of c between
1935 and 1941.

Measurement	Date	$c(10^8 \text{ m s}^{-1})$	
Michelson *et al*	1935	2.997 74	(4)
Anderson	1937	2.997 71	(10)
Hüttel	1940	2.997 71	(10)
Anderson	1941	2.997 76	(6)

(b) Three review values of the period 1935–41 compared with the present day value (after Sanders 1961).

Review	Date	$c(10^8 \text{ m s}^{-1})$	
Birge	1941	2.997 76	(4)
Stille	1943	2.997 77	(20)
Dorsey	1945	2.997 73	(10)
CCDM	1971	2.997 924 58	(1.2)

effects; indeed ever-helpful colleagues delight in pointing them out! It is easy to avoid being influenced by contemporary work during much of the experiment, for the measurements are performed in terms of local 'Zanzibar' units, in terms of various meter readings, etc, and this is sustained until the final conversion to SI units is made.

In principle, the experiment is completed at this point but unfortunately, although the computations in terms of the local units may have been effected many times, the metrologist is much less expert at making the conversion to absolute units than one might at first assume. The reviewers have uncovered a number of mistakes by experimenters in making these final conversions and corrections: for example, Birge found that the factor 300 instead of $c/(10^8 \text{ cm s}^{-1})$ had been used in converting from electrostatic voltage units to the cgs voltage unit. A re-evaluation of the length of the length standards after the initial result had been computed was a frequent cause of corrections in the velocity of light determinations.

Aside from the discovery of mistakes, the near completion of the experiment brings more frequent and stimulating discussion with interested colleagues and the preliminaries to writing up the work add a fresh perspective. All of these circumstances combine to prevent what was intended to be 'the final result' from being so in practice, and consequently the accusation that one is most likely to stop worrying about corrections when the value is closest to other results is easy to make and difficult to refute. Corrections which are applied at a later date present a further difficulty, particularly those which affect the internal relationships between the experimental results. It might well be that had these been known at the time further experiments would have been undertaken or the systematic uncertainties assigned differently.

9.3 Evaluation of the 'best values'

The methods of obtaining the best values of the fundamental constants have gradually been refined over the years. Of course, no amount of statistical sophistication can make bad results any better and a continuing task of the reviewers is to sort out which data to include in their evaluation. The advent of computers has made it possible to perform more sophisticated analyses than hitherto, which makes assessing the overall situation much simpler. However, the relatively simple methods of the past were quite effective. Thus the Birge–Bond diagram, which was proposed by Bond (1930, 1931), was extensively used by Birge (1932, 1941) in his evaluations in both the three- and two-dimensional forms.

An example of the method applied by Birge to the data available in 1941 is

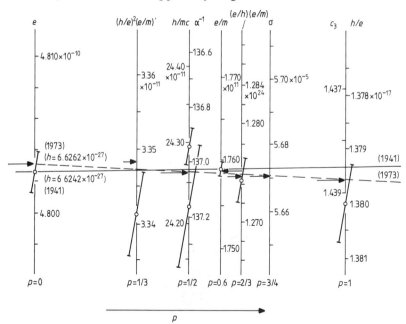

Figure 9.3 The method used by Birge (1941) to display the relationships between the experimental constants as a nomogram with the measurements spaced horizontally according to the value of the parameter p, expressed as a function $\Phi(h^p e)$. The dashed line joins the Cohen and Taylor (1973) best values of e and h/e, and the other 1973 values are indicated by arrows. This provides a tribute to the quality of the work of the metrologists of that time, and to R T Birge in his choice of auxiliary constants. (After Cohen *et al* 1957.)

shown in figure 9.3. The experimental determination E_μ expressed in the form

$$E_\mu = e^p f(e_0, h_0, m_{e,0})$$

was plotted vertically as the ordinate and the value of p as the abscissa, where e_0, h_0, $m_{e,0}$ were assumed origin values. If the chosen value of h_0 was only slightly wrong (which could in any case be achieved by an iterative procedure) the data was compatible with a straight line whose slope gave the best value of $h_0(h_0 + \delta h_0)$ and whose intercept yielded the best value of e.

This Birge–Bond diagram greatly simplified the analysis and allowed the results of different experiments to be intercompared very readily. The solid line in the figure is the one from which Birge obtained his output values for e and h, and the dashed line is that corresponding to present day values. It is remarkable how well the experimental results agree with the latter; indeed, it appears that rather too many of them are within one standard deviation, suggesting that the uncertainties were perhaps a trifle pessimistic.

The three-dimensional representation was not used for detailed analysis, but it was used to show that the 1941 data placed heavy reliance on the values of the Rydberg constant and on the oil-drop determination of the electronic charge. An additional method of displaying the results was in the form of an isometric consistency chart suggested by Bethe (1938). This allowed a large number of different experiments to be shown on a single plot and was a good way of drawing attention to discrepant data. However, it was not capable of yielding a solution by fitting a least squares straight line in the way that the Birge–Bond diagram permitted.

The above methods were not continued much beyond 1950, partly because the increased accuracy and multiplicity of the available experimental results meant that many of the hitherto auxiliary constants had to be evaluated as well (see Cohen *et al* 1957) and the situation was more complex. However, as we will discuss further, the data available in 1982 permitted representation as a two-dimensional plot (figure 9.7). This permits an illustration of how the adjustment procedure works, for we can imagine each result pulling or pushing in a particular direction; the process being rather like a spring, with the reciprocal of the uncertainty determining the strength of the spring. The adjusted output values are represented by the equilibrium position of a number of such springs. If the equilibrium position changes considerably when a particular 'spring' is omitted then that result may well be suspect.

9.3.1 The reviewers' task

The task of the reviewer, in the face of the problems which have been discussed is to:

(i) check on the validity of the theory
(ii) check the computation of the experimental results, seeking clarification where necessary

(iii)   to ensure that the corrections and conversions have been correctly applied

(iv)   to decide whether the uncertainty estimates for the systematic effects are of reasonable magnitude and whether all effects have been adequately allowed for, and finally

(v)   to decide whether the work is sufficiently reliable experimentally and theoretically to be included in the adjustment.

Sometimes work has been excluded because there was insufficient information available about the determination; for example the measurements of μ'_p/μ_N by Mamyrin and Frantsuzov was omitted (incorrectly as it turned out) from the 1965 and 1969 evaluations of the fundamental constants. There is little point either in including results which have too large an uncertainty, even when there are many such results, the reason for this being that experimenters may well have neglected systematic effects which were smaller than say $\sim 10\%$ of the overall uncertainty (possibly by lumping them together under one heading) and there may be rounding errors as well.

9.3.2 The purpose of the reviews

Reviews providing the best values have two purposes. The first is to provide the scientific community as a whole with the most up to date and correct set of units and conversion factors. Happily, the vast majority of these users are not interested in using the values at the full precision and are unaffected by quite large changes such as occurred following the 1969 and 1973 reviews (of course they derive considerable comfort whenever the experts at precision measurement prove to be fallible!). However, there are scientists who have a need for the results at the full precision. The consequences of the comparison between their results and those from the evaluation may have a wide impact on the scientific community: the validity of QED, possible time variations of G, a new method of defining a base SI unit, and so on. This latter group of scientists provides the second purpose of the reviews and also generates the main pressure for fresh evaluations.

A possible consequence of these disparate needs could be that two sets of values would be brought into existence, an official set for the bulk of the users and a more accurate one for the minority. The values quoted in books are often truncated to perhaps four significant figures, since this suffices for the reader. It is obviously desirable, if possible, to avoid having two such official sets and the compromise has been to have an evaluation roughly once per decade. In this time, the measurement frontier will, on the average, have advanced by approaching an order of magnitude. Of course, even one evaluation per decade may be too frequent for many users, for the sources which they consult may not contain the latest values. Thus Taylor (1970) has reported that a well known physics handbook still included the Birge (1941) values (as well as later ones) in the 1956–7 edition. This propagation delay in the scientific literature underlines the necessity for the evaluations to produce values whose precision is well in advance of that required by the majority of the users.

9.3.4 The basic theory of the least squares method

The procedure for evaluating the fundamental constants has been considerably refined over the years and the techniques have been discussed by Bearden and Thomsen (1952), Cohen et al (1957) Cohen and DuMond (1965), Taylor et al (1969), Cohen (1980), Taylor (1981) and by Tuninsky and Kholin (1975). A detailed discussion would be out of place here, but we discuss the method used to obtain a unique set of values from the experiments which have been discussed earlier and also indicate how the output values and uncertainties may be used to calculate other combinations of the constants.

The experimental data comprise direct and indirect information about the fundamental constants and the first task is to decide on which constants to use as the unknowns to be adjusted. The experimental observations are then expressed in terms of these unknowns by combining them with constants which are known with at least an order of magnitude greater precision. Since

$$\sigma^2 = (\sigma_i^2 + \sigma_A^2)^{1/2}$$

and hence, if

$$\sigma_A \lesssim 0.1\sigma_i$$

so that the uncertainty in the auxilliary constant may safely be neglected, $\sigma \sim \sigma_i$.

In the 1973 adjustment, which we consider for purposes of illustration, the unknowns chosen for adjustment were $\bar{R}$, K, α, N_A, Λ and μ. The observations were expressed in terms of these as shown in table 9.4. The relations between the constants $\bar{R}$, $K\alpha$... (m in number), and the n experimental data X_1, $X_2, \ldots, X_n$, combined with auxiliary constants a_i, were of the form

$$\bar{R}^{i_n} K^{j_n} \alpha^{k_n} \cdots = a_n X_\eta \tag{9.1}$$

where $i, j, \ldots$ were the powers to which $\bar{R}$, K etc were raised, $n > m$ and $1 \leq \eta \leq n$.

The relations (9.1) may be reduced to the linear approximation of a Taylor expansion by expressing the unknowns as

$$\begin{aligned}
\bar{R} &= \bar{R}_0(1 + \delta\bar{R} + \cdots) \\
K &= K_0(1 + \delta K + \cdots) \\
\alpha &= \alpha_0(1 + \delta\alpha + \cdots)
\end{aligned} \tag{9.2}$$

where $\bar{R}_0$, K_0, $\alpha_0, \ldots$ etc are the chosen origin values so that the fractional terms are all small compared with unity. Equations (9.1) now become

$$i\delta\bar{R} + j\delta K + \cdots = a_i X_\eta - (\bar{R}_0^{l_n} K_0^{m_n} \alpha_0^{n_n}). \tag{9.3}$$

Table 9.4 The relationships between the evaluated quantities and the numerical values of the measured quantities { } in the 1973 evaluation by Cohen and Taylor.

Measured quantity	Operational equation
$\Omega_{B169}/\Omega \equiv \bar{R}$	$\bar{R} = \{\bar{R}\}$
$A_{B169}/A \equiv K$	$K = \{K\}$
F	$\alpha N_A \bar{R} = (\mu_0/4)c(2e/h)_{B169}\{F_{B169}\}$
γ'_p (low field)	$\alpha^2 \bar{R}^{-1} = \dfrac{4R_\infty}{c(\mu'_p/\mu_B)(2e/h)_{B169}}\{\gamma'_p(\text{low})_{B169}\}$
γ'_p (high field)	$\alpha^2 K^{-2}\bar{R}^{-1} = \dfrac{4R_\infty}{c(\mu'_p/\mu_B)(2e/h)_{B169}}\{\gamma'_p(\text{high})_{B169}\}$
μ'_p/μ_N	$\alpha K^{-2}N_A^{-1}\bar{R}^{-2} = \dfrac{16R_\infty}{M_p\mu_0(\mu'_p/\mu_B)c^2(2e/h)_{B169}^2}\{\mu'_p/\mu_N\}$
Λ	$\Lambda = \{\Lambda\}$
$N_A\Lambda^3$	$N_A\Lambda^3 = \{N_A\Lambda^3\}$
λ_C	$\alpha^2\Lambda^{-1} = 2 \times 10^{-10} R_\infty\{\lambda_C\}$
α^{-1}	$\alpha^{-1} = \{\alpha^{-1}\}$
$\mu_\mu/\mu_p \equiv \mu$	$\mu = \{\mu\}$
ν_{Mhfs}	$\alpha^2\mu = \dfrac{3(1 + m_e/m_\mu)^3(1.000\,197\,21(201))}{16R_\infty c(\mu_e/\mu_B)(\mu_p/\mu_B)}\{\nu_{\text{Mhfs}}\}$

If we knew the true values of $\bar{R}$, K, etc, we could substitute them into the right-hand side of equation (9.3), but we would not, in general, obtain the values of C_η, but instead the quantity $C_\eta(1 - r_\eta)$, where the quantity r_η is the relative fractional error in the measured numeric C_η. Since we do not know the true values of the primary variables $\bar{R}$, K, ..., we cannot compute r_η, from equation (9.2). However we do have an estimate of some of the parameters of its probability distribution.

In introducing the parameters r_η we have changed from a set of over-determined equations to one in which there are n equations and $n + k$ unknowns and so, in order to obtain a solution, it is necessary to introduce additional conditions. There are several possibilities and these are discussed in the next section.

9.3.5 Uncertainties in the output values

The generalised form of equation (9.1) for the experimental data, after being combined with the appropriate combination of the auxiliary constants a_η is

$$\prod_{j=1}^{m} Z_j^{\Upsilon \eta_j} = X_\eta$$

where η_j denotes the ith observational equation, η a number, Z_j is the jth adjustable constant, m a number and Y_{nj} is the exponent of the jth adjustable constant in the ηth observational equation (note that Y_{nj} is an integer or combination of integers, which may be positive, negative or zero). The generalised linearised form of equation (9.3) with the residuals conveniently expressed in parts per million, is

$$\sum_{j=1}^{n} Y_{j\eta} z_j = B_\eta$$

where $z_j = [(Z_j - Z_{j0})/Z_{j0}] \times 10^6$ and $B_\eta = [(X_\eta - X_0)/X_0] \times 10^6$. The normalised residuals r_η are given by

$$r_\eta = (1/\sigma_\eta) \sum_{j=1}^{n} (Y_{j\eta} z_j - B_\eta)$$

where the uncertainties σ_η are expressed in parts per million. The least squares solution is obtained by minimising the sum of the squares of the normalised residuals with respect to z_j. That is

$$\partial \left(\sum_{\eta=1}^{n} r_\eta^2 \right) \Big/ \partial z_j = 0.$$

This leads to a set of j linear equations, known as the normal equations, which are given by

$$\sum_{k=1}^{j} G_{jk} z_k = D_j \qquad j = 1, \dots, J$$

where

$$G_{jk} = \sum_{1}^{n} (Y_{j\eta} B_\eta)/\sigma_\eta^2$$

and

$$D_j = \sum_{1}^{n} (Y_{j\eta} B_\eta/\sigma_\eta^2).$$

These equations are then solved for the j unknowns Z_k and the adjusted values of the original variables are obtained from

$$Z_k = Z_{k0}(1 + Z_k \times 10^6).$$

It can be shown that the standard deviation uncertainty σ_k in Z_k is related to the matrix G by the elements of G^{-1}

$$\sigma_k^2 = (G^{-1})_{kk}.$$

The matrix G^{-1} is called the error matrix. Note too that the generalised expression for χ^2, having an expectation value $n - m$, is

$$\chi^2 = \sum_{\eta=1}^{n} \left[\left(\prod_{j=1}^{J} Z_j - A_\eta \right)^2 \sigma_\eta^{-2} \right].$$

In general, the adjusted constants Z_j will be correlated even when, as is usually the case, the uncertainties in the input values are not correlated. These correlations were evident in the changes in the 'best values' that were shown in table 1.3 and are perhaps one of the disadvantages of the method of least squares. It is necessary to use the generalised law of error propagation in computing the uncertainties of any output constants which involve two or more of the adjusted constants. This requires using the variances of the individually adjusted constants together with the covariances which are associated with pairs of the adjusted constants. The variance v_{kk} of the kth adjustable constant is of course the square of the standard deviation; thus

$$v_{kk} = \sigma_k^2 = (G^{-1})_{kk}$$

and the covariance of the jth and kth adjusted constants is

$$v_{jk} = (G^{-1})_{jk}.$$

It is apparent therefore that both the diagonal elements (variances) and off-diagonal elements (covariances) of the matrix G^{-1} are required. A useful property of this matrix is that it is symmetrical, i.e.

$$v_{ij} = v_{ji}.$$

9.2.6 Use for estimating the uncertainties of combinations of the output values

The above matrix elements are used for evaluating the uncertainty in a quantity Q which is a combination of the adjusted quantities Z_k as follows. Suppose that Q depends on N statistically correlated quantities, as given by the equation

$$Q = Q(x_1, x_2, \ldots, x_N)$$

then the variance of Q is given by σ_Q^2 where

$$\sigma_Q^2 = \sum_{i=1}^{N} \sum_{j=1}^{N} (\partial Q/\partial x_i)(\partial Q/\partial x_j) v_{ij}$$

where v_{ij} is the covariance between x_i and x_j. For most examples involving the fundamental constants, the quantity Q will depend on the values of z_j as

$$Q = q \prod_{j=1}^{J} Z_j^{Y_{Qj}}$$

where q is a combination of a numerical value and possibly some auxiliary constants. The expression for σ_Q reduces to

$$\sigma_Q^2 = \sum_{i=1}^{J} \sum_{j=1}^{J} (Y_{Qi} Y_{Qj} v_{ij})$$

where v_{ij} are conveniently expressed in parts per million for recent evaluations. It will also be recalled that Y_{Qi} and Y_{Qj} are usually integers or simple combinations of integers.

9.3.7 Correlation coefficients

An alternative expression for σ_Q is in terms of correlation coefficients which are defined by

$$r_{ij} = v_{ij}/(v_{ij}v_{jj})^{1/2} = v_{ij}/\sigma_i\sigma_j$$

which lead to

$$\sigma_Q^2 = \sum_{i=1}^{N} (\partial Q/\partial x_i)^2 \sigma_i^2 + \sum_{i=1}^{N} r_{ij}\sigma_i\sigma_j(\partial Q/\partial x_i)(\partial Q/\partial x_j)$$

and hence as above we obtain

$$\sigma_Q^2 = \sum_{i=1}^{N} Y_{Qi}^2\sigma_i^2 + \sum_{i=1}^{N} r_{ij}\sigma_i\sigma_j Y_{Qi}Y_{Qj}.$$

It is evident that if $r_{ij} = 0$, i.e. there is no correlation between the variances, the equations reduce to the usual equations given by the law of propagation of uncertainties.

The property that $r_{ij} = r_{ji}$ and $r_{ii} = 1$, together with $v_{ij} = v_{ji}$, allows both the correlation coefficients (given below the diagonal in bold type) and the variance – covariance terms (on and above the diagonal) to be given in the same table (as just indicated) instead of separately.

It is not necessary to restrict this table to just those quantities which have been adjusted. Thus, in terms of the two quantities Q and R, where Q was defined above, and

$$R = R(x_i, x_2, \ldots, x_N)$$

or in terms of the constants Z_j,

$$R = r \sum_{j=1}^{J} Z_j^{Y_{Ri}}$$

the combined covariance of the two quantities, V_{QR}, is just

$$V_{QR} = \sum_{i=1}^{N} \sum_{j=1}^{N} (\partial Q/\partial x_i)(\partial R/\partial x_j)v_{ij}.$$

If as before v_{ij} are expressed in (parts per million)^2 this expression may be written as

$$v_{QR} = \sum_{i=1}^{J} \sum_{j=1}^{J} (Y_{Qi}Y_{Rj}v_{ij}).$$

9.4 The Birge ratio and recent developments

Birge (1932) emphasised that there were really two ways of assigning an uncertainty to the weighted mean of a quantity, that obtained by the criterion of internal consistency and that by external consistency. The uncertainty determined by internal consistency σ_I is the usual uncertainty given by

$$\sigma_I^2 = \sigma_m^2 = \left(\sum_{i=1}^{m} (1/\sigma_i^2) \right)^{-1}.$$

The error determined by external consistency σ_E is defined as

$$\sigma_E^2 = \left(\sum_{i=1}^{n} (X_i - \bar{X})^2/\sigma_i^2 \right) \left((n-1) \sum_{i=1}^{n} (1/\sigma_i^2) \right)^{-1}$$

where $\bar{X}$ is the weighted average computed from X_i and σ_i in the usual way. Thus σ_I is the expected uncertainty in the mean as given by the individual *a priori* uncertainties, while σ_E is the uncertainty based on the amount by which each X_i deviates from the mean as a fraction of its uncertainty σ_i. The Birge ratio R_B is simply

$$R_B = \sigma_E/\sigma_I$$

and should be consistent with unity if the data are actually distributed as expected from the individual *a priori* uncertainties. If R_B is significantly greater than unity it implies that some of the measurements are suspect; that is the particular σ_i may have been underestimated. On the other hand, a value of R_B which is significantly less than unity indicates that the data are rather too highly compatible, which in turn implies that some at least of the uncertainties have been overestimated. Since the Birge ratio is related to χ^2 by

$$R_B^2 = \chi^2/(N-1)$$
$$= \chi^2/(n-m)$$

it has a distribution which may be inferred from the χ^2 distribution.

The expectation value of χ^2 is equal to the number of degrees of freedom, and the extent to which the observations yield this value is a measure of the consistency of the data. In practice, the number of degrees of freedom is not very great and the χ^2 probability distribution is quite broad and skew for small $n - m$. If the deviation of $\chi^2/(n - m)$ from unity is the result of only one of the measurements, then this may give sufficient reason for excluding that particular measurement from the data. One adjustment procedure is to exclude a few of the results from the adjustment in turn and to examine the consequent values of the Birge ratio as a test of the reliability of the excluded values. This provides a guide as to whether the results should be omitted altogether. This procedure is equivalent to multiplying the weights by an additional factor which either takes the values zero or unity. This 'all or nothing' approach was used in earlier evaluations and has given considerable cause for disquiet over the years. Alternative methods have been sought which are soundly in accord with statistical methods.

It has always been assumed that the value obtained for X_i has an associated uncertainty, but there has been little discussion until recently of the uncertainty associated with the assigned uncertainty. This is rather surprising in view of the use of the uncertainty as a weighting factor in evaluating the best values. A random component has an uncertainty determined by the number of degrees of freedom, which is essentially the number of observations less the number of systematic effects that have been investigated. Estimating the uncertainty in the systematic component is difficult and many scientists have been unhappy about combining the random and systematic uncertainties. The group at VNIIM in Leningrad have attempted to overcome this objection by applying an additional criterion which enables the assigned weights to be expanded in order to obtain a Birge ratio of unity. A disadvantage of this method is that it treats all metrologists as being optimists, by always expanding the errors (it has to be admitted that there is ample historical precedent to support this thought!). Clearly, in contracting the overall uncertainty one might encounter a difficulty if the contracted value was comparable with the random component and a more sophisticated treatment would be required. The suggested methods considered for overcoming the problem have been (Cohen 1976) to take one of the following criteria

$$\sum_i (1 - \sigma_i'^2/\sigma_i^2) = \min$$

$$\sum_i (1 - \sigma_i^2/\sigma_i'^2) = \min$$

$$\sum_i \ln^2(\sigma_i'^2/\sigma_i^2) = \min$$

$$\sigma_i'^2 = \sigma_r^2 + R_{int}^2 \sigma_{sy}^2$$

where σ_i is the initial standard deviation of the input data, σ_r is the random and σ_{sy} the systematic component of the initial standard deviations, R_{int} is the internal Birge ratio introduced in (§9.5) and σ_i' are the standard deviations satisfying the condition of consistency.

$$\chi_{cal}^2(\sigma_i'^2) = F.$$

All of the above methods for handling inconsistent data result in the expansion of *a priori* variances by a factor

$$R_i^2 = \sigma_i'^2/\sigma_i^2.$$

Ideally, such factors would improve the variances of those results whose uncertainties estimates were erroneous without distorting the remainder. A disadvantage of this type of method would appear to be that once one starts to adjust the value of χ^2 there is no well defined value other than $\chi^2_{cal} = F$ at which to stop. Fortunately the question of which is the best statistical method may well prove to be rather an academic one. Thus Taylor (1981) analysed a number of different methods of treating the experimental data that was available in 1973 and concluded, as one might hope, that the particular statistical method employed is less important than the decision to include or exclude particular results.

One difficulty in testing these more sophisticated techniques is that any method applied retrospectively to the 1965 and 1969 evaluations, which allows the data that was given zero weight in these evaluations to be included (namely some of the data on the fine structure constant and μ'_p/μ_N respectively), will lead to a set of output values which are closer to the more recent values, and hence will appear to be better. Of course, if it becomes apparent in the future that, as with earlier evaluations, the present set of data also has unsuspected systematic effects our ideas might have to be modified further—which adds to the interest in the subject.

It seems likely that the discussions about which algorithm to use will continue for some time to come and it may be that statistical theory will not provide a unique 'best choice'. The results from most of the algorithms will probably agree with one another within about one standard deviation of the output uncertainties. It is, of course, quite possible that when applied to the data which were available for the relatively few past evaluations, or will become available in the future, one method might appear to be better than the others. Whatever method is finally adopted, it is likely to be one which makes minimum changes to the input values. It must be apparent that when it comes to changing other people's assigned uncertainties, or omitting results altogether, the reviewer's task can be a lonely one at times!

9.5 Recent evaluations of the fundamental constants

The evaluation of the 'best values' of the fundamental physical constants proceeds in well characterised stages. The first stage is to take an overall look at the data available, from the point of view of the reliability of the experimental work and of the underlying physical theory. Next, the data are expressed, as we have discussed, in terms of combinations of the fundamental constants so that the constants that are to be subject to evaluation can be decided upon. After deciding on the statistical algorithm the output values can finally be computed.

The constants c, k and G are essentially obtained as individual measurements rather than in combination with constants having a comparable measurement uncertainty. The remaining constants N_A, m_p, m_e, h and e must be evaluated from the experimental data. For the evaluations prior to 1969 it was not necessary to include the ampere conversion constant K in the evaluation and more recently the ohm conversion constant $\bar{R}$ has been included as well (these are the conversion constants of the ampere and ohm respectively from maintained to SI units). Latterly both K and $\bar{R}$ have changed from being auxiliary to evaluated constants in the evaluations. Although the bulk of the information on $\bar{R}$ comes from the calculable capacitor absolute realisations of the SI ohm, the measurements of $\gamma'_{p(low)}$ and of R_H, the quantised Hall resistance, are changing the situation markedly.

For the 1973 evaluation Cohen and Taylor took α, K, $\bar{R}$, Λ, μ and N_A as the constants to be evaluated. The available experimental results were then expressed in terms of these by making use of the auxiliary constants as shown in the table 9.4. The accuracy of these auxiliary constants may be gauged from figure 9.4. The accuracy of the available experimental data is similarly displayed in figure 9.5. In the Taylor *et al* evaluation (1969) their unknown were α, e, K, Λ and N_A. Their reason for this was that they were seeking to demonstrate that the Josephson effect E_J was indeed equal to $2e/h$ when they were both expressed in terms of the same electrical units.

By 1973, the precision of the Josephson effect measurements had so surpassed the information concerning $2e/h$ from the rest of physics that it became entirely reasonable to take the Josephson effect $2e/h$ as an auxiliary constant. That is it was assumed that the measurements determined a constant that was identically $2e/h$, as per the theory of the effects.

One result of the 1969 evaluation was to spur a number of different groups to undertake and complete experiments pertaining to the fine structure

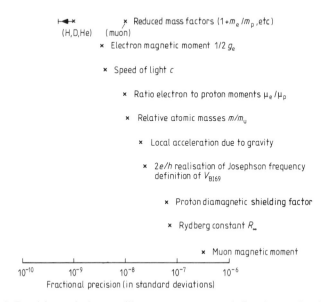

Figure 9.4 Precision of the auxiliary constants used in the evaluation of the fundamental constants by Cohen and Taylor (1973).

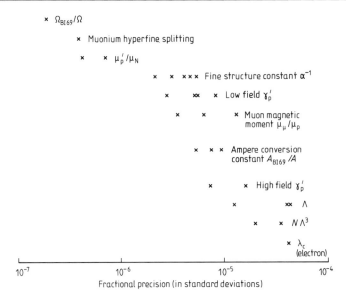

Figure 9.5 Precision of the observational data contributing to the 1973 evaluation of the fundamental constants.

constant (the first Precision Measurement and Fundamental Constants Conference held at the NBS in 1970 helped this process considerably). These measurements included results from spectroscopy as well as the muon and electron $(g-2)$ measurements. In addition, between 1969 and 1973, it became more apparent that Taylor *et al*, in rejecting the Leningrad determination of μ'_p/μ_N (as had DuMond and Cohen earlier), had rejected a very good measurement. The major change in the output values between 1969 and 1973 was largely attributable to these two factors.

Between 1973 and the present there have been a number of advances, many of which were foreshadowed by the above-mentioned Gaithersberg Conference. Thus the Rydberg constant is now known with more than an order of magnitude improved accuracy as too are the $g_e - 2$ measurements.

The Avogadro constant too has been measured with part per million precision, making it no longer necessary to evaluate the conversion constant from x units to nanometres. The gyromagnetic ratio of the proton in weak magnetic fields has also been measured with sub-ppm precision in a number of laboratories since 1973, notably at the NBS, as has been discussed earlier. There are measurements of the quantised Hall effect resistance and the Faraday, as well as improved measurements of $\gamma'_{p(high)}$. Many countries too have stabilised their maintained volt with the aid of the Josephson effects. A number of them have fixed their volt so as to give E_J a value of 483 594.0 GHz per maintained volt (other countries including France, the USA and USSR are offset by a known amount from this value: all of the values will be unified when

K is regarded as being known with sufficiently high precision to enable a 'once and for all' change in the nationally maintained and disseminated volts to be made). Thus we have

$$E_J = K\bar{R}(2e/h)$$

$$= 483\,594 \times 10^9 \text{ Hz V}_{\text{lab}}^{-1}$$

as an exact auxiliary constant (note V_J is realised in practice to about 3 parts in 10^8 and this accuracy could easily be improved if the need arose).

The accurate measurements of m_p/m_e at the University of Washington which are supported by those at Mainz, as well as by the μ_p'/μ_N determinations, enable m_p/m_e to be evaluated separately. The resulting value can then be used as an auxiliary constant with the other data. The value obtained by Cohen (1983) was $m_p/m_e = 1836.152\,83(23)$. Similarly the absolute ohm measurements may be evaluated to give $\bar{R} = 1 - 0.82(19) \times 10^{-6}$.

Cohen (1983) pointed out that the available data enable the number of unknowns that must be evaluated to be reduced to two, namely α and K, as shown in table 9.5. The data available to the beginning of 1983 yield as output values (Cohen 1982):

$$\alpha^{-1} = 137.036\,004(32) \qquad \text{or } 2.4 \text{ parts in } 10^7$$

and

$$K = 1.000\,005\,9(14) \qquad \text{or } 1.4 \text{ parts in } 10^6.$$

Table 9.5 The observational equations relating the experimentally determined values of $\gamma_{p(\text{low})}'$, $\gamma_{p(\text{high})}'$, F_{lab} and N_A with α and K, M_p is the molar mass of the proton (after Cohen 1983) and $\{\quad\}$ denotes the numerical value of the measured quantity.

Observed quantity	Relationship
$\gamma_{p(\text{low})}'$	$\alpha^2 = [4\bar{R}R_\infty/(\mu_p'/\mu_B)cE_J]\{\gamma_{p(\text{low})}'\}$
$\gamma_{p(\text{high})}'$	$\alpha^2 K^{-2} = [4\bar{R}R_\infty/(\mu_p'/\mu_B)cE_J]\{\gamma_{p(\text{high})}'\}$
F_{lab}	$\alpha^2 K^{-2} = [4\bar{R}R_\infty(\mu_p'/\mu_N)/(\mu_p'/\mu_B)M_p(10^{-3} \text{ kg mol}^{-1})]\{F_{\text{lab}}\}$
N_A	$\alpha K^{-2} = [16\bar{R}^2R_\infty(m_p/m_e)/M_p(10^{-3} \text{ kg mol}^{-1})\mu_0c^2E^2]\{N_A\}$

There are considerable discrepancies between the data available having a bearing on K. The Birge ratio is high, indicating a high degree of inconsistency between the experimental results. This can be shown by plotting the results on an isometric consistency chart as shown in figure 9.6 which is based on that given by Cohen (1982). The axes are essentially K and F so that measurements of the Avogadro constant N_A define a curve KF, and those of the fine structure constant α define a line K/F. The measurements included on the chart are the low-field determinations of the gyromagnetic ratio of the proton, which are:

1   that of Williams and Olsen (1979) at the NBS
2   those at NIM–PRC in China by Wang (1982)
3   those at VNIIM reported by Tarbeyev (1982)
4   the NPL measurement of Vigoureux (1976).

The two measurements of $\gamma'_{p(high)}$ are:

5   those of Kibble and Hunt (1976) at the NPL and
6   Wang *et al* at NIM–PRC.

The fine structure constant determinations comprise:

7   the Washington measurement of $g_e/2$
8   the NBS–NRL–BTL measurement of the quantised Hall resistance (which are essentially confirmed by the other measurements in Japan, the PTB and the NPL)
9   the helium fine structure measurements at Yale by Hughes *et al.*

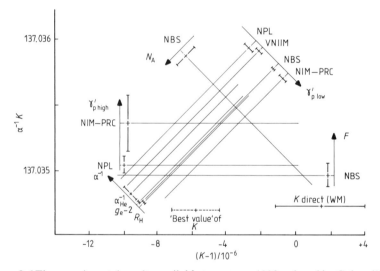

Figure 9.6 The experimental results available to autumn 1983 reduced by Cohen (1983) to a two-dimensional plot by expressing them in terms of auxiliary constants and the unknowns $\alpha^{-1}K$ and $K-1$. The results are identified in terms of the metrological laboratory in which they were obtained, together with the numbers assigned in the text (after Cohen 1983). Note: At the time of writing the exact relationship between the quantised Hall effect values and the others is still being evaluated, since the absolute realisation of the SI ohm is involved as well as the relationships between the ohms maintained in different countries and so we have only shown the value given in Cohen (1983). This value may be offset too much from the other results. The relationships between the R_H results and with the other α-values will, no doubt, be resolved in the forthcoming evaluation by Cohen and Taylor. It is a tribute to the rapid progress with the measurements of the quantised Hall resistance that the discrepancies to be resolved are already smaller than a part in a million.

The Avogadro constant measurement

   10   by Deslattes *et al* at NBS.

   11   Finally, the Faraday constant by Bower and David at NBS completes the essential list.

Since the experiments simply determine combinations of α and K, the plots may be linearised by expressing them in terms of an assumed origin value that differs from the output value by only a few parts per million. If all of the data were consistent the lines would cross at a single point, or rather enclose an area according to the assigned uncertainties. The PTB lattice parameter measurement differs from that of Deslattes and if it is confirmed by a density measurement to give a higher value for the Avogadro constant, then it will strongly suggest that the NBS measurement of N_A might have an unsuspected systematic error and the picture will look rather tidier.

The measurements pertaining to the fine structure constant are in quite good agreement (these include such measurements as the electron $g_e - 2$, the proton low field γ'_p, the spectroscopic measurements in helium and the quantised Hall effect measurements). The high field γ'_p and Faraday measurements are discrepant, as too are the direct measurements of the ampere conversion constant K.

It is once again very evident that there is rarely a 'best time' at which to evaluate the 'best values' for there are always some unanswered questions and the picture only becomes clear with the advantage of hindsight. Given the evaluated values of α^{-1} and K, the values of the other fundamental physical constants may be evaluated from the equations below, using the full variance and covariance matrix

$$e = (4/\mu_0 E_J)\alpha K \bar{R}$$

$$h = \frac{8}{\mu_0 c}\left(\frac{K\bar{R}}{E_J}\right)^2$$

and

$$m_e = \frac{16 R_\infty}{\mu_0 \alpha}\left(\frac{K\bar{R}}{cE_J}\right)^2.$$

(NB   At the time of writing it is not clear whether the latest evaluation will have been published and internationally adopted by IUPAC, IUPAP, etc, before this book is published (see appendix) and the above procedure described by Cohen (1982) should be taken as illustrating the discrepancies between the data and how they may be evaluated.)

9.6 The pace of metrology

One of the salutary experiences in metrology is to realise the transient nature of many of the measurements. The classic determinations survive in the textbooks, of course, but the actual result, except where it leads to a new discovery, has little of lasting interest. Many of the measurements of the velocity of light perhaps illustrate this, and it may be that for many of them it was a constant that was an incidental measurement to prove an apparatus rather than a prime reason for the work.

The frontier of accurate measurement advances by roughly a factor of ten every fifteen to fifty years and the pace varies from one constant to another. Both the velocity of light and the universal gravitational constant have been measured for centuries and yet the former is known to parts in 10^{10} and the latter to little better than 0.01%. This underlines the important point that in metrology measurements to $\sim 1\%$ can be just as exacting as measurements to the nth decimal place. The pace of advance can change as well; for example in time measurement, the move to atomic standards from those depending essentially on the stability of a length (be it the radius of the Earth, the length of a pendulum, or the thickness of a quartz crystal) to atomic frequency standards saw a greatly accelerated progress—almost a factor of ten every five to ten years.

The last twenty years have certainly seen some exciting progress in the measurement of the fundamental constants, particularly from the application of lasers and of cryogenic methods of measurement. Whether this rate of advance will continue indefinitely, or at the same rate, is impossible to forecast. There may well be some ultimate quantum limits which will prove impossible to circumvent, but these seem likely to fall well beyond the likely progress during the latter part of this century. The author has attempted to forecast, on the basis of previous progress, the likely accuracy with which many of the fundamental constants will be known by about the turn of the century (figure

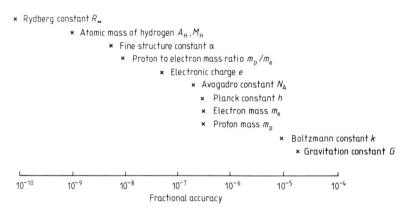

Figure 9.7 The accuracies that are likely to be achieved for many of the fundamental constants by the year 2000 on the basis of extrapolation of the earlier rates of progress. One might hope to do rather better than shown on h, m_e and m_p.

9.7). (He will of course be very pleased if experimenters prove him to be a pessimist!)

9.7 Conclusion

One of the difficulties in metrology stems from the fact that there is some loss of accuracy as one gets away from the base unit: we cannot measure 1 g or 100 kg with the same relative accuracy as we can compare 1 kg masses. Neither can we measure the microsecond, nor the century, with quite the accuracy that we can realise the second. Interestingly enough, much of the improved accuracy of realisation of the second comes from the increasing time that we take to realise it: averaging times of thousands of seconds are required for the highest accuracy.

The measurements of the fundamental constants leapfrog this type of problem. We know the mass of the proton, for example, with almost the same accuracy as we know the mass of a 1 g standard. A further difficulty that can be overcome stems from the fact that a fundamental physical constant provides the metrologist with something to measure that is known to be absolutely stable. The situation often arises in high-precision measurements where it is impossible to say with certainty whether the measuring equipment or the standard has changed. This is a very real problem in the national standards laboratories, for all other users can refer the problem of whether or not their standard has changed by comparing it with another that is one level back in the dissemination chain. The advent of the Josephson effect voltage standard that was known to be absolutely stable led rapidly to improvements in the manufacture of standard cells, for it became possible to say which were stable and which were not (some of the most stable cells were actually rejected as drifting upwards in the past because the maintained volt was drifting downwards!).

Measurements of the fundamental constants also provide the motivation for the development of new measurement techniques—lending truth to the proverb 'necessity is the mother of invention'. As has been seen from the content of this book, the measurement of quantities with the highest possible accuracy can take one from the exciting areas of cosmology and fundamental physics, through to the equally important regions of technological applications and that of enabling one scientist or technologist to communicate the results of their experiments to another in a different laboratory—perhaps many years later.

Metrology is one of the most exacting of the scientific disciplines, yet nowhere is it more true that the regularly cited reference of today becomes the forgotten work of tomorrow. Equally, if we progress and see a little further, it is also because as Newton, and others before and since have recognised, 'although we may be dwarfs we have stood on the shoulders of giants'. It is salutary to read again the inspiring and prescient views of Maxwell:

> This characteristic of modern experiments—that they consist principally of measurements,—is so prominent, that the opinion seems to have got abroad, that in a few years all the great physical constants will have been approximately estimated, and that the only occupation which will then be left to men of science will be to carry on these measurements to another place of decimals.
>
> ... But the history of science shows that even during that phase of her progress in which she devotes herself to improving the accuracy of the numerical measurement of quantities with which she has long been familiar, she is preparing the materials for the subjugation of new regions, which would have remained unknown if she had been contented with the rough methods of her early pioneers.[†]

Precision measurements have always played an important part in testing theory to the limit and we have seen that there is every sign that this process will continue. It is probable that, even now, there is a precision measurement in progress whose results will revolutionise our view of physics—or perhaps the reader will be encouraged by this book to think of one!

[†] Taken from *The Scientific Papers of James Clark Maxwell* vol II (1890) (Cambridge: Cambridge University Press). The sentence concerning the 'great physical constants' is the most frequently quoted part. Reproduced by permission of Cambridge University Press.

Appendix

Table A.1 Fundamental physical constants and energy conversion factors: the recommended 'best values' by Cohen and Taylor (1973). The figures in parentheses are the standard deviation uncertainties in the last digit quoted.

	Quantity	Symbol	Value	(σ)	Units	Uncertainty (parts in 10^6)
General	speed of light in vacuum	c	2.997 924 580	(12)	$10^8\ \mathrm{m\,s^{-1}}$	0.004
	permeability of vacuum	μ_0	4π exactly		$10^{-7}\ \mathrm{H\,m^{-1}}$	—
	permittivity of vacuum $1/\mu_0 c^2$	ε_0	8.854 187 818	(71)	$10^{-12}\ \mathrm{F\,m^{-1}}$	0.008
	Planck constant	h	6.626 176	(36)	$10^{-34}\ \mathrm{J\,Hz^{-1}}$	5.4
			4.135 701	(11)	$10^{-15}\ \mathrm{eV\,Hz^{-1}}$	2.6
	$h/2\pi$	$\hbar$	1.054 588 7	(57)	$10^{-34}\ \mathrm{J\,s}$	5.4
			6.582 173	(17)	$10^{-16}\ \mathrm{eV\,s}$	2.6
	elementary charge	e	1.602 189 2	(46)	$10^{-19}\ \mathrm{C}$	2.9
			$\{$4.803 242	(14)	$10^{-10}\ \mathrm{esu}$	2.9$\}$
	Avogadro constant	N_A	6.022 045	(31)	$10^{23}\ \mathrm{mol^{-1}}$	5.1
	atomic mass unit, $10^{-3}\ \mathrm{kg\,mol^{-1}}\,N_A^{-1}$	u	1.660 565 5	(86)	$10^{-27}\ \mathrm{kg}$	5.1
			931.501 6	(26)	$10^6\ \mathrm{eV}$	2.8
	MeV to kilogram, $10^6\,ec^2$	MeV	1.782 675 9	(51)	$10^{-30}\ \mathrm{kg}$	2.9
	Faraday constant of electrolysis $N_A e$	F	9.648 456	(27)	$10^4\ \mathrm{C\,mol^{-1}}$	2.8
			$\{$2.892 534 2	(82)	$10^{14}\ \mathrm{esu\,mol^{-1}}$	2.8$\}$
	gravitational constant	G	6.672 0	(41)	$10^{-11}\ \mathrm{N\,m^2\,kg^{-2}}$	615

Quantity	Symbol	Value	(σ)	Units	Uncertainty (parts in 10^6)
Magnetic					
Bohr magneton $[c](eh/2m_ec)$	μ_B	9.274 078	(36)	10^{-24} J T^{-1}	3.9
		5.788 378 5	(95)	10^{-5} eV T^{-1}	1.6
		1.399 612 3	(39)	10^{10} Hz T^{-1}	2.8
		46.686 04	(13)	m^{-1} T^{-1}	2.8
		0.671 712	(21)	K T^{-1}	31
magnetic flux quantum $[c]^{-1}(hc/2e)$	Φ_0	2.067 850 6	(54)	10^{-15} Wb	2.6
	h/e	4.135 701	(11)	10^{-15} J s C^{-1}	2.6
		{1.379 521 5	(36)	10^{-17} ergs esu^{-1}	2.6}
nuclear magneton $[c](eh/2m_pc)$	μ_N	5.050 824	(20)	10^{-27} J T^{-1}	3.9
		3.152 451 5	(53)	10^{-8} eV T^{-1}	1.7
		7.622 532	(22)	10^6 Hz T^{-1}	2.8
		2.542 603 0	(72)	10^{-2} m^{-1} T^{-1}	2.8
		3.658 26	(12)	10^{-4} K T^{-1}	31
Particles					
Electron					
Bohr radius $[\mu_0 c^2/4\pi]^{-1}(\hbar^2/m_e e^2)=\alpha/4\pi R_\infty$	a_0	5.291 770 6	(44)	10^{-11} m	0.82
charge to mass ratio	e/m_e	1.758 804 7	(49)	10^{11} C kg^{-1}	2.8
		{5.272 764	(15)	10^{17} esu g^{-1}	2.8}
classical electron radius $[\mu_0 c^2/4\pi](e^2 m_e c^2)=\alpha^3/4\pi R_\infty$ $=\alpha\lambda\!\!\!/_c$	r_e	2.817 938 0	(70)	10^{-15} m	2.5
Compton wavelength of the electron $(h/m_e c)=\alpha^2/2R_\infty$	λ_C	2.426 308 9	(40)	10^{-12} m	1.6
$\lambda_C/2\pi=\alpha a_0$	$\lambda\!\!\!/_C$	3.861 590 5	(64)	10^{-13} m	1.6

Quantity	Symbol	Value		Units	Uncertainty
magnetic moment of the electron	μ_e	9.284 832	(36)	10^{-24} J T^{-1}	3.9
magnetic moment in Bohr magnetons (g-factor, g_e)	$\mu_e/\mu_B = g_e/2$	1.001 159 656 7	(35)	—	0.0035
ratio of electron and proton magnetic moments	μ_e/μ_p	658.210 688 0	(66)	—	0.010
mass of the electron at rest	m_e	9.109 534	(47)	10^{-31} kg	5.1
		5.485 802 6	(21)	10^{-4} u	0.38
		0.511 003 4	(14)	MeV	2.8
1 electron volt	eV	1.602 189 2	(46)	10^{-19} J	2.9
in frequency units e/h		2.417 969 6	(63)	10^{14} Hz	2.6
in wavenumber units e/hc		8.065 479	(21)	10^5 m^{-1}	2.6
in temperature units e/k		1.160 450	(36)	10^4 K	31
Thomson cross section $(8/3)\pi r_e^2$	σ_e	0.665 244 8	(33)	10^{-28} m^2	4.9
Compton wavelength of the proton					
$h/m_p c$	$\lambda_{C,p}$	1.321 409 9	(22)	10^{-15} m	1.7
$\lambda_{C,p}/2\pi$	$\lambdabar_{C,p}$	2.103 089 2	(36)	10^{-16} m	1.7
gyromagnetic ratio of free proton	γ_p	2.675 198 7	(75)	10^8 rad s^{-1} T^{-1}	2.8
	$\gamma_p/2\pi$	4.257 711	(12)	10^7 Hz T^{-1}	2.8
gyromagnetic ratio of protons in H$_2$O	γ'_p	2.675 130 1	(75)	10^8 rad s^{-1} T^{-1}	2.8
	$\gamma'_p/2\pi$	4.257 602	(12)	10^7 Hz T^{-1}	2.8
magnetic moment of free proton	μ_p	1.410 617 1	(55)	10^{-26} J T^{-1}	3.9
in Bohr magnetons	μ_p/μ_B	1.521 032 209	(16)	10^{-3}	0.011
in nuclear magnetons	μ_p/μ_N	2.792 845 6	(11)	—	0.38
magnetic moment of protons in H$_2$O,					
in Bohr magnetons	μ'_p/μ_B	1.520 993 22	(10)	10^{-3}	0.066
in nuclear magnetons	μ'_p/μ_N	2.792 7740	(11)	—	0.38
mass of proton at rest	m_p	1.672 648 5	(86)	10^{-27} kg	5.1
		1.007 276 470	(11)	u	0.011
		938.2796	(27)	MeV	2.8

Proton

	Quantity	Symbol	Value	(σ)	Units	Uncertainty (parts in 10^6)
Neutron	ratio of proton mass to electron mass	m_p/m_e	1836.151 52	(70)	—	0.38
	Compton wavelength of neutron					
	$h/m_n c$	$\lambda_{C,n}$	1.319 5909	(22)	10^{-15} m	1.7
	$\lambda_{C,n}/2\pi$	$\lambdabar_{C,n}$	2.100 1941	(35)	10^{-16} m	1.7
	mass of neutron at rest	m_n	1.674 9543	(86)	10^{-27} kg	5.1
			1.008 665 012	(37)	u	0.037
			939.573 1	(27)	MeV	2.8
Muon	mass at rest	m_μ	1.883 566	(11)	10^{-28} kg	5.6
			0.113 429 20	(26)	u	2.3
			105.659 48	(35)	MeV	3.3
	mass to that of electron	m_μ/m_e	206.768 65	(47)	—	2.3
	magnetic moment in terms of proton magnetic moment	μ_μ	4.490 474	(18)	10^{-26} JT^{-1}	3.9
	free muon g-factor or muon magnetic moment	μ_μ/μ_p	3.183 3402	(72)	—	2.3
	moment in units of $[c](eh/2m_\mu c)$	$g_\mu/2$	1.001 166 16	(31)	—	0.31
Spectroscopic	fine structure constant $[\mu_0 c^2/4\pi][e^2/\hbar c]$	α	7.297 3506	(60)	10^{-3}	0.82
		α^{-1}	137.036 04	(11)	—	0.82
	Rydberg constant (fixed nucleus) $[\mu_0 c^2/4\pi]^2 (m_e e^4/4\pi\hbar^3 c)$	R_∞	1.097 373 177	(83)	10^7 m^{-1}	0.075
			2.179 907	(12)	10^{-18} J	5.4
			13.605 804	(36)	eV	2.6
			3.289 842 00	(25)	10^{15} Hz	0.075
			1.578 885	(49)	10^5 K	31

	Symbol	Value		Units	
energy × wavelength (or energy ÷ wavenumber)		1.239 852 0	(32)	10^{-6} eV m	2.6
		1.986 447 6	(107)	10^{-25} J m	5.4
energy ÷ frequency		4.135 701	(11)	10^{-15} eV Hz^{-1}	2.6
frequency ÷ energy		2.417 969 6	(63)	10^{11} Hz eV^{-1}	2.6
ratio, kx unit to angström, $\Lambda = \lambda(A)$ ÷ $\lambda(kxu)$; $\lambda(CuK\alpha_1) = 1.537 400$ kxu	Λ	1.002 007 72	(54)	—	5.3
ratio Å* to angström, $\Lambda^* = \lambda(Å)/$ $\lambda(Å^*)$; $\lambda(WK\alpha_1) = 0.209 0100$ Å*	Λ^*	1.000 020 5	(56)	—	5.6
Thermal Boltzman constant R/N_A	k	1.380 662	(44)	10^{-23} J K^{-1}	32
		0.861 735	(28)	10^{-4} eV K^{-1}	32
molar gas constant, $p_0 V_m/T_0$	R	8.314 41	(26)	J mol^{-1} K^{-1}	31
molar volume of ideal gas at STP ($T_0 = 273.15$ K; $p_0 = 101 325$ Pa = 1 atm)	V_m	22.413 83	(70)	10^{-3} m^3 mol^{-1}	31
first radiation constant $2\pi hc^2$	c_1	3.741 832	(20)	10^{-16} W m^2	5.4
second radiation constant hc/k	c_2	1.438 786	(45)	10^{-2} m K	31
Stefan–Boltzmann constant $2\pi^5 k^4/15 h^3 c^2$	σ	5.670 32	(71)	10^{-8} W m^{-2} K^{-4}	125

References

Alasia F, Cannizzo L, Cerutti G and Marson I 1982 *Metrologia* **18** 221

Alpher R A 1973 *Am. Sci.* **61** 52

Altarev I S, Borisov U V, Borovikova N V, Brandin A B, Egorov A I, Ezhov V F, Ivanov S N, Lobashev V M, Nazarenko V A, Ryabov V L, Serebrov A P and Taldaev R R 1981 *Phys. Lett.* **102B** 13

Alvarez L W and Bloch F 1940 *Phys. Rev.* **57** 111

Alvarger T, Bailey J M, Farley F J M, Kjellman J and Wallin I 1966 *Ark. Fys.* **31** 145

Alvarger T, Farley F J M, Kjellman J and Wallin I 1964 *Phys. Rev. Lett.* **12** 260

Amin S R, Caldwell C D and Lichten W 1981 *Phys. Rev. Lett.* **47** 1234

Anderson W C 1937 *Rev. Sci. Instrum.* **8** 239

—— 1941 *J. Opt. Soc. Am.* **31** 187

Ando M, Bailey D and Hart M 1978 *Acta Crystallogr.* **A34** 484

Andrew E R 1958 *Nuclear Magnetic Resonance* (Cambridge: Cambridge University Press)

Andrews D A and Newton G 1976 *Phys. Rev. Lett.* **37** 1214

Aplin P S 1972 *J. Gen. Rel. Grav.* **3** 111

Ashby N and Allan D W 1979 *Radio Sci.* **14** 649

Aslakson C I 1949 *Nature* **64** 71

——1951 *Trans. Am. Geophys. Union.* **32** 813

Aspden H and Eagles D M 1972 *Phys. Lett.* **41A** 423

Atkinson R d'E 1968 *Phys. Rev.* **170** 1193

Audoin C 1980 in *Metrology and Fundamental Constants: Proceedings of the International School of Physics 'Enrico Fermi', Course 68* ed. Ferro Milone A Giacomo P and Leschuitta S (New York: North-Holland) pp 223–55

Audoin C and Vanier J 1976 *J. Phys. E: Sci. Instrum.* **9** 697

Backlin E and Flemberg H 1936 *Nature* **137** 655

Bahcall J N and Salpeter E E 1965 *Astrophys. J.* **142** 1677

Bachall J N, Sargent W L W and Schmidt M 1967 *Astrophys. J.* **149** L111

Bahcall J N and Schmidt M 1967 *Phys. Rev. Lett.* **19** 129

Bailey F 1843 *Mon. Not. R. Astron. Soc.* **5** 197

Bailey J, Borer K, Combley F, Drumm H, Farley F J M, Field J H, Flegel W, Hattersley P M, Krienen F, Lange F, Picasso E and von Ruden W 1977 *Phys. Lett.* **68B** 191

Bailey J, Borer K, Combley F, Drumm H, Krienen F, Lange F, Picasso E, von Ruden W, Farley F J M, Field J H, Flegel W and Hattersley P M 1977 *Nature* **268** 301

Baird J C, Brandenberger J, Gondaira K-I and Metcalf H 1972 *Phys. Rev.* A5 564

Baird K M, Smith D S and Witford B G 1980 *Opt. Commun.* **31** 367

Baird R C 1967 *Proc. IEEE* **55** 1032

Baklankov E V and Chebotaev V P 1974 *Opt. Commun.* **12** 312

Bardeen J, Cooper L N and Schrieffer J R 1957 *Phys. Rev.* **108** 1175

Barger R L, English T C and West J B 1976 *Proceedings of the Fifth International Conference on Atomic Masses and Fundamental Constants* ed. Sanders J H and Wapstra W H (New York: Plenum) pp 565–8

Barker W A and Glover F N 1955 *Phys. Rev.* **99** 317

Barnes I L, Moore L J, Machlan L A, Murphy T S and Shields W R 1975 *J. Res. NBS* **79A** 727

Barrow J D 1978 *Mon. Not. R. Astron. Soc.* **184** 677

Bartelt J, Courant H, Heller K, Joyce T, Marshak M, Peterson E, Ruddick K, Shupe M, Ayres D S, Dawson J W, Fields T H, May E N and Price L E 1983a *Phys. Rev. Lett.* **50** 651

—— 1983b *Phys. Rev. Lett.* **50** 655

Bartlett D F, Goldhagen P E and Phillips E A 1970 *Phys. Rev.* D2 483

Bass L and Schrödinger E 1955 *Proc. R. Soc.* A232 1

Bates S J and Vinal G W 1914 *J. Am. Chem. Soc.* **36** 916

Batuecas T 1964 *Proceedings of the Second International Conference on Nuclidic Masses* ed. Johnson W H (Vienna: Springer) pp 139ff

—— 1972 in *Atomic Masses and Fundamental Constants 4* ed. Sanders J H and Wapstra A H (New York: Plenum) p 534

Baum W A and Nielson R F 1976 *Astrophys. J.* **209** 319

Bay Z 1972 in *Atomic Masses and Fundamental Constants 4* ed. Sanders J H and Wapstra A H (New York: Plenum) pp 323ff

Bay Z and Luther G G 1971 in *Precision Measurement and Fundamental Constants* (NBS Special Publication 343) ed. Langenberg D N and Taylor B N (Washington DC: US Government Printing Office) p 59

Bay Z, Luther G G and White J A 1972 *Phys. Rev. Lett.* **29** 189

Bay Z and White J A 1972 *Phys. Rev.* D5 796

Beams J W 1964 *Proc. Virginia J. Sci.* **15** 269

—— 1971 *Phys. Tod.* **24** 34

Beams J W, Hulburt C W, Lotz W E Jr and Montague R M Jr 1955 *Rev. Sci. Instrum.* **26** 1181

Bearden J A 1931 *Phys. Rev.* **38** 2089

—— 1938 *Phys. Rev.* **54** 698

—— 1965 *Phys. Rev.* **137B** 455

Bearden J A, Johnson F T and Watts H M 1951 *Phys. Rev.* **81** 70

Bearden J A and Schwarz G 1941 *Phys. Rev.* **59** 934

Bearden J A and Thomsen J S 1957 *Suppl. Nuovo Cimento.* **5** 267

—— 1959 *Am. J. Phys.* **27** 569

Beck G, Bethe H and Reizler G 1931 *Naturwiss* **19** 29

Becker P, Dorenwendt K, Ebeling G, Lauer R, Lucas W, Probst G, Radmacher H-J, Reim G, Seyfried P and Seigert H 1981 *Phys. Rev. Lett.* **46** 1540

Beg M A, Lee B W and Pais A 1964 *Phys. Rev. Lett.* **13** 514

Bekenstein J D 1980 *Comment. Astrophys.* **8** 89

Bergstrand E 1952 in *Recent Developments and Techniques in the Maintenance of Standards* (London: HMSO)

—— 1957 *Ann. Fr. Chronom.* **11** 97

Beringer R and Heald M A 1954 *Phys. Rev.* **95** 1474

Berthold J W III, Jacobs S F and Norton M A 1976 *Appl. Opt.* **15** 1898

Bessel F W 1832 *Pogg. Ann. Phys.* **25** 401

Bethe R A 1938 *Phys. Rev.* **53** 681

Birge R T 1929 *Rev. Mod. Phys.* **1** 1

—— 1932 *Phys. Rev.* **40** 228

—— 1941 *Rep. Prog. Phys.* **8** 90

—— 1945 *Am. J. Phys.* **13** 16

Blackman J A and Series G W 1973 *J. Phys. B: At. Mol. Phys.* **6** 1090

Blake G M 1977 *Mon. Not. R. Astron. Soc.* **178** 41

Blaney T G, Bradley C C, Edwards G J, Jolliffe B W, Knight D J E, Rowley W R C and Shotton K C 1977 *Proc. R. Soc.* **61** 89

Bloch F 1953 *Physica* **19** 821

Bloch F and Jeffries C D 1950 *Phys. Rev.* **80** 305

Bloch F, Nicodemus D B and Staub H R 1949 *Phys. Rev.* **74** 1025

Bodwin G T and Yennie D R 1978 *Phys. Rev.* **43** 267

de Boer H 1984 in *Precision Measurement and Fundamental Constants 2* ed. Taylor B N and Phillips W D (NBS Special Publication 635) (Washington DC: US Government Printing Office) p 561

Bollinger J J, Lundeen S R and Pipkin F M 1984 in *Precision Measurement and Fundamental Constants 2* ed. Taylor B N and Phillips W D (NBS Special Publication 635) (Washington DC: US Government Printing Office) p 141

Bond W N 1930 *Phil. Mag.* **10** 994

—— 1931 *Phil. Mag.* **12** 632

Bondi H and Gold T 1948 *Mon. Not. R. Astron. Soc.* **108** 252

Bondi H and Lyttleton R A 1959 *Nature* **184** 974

Bonse U and Hart M 1965 *Appl. Phys. Lett.* **6** 155

Borie E 1981 *Phys. Rev. Lett.* **47** 568

Bower V E 1971 in *Atomic Masses and Fundamental Constants 4* ed. Sanders J H and Wapstra A H (New York: Plenum) pp 516–20

Bower V E and Davis R S 1980 *J. Res. NBS* **85** 175

Boys C V 1895 *Phil. Trans. R. Soc.* **A1** 1

Boyne H S and Franken P A 1961 *Phys. Rev.* **123** 242

Bragg Sir Lawrence 1947 *J. Sci. Instrum.* **24** 27

—— 1948 *Acta Crystallogr.* **1** 46

Bragg W 1913 *Proc. R. Soc.* **88A** 428

—— *Proc. R. Soc.* **89A** 246, 430

Braginsky V B, Kornieko L S and Polskov S S 1970 *Phys. Rev. Lett.* **33B** 613

Braginsky V B and Manukin A B 1977 in *Measurement of Weak Forces in Physics Experiments* ed. Douglas D H (Chicago: University of Chicago Press)

Braginsky V B and Panov V I 1972 *Sov. Phys.–JETP* **34** 463

Braun C 1897 *Naturwiss. Rundsch.* **12** 273

Braun E, Staben E and von Klitzing K 1980 *P.T.B. Mitteil.* **90** 350

Brecher K 1977 *Phys. Rev. Lett.* **39** 1051, 1236(E)

Breit G and Rabi I 1931 *Phys. Rev.* **38** 2082

Breitenberger E 1971 *Nuovo Cimento. Ser. II* **1B** 1

Briand J P, Tavernier M, Indelicato P, Marrus R and Gould H 1983 *Phys. Rev. Lett.* **50** 832

Brillet A and Hall J A 1979 *Phys. Rev. Lett.* **42** 549

de Broglie L (1954) *Théory Générale des Particules à Spin* 2nd edn (Paris: Gauthier-Villars) p 190

de Broglie R 1940 *La Méchanique Ondulatoire du Photon, Une Nouvelle Théorie de la Lumière* **1** (Paris: Herman) p 39

Brown B C 1969 *Nature* **224** 1189

Brown B C, Masek G E, Maung T, Miller E S, Ruderman H and Vernon V W 1973 *Phys. Rev. Lett.* **30** 763

Brown R A and Pipkin F M 1971 *Bull. Am. Phys. Soc. Ser. II* **16** 85

Bucherer A H 1908 *Verh. Dtsch. Phys. Gesch.* **10** 688

—— 1909 *Ann. Phys., Lpz* **28** 513

Burger T J 1978 *Nature* **271** 402

Burgess G K 1902 *Phys. Rev.* **14** 257

Busch H 1922 *Phys. Ztg.* **23** 428

Byrne J 1983 in *The Neutron and its Applications* Inst. Phys. Conf. Ser. 64 (Bristol: The Institute of Physics) pp 15–20

Byrne J, Morse J, Smith K F, Shaikh F, Greene K and Greene G L 1980 *Phys. Lett.* **92B** 274

Cabrera B 1982 *Phys. Rev. Lett.* **48** 1378

Cabrera B, Taber M, Gardner R and Bourg J 1983 *Phys. Rev. Lett.* **51** 1933

Cage M E, Dziuba R F, Field B F, Williams E R, Garvin S M, Gossard A G, Tsui D C and Wagner R J 1983 *Phys. Rev. Lett.* **51** 1374

Campbell A 1907 *Phil. Trans. R. Soc.* **79A** 428

Canuto V 1981 *Nature* **290** 739

Canuto V and Hsieh S H 1978 *Astrophys. J.* **224** 302

Canuto V and Lodenquai J 1977 *Astrophys. J.* **211** 342

Cap F 1953 *J. Phys. Radiat.* **14** 213

Capptuller H 1964 in *Proceedings of the Second International Conference on Nuclidic Masses* ed. Johnson W H Jr (Vienna: Springer) p 105

Caputo M 1962 *Attiv. R. Accad. Lincei Rend.* **32** 599–615

Carr B J and Rees M J 1979 *Nature* **278** 605

Carrigan R A and Trower W P 1983 *Nature* **305** 673

Carter B 1971 *Phys. Rev. Lett.* **26** 33

—— 1974 *Phys. Rev. Lett.* **33** 558

Casperson D E, Crane T W, Denison A B, Egan P O, Hughes V W, Mariam F G, Orth H. Reist H W, Souder P A, Stambaugh R D, Thompson P A and zu Pulitz G 1977 *Phys. Rev. Lett.* **38** 956

Caswell W E, Lepage G P and Sapirstein J 1977 *Phys. Rev. Lett.* **38** 488

Cavendish H 1798 *Phil Trans. R. Soc.* **88** 469

—— 1897 *The Electrical Researches of the Honourable Henry Cavendish, F R S* ed. Maxwell J C (Cambridge: Cambridge University Press; reprinted 1967, London: Frank Cass) pp 104–13

CCDM 1973 *P.-V. Seances Com. Int. Poids Mes. Ser.* 2 **41** 113

Cerutti G, Carnizzo L, Sakuma A and Hostache J 1974 *VDI Berichte* 212 49

Charpak G, Farley F J M, Garwin R L, Muller T, Sens J C, Telegdi V L and Zichi A 1961 *Phys. Rev. Lett.* **6** 128

Charpak G, Farley F J M, Garwin R L, Muller T, Sens J C and Zichi A 1965 *Nuovo Cimento* **37** 1241

Chitre S M and Pal Y 1968 *Phys. Rev. Lett.* **20** 278

Christensen C J, Nielsen A, Bafasen A, Brown W K and Rustad B M 1972 *Phys. Rev.* D**5** 1628

Christmas P and Cross P 1978 *Metrologia* **14** 147

Chu D Y 1939 *Phys. Rev.* **55** 175

Classen J 1908 *Phys. Ztg.* **9** 762

—— *Verh. Dtsch. Phys. Gesch.* **10** 700

Clothier W G 1965 *Metrologia* **1** 181

Cochran G D and Franken P A 1968 *Bull. Am. Phys. Soc.* **13** 1379

Cohen E R 1952 *Phys. Rev.* **88** 353

—— 1953 *Rev. Mod. Phys.* **25** 709

Cohen E R, Crowe K M and DuMond J W M 1957 *Fundamental Constants of Physics* (New York: Wiley Interscience)

Cohen E R and DuMond J W M 1965 *Rev. Mod. Phys.* **37** 537–94

Cohen E R and Taylor B N 1973 *J. Chem. Ref. Data* **2** 663

—— 1984 *Review of the 'best values'* (to be published)

Cohen E R and Wapstra A H 1983 *Nucl. Instrum. Methods Phys. Res.* **211** 153

Cohen S A and King J G 1969 *Nature* **222** 1158

Cohen V W, Corngold N R and Ramsey N F 1956 *Phys. Rev.* **104** 283

Colclough A R 1979 *Proc. R. Soc.* **365A** 349

—— 1981 *Precision Measurement and Fundamental Constants 2* ed. Taylor B N and Phillips W D (NBS Special Publication 635) (Washington DC: US Government Printing Office) p 263

Colclough A R, Quinn T J and Chandler T D 1979 *Proc. R. Soc.* **368A** 1732

Collington D J, Dellis A N, Sanders J H and Turberfield K C 1955 *Phys. Rev.* **99** 1622

Combley F, Farley F J M, Field J H and Picasso E 1979 *Phys. Rev. Lett.* **42** 1383

Condon E U and Shortley C H 1935 *The Theory of Atomic Spectra* (Cambridge: Cambridge University Press)

Cook A H 1957 *Nature* **139** 323

—— 1967 *Phil. Trans. R. Soc.* **261** 211

—— 1968 *Contemp. Phys.* **9** 227

—— 1970 in *Precision Measurement and Fundamental Constants* ed. Taylor B N and Langenberg D N (NBS Special Publication 343) (Washington DC: US Government Printing Office) pp 457ff

—— 1972 *Rep. Prog. Phys.* **35** 463

Cooper P S, Alguard M J, Ehrlich R D, Hughes V W, Kobayakawa H, Ladish J S, Lubell M S, Sasao N, Schuler K P, Souder P A, Coward D H, Miller R H, Prescott C Y, Sherden D J, Sinclair C K, Baum G, Raith W and Kondo K 1979 *Phys. Rev. Lett.* **42** 1386

Cosens B L 1968 *Phys. Rev.* **173** 49

Cosens B L and Vorburger T V 1970 *Phys. Rev.* **A2** 16

Coulomb C A 1788 *Acad. R. Soc. Paris Histoire* 1785 pp 569, 579 (Transl. in part into English in Migie W F 1965 *A Source Book in Physics* (Cambridge MA: Harvard

University Press) pp 408–20)

Cranshaw T E and Schiffer J P 1960 *Nature* **185** 653

Craig D N, Hoffman J I, Law C A and Hamer W J 1960 *J. Res. NBS* A**64** 381

Creer K M 1965 *Nature* **205** 539

Crovini L and Actis A 1978 *Metrologia* **14** 69

Csillag L 1968 *Acta Phys. Acad. Sci. Hung.* **24** 1 (also in *Atomic Masses and Fundamental Constants 4* ed. Sanders J H and Wapstra A H (New York: Plenum) pp 411ff

Curott D R F 1965 *PhD Thesis*, Princeton University (unpublished)

Curtis I, Morgan I G, Hart M and Miln A 1971 in *Precision Measurement and Fundamental Constants* ed. Taylor B N and Langenberg D N L (NBS Special Publication 343) (Washington DC: US Government Printing Office) p 285

Cushing J T 1981 *Am. J. Phys.* **49** 1133

Cvitanov P and Kinoshita T 1974 *Phys. Rev.* D**10** 4007

Davis R S and Bower V 1979 in *Atomic Masses and Fundamental Constants 6* ed. Nolan J A Jr and Benenson W (New York: Plenum)

Dayhoff E S, Triebwasser S and Lamb W E Jr 1953 *Phys. Rev.* **89** 106

Dearborn D S and Schramm D N 1974 *Nature* **247** 441

Deslattes R D 1969 *Appl. Phys. Lett.* **15** 386

—— 1970 in *Precision Measurements and Fundamental Constants* ed. Taylor B N and Langenberg D N (NBS Special Publication 343) (Washington DC: US Government Printing Office) p 279

—— 1980a in *Metrology and Fundamental Constants, Proceedings of the 68th International School 'Enrico Fermi'* ed. Ferro-Milone A, Giacomo P and Leschuitta S (Amsterdam: North-Holland) pp 38ff

—— 1980b *Ann. Rev. Phys. Chem.* **31** 435

Deslattes R D, Henins A, Bowman H A, Schoonover R M, Carroll C L, Barnes I C, Machlan L A, Moore L J and Shields W R 1970 *Phys. Rev. Lett.* **33** 463

Dicke R H 1961 *Nature* **192** 440

—— 1962 *Phys. Rev.* **125** 2163

—— 1963 in *Relativity, Groups and Topology* ed. C and B De Witt (London: Gordon and Breach) pp 165–313.

Dirac P A M 1931 *Proc. R. Soc.* A**133** 60

—— 1937 *Nature* **139** 323

—— 1938 *Proc. R. Soc.* A**165** 199

—— 1979 *Proc. R. Soc.* A**365** 19

Dixit M S, Anderson H L, Hargrove C K, Kessler D, Mes H and Thompson D M 1971 *Phys. Rev. Lett.* **27** 878

Dorsay N E 1944 *Trans. Am. Phil. Soc.* **34** (Part 1) 1

Drake C L and Delauze H 1965 *Ann. Inst. Oceanogr.* **46** 7

Drake G W F, Goldman S P and van Wyngaarden A 1979 *Phys. Rev.* A**20** 1299

Dress W B, Miller P D, Pendlebury J M, Perrin P and Ramsey N F 1977 *Phys. Rev.* D**7** 3417

Drever R W P 1961 *Phil. Mag.* **6** 683

—— 1977 *Q. J. R. Astron. Soc.* **18** 9

Drinkwater J W, Richardson O and Williams W E 1940 *Proc. R. Soc.* A**174** 164

Driscoll R L and Bender P L 1958 *Phys. Rev. Lett.* **1** 413

DuMond J W M 1959 *Proc. Nat. Acad. Sci.* **45** 1052

Duane W, Palmer H and Yeh C S 1921 *J. Opt. Soc. Am.* **5** 376

Dunnington F G 1933 *Phys. Rev.* **43** 404

—— 1937 *Phys. Rev.* **52** 475

Dunnington F G, Hemenway C L and Rough J D 1954 *Phys. Rev.* **94** 592

Dylla H F and King J G 1973 *Phys. Rev.* **A7** 1224

Dymond J H and Smith E B 1969 *The Virial Coefficients of Gases* (Oxford: Clarendon)

Dyson F J 1967 *Phys. Rev. Lett.* **19** 1291

—— 1972 in *Aspects of Quantum Theory* ed. Salam A and Wigner E P (Cambridge: Cambridge University Press) pp 213–36

Eddington A S 1929 *Proc. R. Soc.* **A122** 358

Edelstein W A, Hough J, Pugh J R and Martin W 1978 *J. Phys. E: Sci. Instrum.* **11** 701

Elnekave N 1965 *Proc. Verb. CCE Doc.* **24**

Elnekave N and Fau A 1981 in *Precision Measurement and Fundamental Constants 2* ed. Taylor B N and Phillips W D (NBS Special Publication 635) (Washington DC: US Government Printing Office) p 465

Eötvös R V 1896 *Ann. Phys., Lpz* **59** 354

Eötvös R V, Pekar D and Feteke E 1922 *Ann. Phys., Lpz* **68** 11

Erickson G W 1967 *AEC Research and Development Report UCD-CNL88* contract no AT(11-1) Gen. 10

—— 1971 *Phys. Rev. Lett.* **27** 780

Essen L and Froome K D 1951 *Proc. Phys. Soc.* **B64** 862

Essen L and Gordon-Smith A C 1948 *Proc. R. Soc.* **A194** 348

Essen L, Hope E G, Bangham M J and Donaldson R W 1971 *Nature* **229** 110

Essen L and Parry J V L 1955 *Nature* **176** 280

Evenson K M, Wells J S, Peterson F R, Danielson B L, Day G W, Barger R L and Hall J L 1972 *Phys. Rev. Lett.* **29** 1349

Everitt C W F 1978 *Nature* **272** 737

Fabjan C W and Pipkin F M 1972 *Phys. Rev.* **A6** 556

Fabjan C W, Pipkin F M and Silverman M 1971 *Phys. Rev. Lett.* **26** 347

Facy L and Pontikis C 1970 *C.R. Acad. Sci., Paris* **270** 15

Feinberg G 1969 *Science* **166** 552

Feinberg G and Goldhaber M 1959 *Proc. Nat. Acad. Sci.* **45** 130

Field J C and Series G W 1959 *Proc. Symposium on Interferometry at the NPL* (London: HMSO)

Florman E F 1955 *J. Res. NBS* **54** 335

Frank J and Hertz G 1914 *Verh. Dtsch. Phys. Gesch.* **16** 10

Franken P A and Ampulski G W 1971 *Phys. Rev. Lett.* **26** 115

Franken P A and Liebes S 1956a *Phys. Rev.* **104** 1197

—— 1956b *Phys. Rev.* **116** 633

Fraser L J, Carlson E R and Hughes V W 1968 *Bull. Am. Phys. Soc.* **13** 636

Froome K D 1952 *Proc. R. Soc.* **A213** 123

—— 1958 *Proc. R. Soc.* **A247** 109

Froome K D and Essen L 1969 *The Velocity of Light and Radio Waves* (London: Academic)

Fried H M and Yennie D R 1960 *Phys. Rev. Lett.* **4** 583

Frisch O R and Stern E 1933 *Z. Phys.* **85** 4
Fujii Y 1971 *Nat. Phys. Sci.* **234** 5
Fystrom D O 1970 *Phys. Rev. Lett.* **25** 1469
Fystrom D O, Petley B W and Taylor B N 1970 in *Precision Measurement and Fundamental Constants* ed. Taylor B N and Langenberg D N (NBS Special Publication 343) (Washington DC: US Government Printing Office) p 187

Gahler R, Kalus J and Mampe W 1982 *Phys. Rev.* **D25** 2887
Gallinaro G, Marinelli M and Morpurgo G 1977 *Phys. Rev. Lett.* **38** 1255
Gallinaro G and Morpurgo G 1966 *Phys. Rev. Lett.* **23** 609
Gallop J C and Petley B W 1976 *J. Phys. E: Sci. Instrum.* **9** 417
—— 1982 *Nature* **303** 53
Gamov G 1967 *Phys. Rev. Lett.* **19** 759
Garcia J D and Mack J E 1965 *J. Opt. Soc. Am.* **55** 654
Gardner J H and Purcell E M 1951 *Phys. Rev.* **83** 996
Gartner G and Klempt E 1978 *Z. Phys.* **A287** 1
Gibbons G W and Whiting B F 1981 *Nature* **291** 636
Gidley D W, Rich A, Zitzewitz P W and Paul D A L 1978 *Phys. Rev. Lett.* **40** 737
Gidley D W and Zitzewitz P W 1978 *Phys. Lett.* **69A** 97
Gillies G T 1982 *Rapport BPM*-82/9 (Sèvres: BIPM)
Gittus J H 1976 *Proc. R. Soc.* **A348** 95
Glashow S L and Nanopoulos D V 1979 *Nature* **281** 464
Gold T 1968 *Nature* **218** 731
Goldenberg H M, Kleppner D and Ramsey N F 1960 *Phys. Rev. Lett.* **5** 361
Goldhaber A S and Nieto M M 1968 *Phys. Rev. Lett.* **21** 576
—— 1971 *Rev. Mod. Phys.* **43** 277
Goldsmith J E M, Weber E W and Hänsch T W 1978 *Phys. Rev. Lett.* **41** 1525
Gould H and Marrus R 1978 *Phys. Rev. Lett.* **41** 1457
Graff G, Kalinowsky H and Traut J 1980 *Z. Phys.* **A297** 35
Granger J and Ford G W 1972 *Phys. Rev. Lett.* **28** 1479
Greene G L 1977 in *Fundamental Physics with Reactor Neutrons and Neutrinos* Inst. Phys. Conf. Ser. 42 ed. von Egidy T (Bristol: The Institute of Physics) pp 5–14
——1981 *Metrologia* **17** 83
Greene G L, Ramsey N F, Mampe W, Pendlebury J M, Smith K, Dress W B, Miller P D and Perrin P 1979 *Phys. Rev.* **D20** 2139
Grodkins L, Engelberg D and Bertozzi W 1961 *Bull. Am. Phys. Soc.* **6** 63
Grotch H and Hegstrom R A 1971 *Phys. Rev.* **A4** 59
Grotch H and Yennie D R 1969 *Rev. Mod. Phys.* **41** 350
Gubler H, Munch S and Staub H H 1974 *Helv. Phys. Acta* **46** 722
Gutowsky H S and McClure R E 1951 *Phys. Rev.* **81** 276

Hafele J C and Keating R E 1972 *Science* **177** 166, 168
Hammer W 1914 *Ann. Phys., Lpz.* **43** 653
Hamon B V 1954 *J. Sci. Instrum.* **31** 450
Handschy M A 1982 *Am. J. Phys.* **50** 987
Hänsch T W, Neyfeh M H, Lee S A, Curry T W and Shahin A S 1974 *Phys. Rev. Lett.* **32** 1336

Hänsch T W, Shahin I S and Schawlow A L 1972 *Nature* **235** 63

Hara K, Shista F and Kubota T 1981 *Precision Measurement and Fundamental Constants 2* ed. Taylor B N and Phillips W D (NBS Special Publication 635) (Washington DC: US Government Printing Office) p 479

Harrick N J, Barnes R G, Bray P J and Ramsey N F 1953 *Phys. Rev.* **90** 260

Harrington E 1916 *Phys. Rev.* **8** 738

Harris F K 1972 *Proc. Verb. CCE* 13th Session, Annex E6.

Harrison P W and Rayner G H 1967 *Metrologia* **3** 1

Hart M 1969 *Proc. R. Soc.* A**309** 289

Hartree D R 1926 *Proc. Camb. Phil. Soc.* **24** 89

Hellings R W, Adams P J, Anderson J D, Keesey M S, Lau E L and Standish E M 1983 *Phys. Rev. Lett.* **51** 1609

Hellwig H 1970 *Metrologia* **6** 56

Hellwig H, Jarvis S, Halford D and Bell H E 1973 *Metrologia* **9** 107

Helmer R G, Gehrke R J and Greenwood R C 1980 *Nucl. Instrum. Methods* **166** 547

Henins I and Bearden J A 1964 *Phys. Rev.* **135A** 890

Herzberg G 1956 *Proc. Phys. Soc.* A**234** 516

Heyl P R 1924 *Sci. Pap. NBS* **19** 307–24

—— 1939 *J. Res. NBS* **5** 1234

Heyl P R and Chrzanowski P 1942 *J. Res. NBS* **29** 1

Hillas A M and Cranshaw T E 1959 *Nature* **184** 892

—— 1960 *Nature* **186** 459

Hindman J C 1966 *J. Chem. Phys.* **44** 4582

Hli F Ba 1959 *J. Br. Interplanet. Sci.* **17–18** 540

Hocker L O, Javin A, Ramachandra Rao D, Frenkel L and Sullivan T 1967 *Appl. Phys. Lett.* **10** 147

Hodges C L, Abrams P, Baden A W, Bland R W, Joyce D C, Royer R P, Walters W, Wilson E G, Wong P G Y and Young K C 1981 *Phys. Rev. Lett.* **47** 1651

Hollweg J W 1974 *Phys. Rev. Lett.* **32** 961

Honerjäger R and Klein E 1962 *Z. Phys.* **169** 32

Hopper V D and Laby T H 1941 *Proc. R. Soc.* A**178** 243

Houston W V 1927 *Phys. Rev.* **30** 608

—— 1937 *Phys. Rev.* **52** 75

Hoyle F and Narlikar J V 1972 *Mon. Not. R. Astron. Soc.* **155** 323

Hughes V W 1957 *Phys. Rev.* **129** 2566

—— 1981 in *Precision Measurement and Fundamental Constants 2* ed. Taylor B N and Phillips W D (NBS Special Publication 635) (Washington DC: US Government Printing Office) p 237

Hughes V W, Robinson H and Beltran-Lopez V 1960 *Phys. Rev. Lett.* **4** 342

Hutchison C A and Johnston H L 1940 *J. Am. Chem. Soc.* **62** 3165

Hutchison D A 1944 *Phys. Rev.* **66** 144

Huttel A 1940 *Ann. Phys., Lpz* **37** 365

Inouye T, Kitsunezaki T, Senda O and Ando K 1971 in *Precision Measurement and Fundamental Constants* ed. Taylor B N and Langenberg D N (NBS Special Publication 343) (Washington DC: US Government Printing Office) p 465

Ishida Y, Fukushima I and Suetsugu T 1937 *Sci. Pap. Inst. Phys. Chem. Tokyo* **32** 57

Ives H E and Stilwell G R 1932 *J. Opt. Soc. Am.* **28** 215

—— 1941 J. Opt. Soc. Am. **31** 369

Jacobs R R, Lea K R and Lambe W E Jr 1971 *Phys. Rev.* **A3** 884

Jaseja T S, Javan A, Murray J and Townes C H 1964 *Phys. Rev.* **A133** 1221

Johnson D P 1974 *J. Res. NBS* **78** 41

Johnston H L and Hutchison D A 1942 *Phys. Rev.* **62** 32

Johnstone-Stoney G 1881 *Phil. Mag.* **11** 381

von Jolly A 1873 *Pop. Sci. Mon.* **21** 565

Jones G R Jr, Ritter R C and Gillies G T 1982 *Metrologia* **18** 209

Jones L W 1977 *Rev. Mod. Phys.* **49** 717

Jones R V and Richards J S 1973 *J. Phys. E: Sci. Instrum.* **7** 589

Joos G 1930 *Ann. Phys., Lpz.* **7** 385

Josephson B D 1962 *Phys. Lett.* **1** 251

Joyce D C, Abrams P C, Bland R W, Johnson R T, Lindgren M A, Savage M H, Scholz M H, Young B A and Hodges C L 1983 *Phys. Rev. Lett.* **51** 731

Karagioz O V, Ismaylov V P, Agafonov N L, Kocheryan E G, Tarakanov Yu A 1976 *Izv. Acad. Sci. USSR Phys. Solid Earth* **12** 351

Kaufman S L, Lamb W E Jr, Lea J R and Leventhal M 1971 *Phys. Rev.* **A4** 2128

Kaufman W 1897 *Ann. Phys., Lpz.* **61** 544

—— 1898 *Ann. Phys., Lpz* **62** 596

Keiser G M and Faller J E 1981 in *Proceedings of the Second Marcel Grossman Meeting on Recent Developments in General Relativity* ed. Ruffini R (Amsterdam: North-Holland)

Keiser P T, Faller J E and McHagan K H 1981 *Precision Measurement and Fundamental Constants 2* ed. Taylor B N and Phillips W D (NBS Special Publication 635) (Washington DC: US Government Printing Office) p 639

Kellog J M B, Rabi I I, Ramsey N F and Zacharias J R 1939 *Phys. Rev.* **55** 318

—— 1940 *Phys. Rev.* **57** 677

Kennedy R J and Thorndike E M 1932 *Phys. Rev.* **42** 400

Kessler E G Jr 1973 *Phys. Rev.* **A7** 408

Kessler E G, Deslattes R D, Sauder W C and Henins A 1979 in *Neutron Capture Gamma-Ray Spectroscopy* ed. Chrein R E and Kane W R (New York: Plenum) p 427

Kibble B P 1976 in *Atomic Masses and Fundamental Constants 2* ed. Sanders J H and Wapstra A H (New York: Plenum)

—— 1982 in *Precision Measurement and Fundamental Constants 2* ed. Taylor B N and Phillips W D (NBS Special Publication 635) (Washington DC: US Government Printing Office) p 461

Kibble B P and Hunt J G 1979 *Metrologia* **15** 5

Kibble B P, Rowley W R C, Series G W and Shawyer R E 1973 *J. Phys. B: At. Mol. Phys.* **6** 1079

Kim Y M and Fedele P D 1982 *Phys. Rev. Lett.* **48** 403

Kind D H, Lucas W, Peier D and Schulz B 1983 *IEEE Trans. Instrum. Meas.* **IM-32** 8

King J G 1960 *Phys. Rev. Lett.* **5** 562

Kinoshita T 1981 in *Quantum Metrology and Fundamental Constants* ed. Cutler P and Lucas A A (New York: Plenum) p 313

Kinsler L E and Houston W V 1934 *Phys. Rev.* **46** 523

Kirchner F 1924 *Phys. Ztg.* **25** 302

—— 1929 *Phys. Ztg.* **30** 773

—— 1932 *Ann. Phys., Lpz.* **12** 503

Kirchner F and Wilhelmy W 1957 *Suppl. Nuovo Cimento* **6** 246

Klein E 1968 *Z. Phys.* **208** 28

Klein H H, Klempt G and Storm L 1979 *Metrologia* **15** 143

Kleinpoppen H 1961 *Z. Phys.* **164** 174

von Klitzing K, Dorda G and Pepper M 1980 *Phys. Rev. Lett.* **45** 494

Koch W F 1979 in *Atomic Masses and Fundamental Constants 6* ed. Nolan J A Jr and Benenson W (New York: Plenum)

Koch W F and Diehl W C 1976 *Talanta* **23** 509

Koenig S H, Prodell A G and Kusch P 1952 *Phys. Rev.* **88** 191

Kohlrausch F and Kohlrausch W 1884 *Wied. Ann.* **27** 1

Kponou A, Hughes V W, Johnson C E, Lewis S A and Pichanick F M J 1971 *Phys. Rev. Lett.* **26** 1613

Kugel H W, Leventhal M, Patel G N, Wood O R II and Murnick D E 1975 *Phys. Rev. Lett.* **35** 647

Kugel H W and Murnick D E 1977 *Rep. Prog. Phys.* **40** 297

Kuhnen F and Furtwangler P 1906 *Veroeff. Konigl. Preuss. Geod. Inst.* **27** 1

Lamb W E Jr 1941 *Phys. Rev.* **60** 817

Lamb W E Jr and Retherford R C 1950 *Phys. Rev.* **79** 549

—— 1951 *Phys. Rev.* **81** 222

—— 1952 *Phys. Rev.* **86** 1014

Lamb W E Jr and Skinner M 1950 *Phys. Rev.* **78** 539

Lambe E B D 1959 *PhD Thesis* Princeton University (unpublished)

Lambe E B D 1969 in *Polarization Matière et Rayonnement* (Paris: Société Française de Physique) p 441

Lampard D G 1957 *Proc. IEEE* **104C** 271

Lampard D G and Cutkosky R D 1960 *Proc. IEEE* **107C** 112

LaRue G S, Fairbank W M and Phillips J D 1979 *Phys. Rev. Lett.* **42** 142

LaRue G S, Phillips J D and Fairbank W M 1981 *Phys. Rev. Lett.* **46** 967

Lawrence E O 1926 *Phys. Rev.* **28** 947

Layzer A J 1960 *Phys. Rev. Lett.* **4** 580

Lea K 1971 in *Atomic Masses and Fundamental Constants 4* ed. Sanders J H and Wapstra A H (New York: Plenum)

Learned J, Reines F and Soni A 1979 *Phys. Rev. Lett.* **43** 907, 1626

Lee K P, Shields F D and Wiley W J 1965 *J. Acoust. Soc. Am.* **37** 724

Lee S A, Wallenstein R and Hänsch T W 1975 *Phys. Rev. Lett.* **35** 1262

LeFloch A, Lenormand J M, Jezequel G and Le Naour R 1981 *Opt. Lett.* **6** 48

Lenz F 1951 *Phys. Rev.* **82** 554

Lepage G P 1977 *Phys. Rev.* **16A** 863

Lepage G P, Yennie D R and Erickson G W 1981 *Phys. Rev. Lett.* **47** 1640

Levine M and Roskies R 1976 *Phys. Rev.* **D14** 2191

J-M Levy-Leblond 1979 in *Problems in the Foundations of Physics LXXII Corso* (Societa Italiana di Fisica Bologna) pp 237–63

Lewis G N and Adams E Q 1914 *Phys. Rev.* **3** 92

Liebes S and Franken P A 1959 *Phys. Rev.* **116** 633

Lipworth E and Novick R 1957 *Phys. Rev.* **108** 1434

Lloyd M G 1909 *Terr. Mag. Atmos. Elect.* **14** 67–71

Long D R 1976 *Nature* **260** 417

—— 1980 *Nuovo Cimento.* **55B** 252

—— 1981 in *Precision Measurement and Fundamental Constants 2* ed. Taylor B N and Phillips W D (NBS Special Publication 635) (Washington DC: US Government Printing Office) p 587

Lowenthal D D 1973 *Phys. Rev.* **D8** 2349

Lawry R A, Towler W R, Parker H M, Kulthau A R and Beams J W 1972 in *Atomic Masses and Fundamental Constants 4* ed. Sanders J H and Wapstra A H (New York: Plenum) pp 521–7

Lukirsky P and Prilezeau S 1928 *Z. Phys.* **49** 236

Lundeen S R and Pipkin F M 1975 *Phys. Rev. Lett.* **34** 1368

—— 1981 *Phys. Rev. Lett.* **46** 232

Luther G G and Towler W R 1982 *Phys. Rev. Lett.* **48** 121

Luther G G, Towler W R, Deslattes R D, Lowry R A and Beams J W 1976 in *Atomic Masses and Fundamental Constants 4* ed. Sanders J H and Wapstra A H (New York: Plenum) p 592

Luxon J L and Rich A 1974 *Phys. Rev. Lett.* **29** 665

Lyttleton R A and Bondi H 1959 *Proc. R. Soc.* **252** 313

McDonald D G, Risley A S, Cupp J D and Evenson K M 1974 *Appl. Phys. Lett.* **24** 335

McElhinny M W, Taylor S R and Stevenson D J 1978 *Nature* **271** 316

McGlashan M L 1971 *Physico-chemical Quantities and Units* (London: Royal Institute of Chemistry)

Mackay A L 1977 *The Harvest of a Quiet Eye* (Bristol: The Institute of Physics) p 31

Mackenzie I C C 1954 *Ord. Surv. Print. Off. Pap. No 19* (London: HMSO)

McSkimin H J 1953 *J. Appl. Phys.* **24** 988

McWeeny R 1973 *Nature* **243** 196

Mader D L, Leventhal M and Lamb W E Jr 1971 *Phys. Rev.* **A3** 1832

Mamyrin B A, Aruyev N N and Alekseenko S A 1972 *Zh. Eksp. Teor. Fiz.* **63** 3 (Engl. Transl. 1977 *Sov. Phys.–JETP* **36** 1)

Mamyrin B A and Frantsuzov A A 1964 *Dokl. Akad. Nauk.* **159** 777 (Engl. Transl. 1965 *Sov. Phys.–Dokl.* **9** 1082)

Mansfield V N 1976 *Nature* **261** 560

Mariam F G, Beer W, Bolton P R, Egan P O, Gardner C J, Hughes V W, Lu D C, Souder P A, Orth H, Vetter J, Moser U and zu Pulitz G 1982 *Phys. Rev. Lett.* **49** 993

Marinelli M and Morpurgo G 1980 *Phys. Lett.* **94B** 433

—— 1982 *Phys. Rep.* **85** 162

Marion J B and Winkler H 1967 *Phys. Rev.* **156** 1062

Martin J C and Quinn T J 1981 in *Precision Measurement and Fundamental Constants 2* ed. Taylor B N and Phillips W D (NBS Special Publication 635) (Washington DC: US Government Printing Office) p 291

Masui T 1971 in *Precision Measurements and Fundamental Constants* ed. Taylor B N and Langenberg D N (NBS Special Publication 343) (Washington DC: US Government Printing Office) pp 83–5

Matrarrese L M and Evenson K M 1970 *Appl. Phys. Lett.* **17** 8

Maxwell J C 1883 *A Treatise on Electricity and Magnetism* (Oxford: Oxford University Press) vol 1 (3rd edn (1891) reprinted 1954 (New York: Dover)) pp 80–86

Merrill P W 1917 *Astrophys. J.* **46** 357

Michelson A A and Morley E W 1887 *Am. J. Sci.*

Michelson A A, Pease F G and Pearson F 1935 *Astrophys. J.* **82** 26

Millikan R A 1910 *Phil. Mag.* **19** 223
—— 1916 *Phys. Rev.* **7** 355
—— 1923 *Phys. Rev.* **22** 1
—— 1935 *Electrons (+ and −), Protons, Neutrons and Cosmic Rays* (Chicago: University of Chicago Press)
Mills A P Jr and Berman G H 1975 *Phys. Rev. Lett.* **34** 246
Mohr P J 1975a *Phys. Rev. Lett.* **34** 1050
—— 1975b *Phys. Rev. Lett.* **42** 1575
Morris D 1971 *Metrologia* **7** 162
Moss G E, Miller L R and Forward R L 1971 *Appl. Opt.* **10** 2495
Mott N F and Massey H S W 1965 *Theory of Atomic Collisions* 3rd edn (Oxford: Clarendon) p 224
Mungall A G 1971 *Metrologia* **7** 49
Myint T D, Kleppner D, Ramsey N F and Robinson H G 1966 *Phys. Rev. Lett.* **17** 405

Narasimham M A and Strombotne R L 1971 *Phys. Rev.* **A4** 14
Newell G F 1950 *Phys. Rev.* **80** 476
Newman D, Ford G W, Rich A and Sweetman E 1978 *Phys. Rev. Lett.* **40** 1355
Newton G, Andrews D A and Unsworth P J 1979 *Phil. Trans. R. Soc.* **290** 373
Newton I 1687 *Philosophiae Naturalis Principia Mathematica* (London)
Newton R R 1975 *Mon. Not. R. Astron. Soc.* **169** 331
Nieto M M and Goldman T 1980 *Phys. Lett.* **79A** 449
Nilsson A 1953 *Ark. Phys.* **6** 513
Noerdlinger P D 1973 *Phys. Rev. Lett.* **30** 761
Nordvedt K 1968 *Phys. Rev.* **169** 1017

Ogorodnikov D D, Samoilov I M and Solntsev A M 1979 *Zh. Eksp. Teor. Fiz.* **76** 1881 (Engl. Transl. 1979 *Sov. Phys.–JETP* **49** 953)
O'Hanlon J 1972 *Phys. Rev. Lett.* **29** 137
Ohlin P 1940 *Ark. Math. Astron. Phys.* **B27** no 10 (also (1942), **A29** no 3; (1944), **B29** no 4; (1946) **A31** no 9)
Olpin A R 1930 *Phys. Rev.* **36** 251
Olsen P T, Cage M E, Phillips W D and Williams E R 1980 *IEEE Trans. Instrum. Meas.* **IM29** 234
Olsen P T and Driscoll R L 1972 in *Atomic Masses and Fundamental Constants 4* ed. Sanders J H and Wapstra A H (New York: Plenum) p 471
Olsen P T, Phillips W D and Williams E R 1981 in *Precision Measurement and Fundamental Constants 2* ed. Taylor B N and Phillips W D (NBS Special Publication 635) (Washington DC: US Government Printing Office) p 475
Olsen P T and Williams E R 1974 *IEEE Trans. Instrum. Meas.* **IM-23** 302
Ono A 1981 in *Precision Measurement and Fundamental Constants 2* ed. Taylor B N and Phillips W D (NBS Special Publication 635) (Washington DC: US Government Printing Office) p 299
Otsuka J, Sekine F, Nakamura A, Hagiwara H and Tanaka K 1982 *Bull. P.M.E.(T.I.T)* 49 p 39

Pagel B E J 1977 *Mon. Not. R. Astron. Soc.* **179** 81
Park D and Williams E 1971 *Phys. Rev. Lett.* **26** 1393
Pegg D T 1977 *Nature* **267** 408
Pendlebury J M, Smith K and Unsworth P 1979 *Rev. Sci. Instrum.* **50** 535

Peres A 1967 *Phys. Rev. Lett.* **19** 1293

Perles J 1928 *Naturwiss.* **16** 1094

Perry C T and Chaffee E L 1930 *Phys. Rev.* **36** 904

Peters H E 1971 *Proceedings of the First Frequency Standards and Metrology Seminar* (Quebec) 29

Petley B W 1979 in *Encycl. Dictionary Phys. Suppl.* Vol 4 ed. Thewls J (Oxford: Pergamon) pp 354–7

—— 1980 in *Metrology and Fundamental Constants* ed. Ferro-Milone A, Giacomo P and Leschuitta S (Amsterdam: North-Holland) pp 358–463

—— 1983 in *Quantum Metrology and Fundamental Constants* ed. Cutler P and Lucas A (New York: Plenum) pp 293–311, 333–51, 383–401

Petley B W and Morris K 1965 *J. Sci. Instrum.* **42** 492

—— 1969 *Phys. Lett.* **29A** 289

—— 1974 *J. Phys. A: Math., Nucl. Gen.* **7** 167

—— 1979 *Nature* **279** 141

Petley B W, Morris K and Shawyer R W 1980 *J. Phys. B: At. Mol. Phys.* **13** 3099

Phillips W D, Cooke W E and Kleppner D 1977 *Metrologia* **13** 179

Piccard A and Kessler E 1925 *Arch. Sci. Phys. Nat.* **7** 340

Plimpton S J and Lawton W E 1936 *Phys. Rev.* **50** 1066

Poincaré 1916 *Science and Method* (London: Nelson) p 237

Pontikis C 1972 *C.R. Acad. Sci., Paris* **B274** 437

Potter H H 1923 *Phys. Rev.* **D8** 2731

Pound R V and Rebka G A 1959 *Phys. Rev. Lett.* **3** 439

Poynting J H 1879 *Proc. R. Soc.* **28** 2

—— 1891 *Nature* **44** 165

Poynting J H and Phillips P 1905 *Proc. R. Soc.* **A76** 445

Primack J R and Sher M A 1980 *Nature* **288** 680

Quinn T J, Colclough A R and Chandler T R D 1976 *Phil. Trans. R. Soc.* **283** 367

Quinn T J and Martin J S 1982 in *Precision Measurement and Fundamental Constants 2* ed. Taylor B N and Phillips W D (NBS Special Publication 635) (Washington DC: US Government Printing Office)

Rabi I I Zacharias J R, Millman S and Kusch P 1938 *Phys. Rev.* **53** 318

Ramsey N F 1950 *Phys. Rev.* **78** 699

—— 1956 *Molecular Beams* (Oxford: Clarendon)

—— 1982 in *The Neutron and its Applications* 1982 (Inst. Phys. Conf. Ser. 64) (Bristol: The Institute of Physics) pp 5–14

Rapport S, Bradt H and Mayer W 1971 *Nature* **229** 40

Lord Rayleigh 1876 *Presidential Address to the British Association*

Lord Rayleigh and Mrs Sidgwick 1884 *Phil. Trans.* **175** 411

Rayner G H 1967 *Metrologia* **3** 12

Reasenberg R D and Shapiro H 1978 in *On the Measurement of Variations of the Gravitational Constant* (Gainsville: University of Florida Press)

Redei L B 1966 *Phys. Rev.* **145** 999

—— 1967 *Phys. Rev.* **162** 1299

Reich F 1838 *Phil. Mag.* **12** 283

—— 1852 *Ann. Phys., Lpz.* **161** 189

Renner J 1935 *Mat. Thermeszettudomanyi Ert.* **53** 542

Renner Ya 1974 in *Determination of gravity constant and measurement of certain fine gravity effects* ed. Boulanger Yu D and Sagitov M (NASA Tech. Transl. F-15 722, Accession code no. N74-30831) (Washington DC: US Government Printing Office) pp 26–31

Rich A 1981 *Rev. Mod. Phys.* **53** 1

Richarz F and Kriger-Menzel O 1898 *Ann. Phys., Lpz.* **66** 177

Ritter R C, Gillies G T, Rood R D and Beams J W 1978 *Nature* **271** 228

Roberts D E and Fortson E N 1973 *Phys. Rev. Lett.* **31** 1539

Robertson B 1971 *Phys. Rev. Lett.* **27** 1545

Robertson H P 1949 *Rev. Mod. Phys.* **21** 378

Robinson C F 1939 *Phys. Rev.* **55** 423

Robiscoe R T and Shyn T W 1970 *Phys. Rev. Lett.* **24** 559

Rodden C J 1972 *Selected Measurement Methods for Plutonium and Uranium in the Nuclear Fuel Cycle* (Washington DC: NBS)

Rogers E H and Staub H R 1949 *Phys. Rev.* **74** 1025

Roll P G, Krotkov R and Dicke R H 1964 *Ann. Phys., NY* **26** 442

Rose R D, Parker H M, Lowry R A, Kuhlthan A R and Beams J W 1969 *Phys. Rev. Lett.* **23** 655

Rosenberg R B, Good I J and Crawford J F 1983 *Nature* **305** 8

Rossini F D 1974 *Fundamental Measures and Constants for Science and Technology* (Cleveland OH: CRC) pp 57–65

Rothman T and Matzner M 1982 *Astrophys. J.* **257** 450

Rowley W R C 1983 in *Precision Measurement and Fundamental Constants 2* ed. Taylor B N and Phillips W D (NBS Special Publication 635) (Washington DC: US Government Printing Office) p 57

Rowlinson J S and Tildesley D J 1977 *Proc. R. Soc.* **368A** 281

Roxburgh I 1981 in *Precision Measurement and Fundamental Constants 2* ed. Taylor B N and Phillips W D (NBS Special Publication 635) (Washington DC: US Government Printing Office) p 1

Safinya K A, Chan K K, Lundeen S R and Pipkin F M 1980 *Phys. Rev. Lett.* **45** 1934

Sagitov M V, Milyukov V R, Monakhov E A, Nazhidinov V S and Tadzhidinov Kh G 1977 *Dokl. Akad. Nauk* **245** 567

Sakita B 1964 *Phys. Rev. Lett.* **13** 643

Sakuma A 1970 in *Precision Measurement and Fundamental Constants* ed. Taylor B N and Langenberg D N (NBS Special Publication 343) (Washington DC: US Government Printing Office) pp 447ff

Sanders J H 1957 *Suppl. Nuovo Cimento.* **6** 242

—— 1965 *The Fundamental Atomic Constants* 2nd edn (Oxford: Oxford University Press)

Sanders J H, Tittle K F and Ward J F 1963 *Proc. R. Soc.* **A272** 103

Sanders J H and Turberfield K C 1963 *Proc. R. Soc.* **A272** 79

Sapirstein J 1980 *Phys. Rev. Lett.* **47** 1723

—— 1983 *Phys. Rev. Lett.* **51** 985

Sauder W C, Huddle J R, Wilson J D and LaVilla R E 1977 *Phys. Lett.* **A63** 313

Schöldstrom 1955 *Determination of Light Velocity on the Oland Base Line in 1955* (Stockholm: Aga)

Schrödinger E 1943 *Proc. R. Ir. Acad.* **A49** 135

Schwinberg P B, Van Dyck R S and Dehmelt H 1981 *Phys. Rev. Lett.* **47** 1679

Series G W 1970 in *Precision Measurement and Fundamental Constants* ed. Taylor B N and Langenberg D N (NBS Special Publication 343) (Washington DC: US Government Printing Office) pp 78–82

—— 1974 *Contemp. Phys.* **15** 49

Shane C D and Spedding F H 1935 *Phys. Rev.* **47** 33

Shapiro I I 1971 *Phys. Rev. Lett.* **26** 27

Shapiro I I, Counselman C C and King R W 1976 *Phys. Rev. Lett.* **36** 555

Shapiro I S and Estulin I V 1957 *Sov. Phys.–JETP* **3** 626

Shapiro I S, Smith W B, Ingalls R P and Pettengill G H 1971 *Phys. Rev. Lett.* **26** 27

Shapiro S L 1971 *Astrophys. J.* **76** 291

Shaw A E 1934 *Phys. Rev.* **44** 1006

—— 1937 *Phys. Rev.* **51** 58

—— 1938 *Phys. Rev.* **54** 193

Shaw P E and Davy N 1922 *Proc. Phys. Soc.* **29** 163

Shull C G, Billman K W and Wedgwood F A 1967 *Phys. Rev.* **153** 1414

Shyn T W, Rebane T, Robiscoe R T and Williams W L 1971 *Phys. Rev.* **A3** 116

Sieghbahn M 1931 *Spectroskopie der Röntgenstrahlen* (Berlin: Springer)

Sim P J and Taylor R J 1968 *Metrologia* **4** 149

Simkin G S, Lukin I V, Sikora S V and Strelenskii V E 1967 *Izmen. Technol.* **8** 92

Simons L 1922 *Phil. Mag.* **43** 138

Sloggett G J, Clothier W K, Benjamin D J, Currey M F and Bairnsfather H 1981 in *Precision Measurement and Fundamental Constants 2* ed. Taylor B N and Phillips W D (NBS Special Publication 635) (Washington DC: US Government Printing Office)

Slyakhter A I 1976 *Nature* **264** 340

Smakala A and Kalnajs J 1957 *Suppl. Nuovo Cimento.* **6** 215

Smith K F 1980 *Contemp. Phys.* **21** 631

Smith L G 1971 *Phys. Rev.* **C4** 22

Smith R L, Shields W R and Tabor T D 1956 *US Atomic Energy Commission R and D Report No. GAT-171REV1* (unpublished)

Smoot G F, Gorenstein M V and Muller R A 1977 *Phys. Rev. Lett.* **39** 898

Sokolov Yu L 1981 in *Precision Measurement and Fundamental Constants 2* ed. Taylor B N and Phillips W D (NBS Special Publication 635) (Washington DC: US Government Printing Office) p 135

Solheim J E, Barnes T G III and Smith H G 1976 *Astrophys. J.* **209** 330

Sommer H, Thomas H A and Hipple J A 1957 *Phys. Rev.* **82** 697

Soto M F Jr 1966 *Phys. Rev. Lett.* **17** 1153

Spero R, Hoskins J K, Newman R, Pellam J and Schultz J 1980 *Phys. Rev. Lett.* **44** 1645

Stacey F D, Tuck G J, Holding S C, Maher A R and Morris D 1981 *Phys. Rev.* **D23** 1683

Stanbury P 1983 *Nature* **305** 11

Stein S R and Turneaure J P 1973 in *Low Temperature Physics LT13* (New York: Plenum) 414ff

Stephenson L M 1967 *Proc. Phys. Soc.* **90** 601

Stevens C M, Schiffer J P and Chapka W 1976 *Phys. Rev.* **D14** 716

Stille U 1943 *Z. Phys.* **121** 133

Stover R W, Moran T I and Trischka J W 1967 *Phys. Rev.* **164** 1599

Straumanis M, Ievans A and Karlsons K 1939 *Z. Phys. Chem.* **B42** 143

Stroscio M A 1975 *Phys. Rev.* **A12** 338

Sullivan J B and Frederick N V 1977 *IEEE Trans. Magn.* **MAG-13** 396

Tarbeyev V 1981 in *Precision Measurement and Fundamental Constants 2* ed. Taylor B N and Phillips W D (NBS Special Publication 635) (Washington DC: US Government Printing Office) p 483

Tate D R 1966 *J. Res. NBS* **70C** 149

Taylor B N 1970 in *Precision Measurement and Fundamental Constants* ed. Taylor B N and Langenberg D N (NBS Special Publication 343) (Washington DC: US Government Printing Office) p 495

Taylor B N, Parker W H and Langenberg D N 1969 *Rev. Mod. Phys.* **41** 375

Taylor R J 1980 *Rep. Prog. Phys.* **43** 17

Teller E 1948 *Phys. Rev.* **73** 801

Terrien J 1973 *Nuovo Rev. Opt.* **4** 215

—— 1976 *Rep. Prog. Phys.* **39** 1067

Thomas H A 1950 *Phys. Rev.* **80** 901

Thomas H A, Driscoll R L and Hipple J A 1950 *Phys. Rev.* **78** 787

Thompson A M and Lampard D E 1956 *Nature* **177** 888

Thompson P A, Crane P, Crane T, Amato J J, Hughes V W, zu Putlitz G and Rothberg J E 1973 *Phys. Rev.* **A8** 86

Thomson J J 1897 *Phil. Mag.* **46** 528

—— 1903 *Phil. Mag.* **5** 354

Topping J 1971 *Errors of Observation and their Treatment* 3rd edn (London: Chapman and Hall)

Towe K M 1975 *Nature* **257** 115

Towler W R *et al* 1971 in *Precision Measurement and Fundamental Constants* ed. Taylor B N and Langenberg D N (NBS Special Publication 343) (Washington DC: US Government Printing Office) p 485

Treibwasser S, Dayhoff E S and Lamb W E Jr 1953 *Phys. Rev.* **89** 98

Trigger K R 1956 *Bull. Am. Phys. Soc.* **1** 220

Trimmer W S N, Baierlein R F, Faller J E and Hill H A 1973 *Phys. Rev.* D**8** 3321'

Tu Y 1932 *Phys. Rev.* **40** 662

Turneaure J P and Stein S R 1974 in *Atomic Masses and Fundamental Constants 4* ed. Sanders J H and Wapstra A' H (New York: Plenum)

Van Atta L C 1931a *Phys. Rev.* **38** 876

—— 1931b *Phys. Rev.* **39** 1012

Van der Pauw L J 1958 *Phillips Res. Rep.* **13** 1

Van Dyck R S and Schwinberg P B 1981 *Phys. Rev. Lett.* **47** 395

Van Dyck R S Jr, Schwinberg P B and Dehmelt H G 1977 *Phys. Rev. Lett.* **38** 310

——1978 in *New Frontiers in High Energy Physics* ed. Mursunoglu B, Perlmutter A and Scott L F (New York: Plenum)

——1979 *Bull. Am. Phys. Soc.* **24** 758

Van Dyck R S Jr, Wineland D, Kstrom P E and Dehmelt H G 1976 *Appl. Phys. Lett.* **28** 446

Van Flandern T C 1975 *Mon. Not. R. Astrom. Soc.* **170** 333

—— 1981 *Astrophys. J.* **248** 813

Vigoureux P 1965 *Metrologia* **1** 3

—— 1971 *Units and Standards of Electromagnetism* (London: T and F Wykeham)

—— 1979 *Report NPL-DES* (Teddington: National Physical Laboratory)

Vigoureux P and Dupuy N 1980 *NPL Rep. DES*59 (Teddington: National Physical Laboratory)

Vinal G W and Bates S J 1914 *Bull. NBS* **10** 425

Volet Ch 1946 *C.R. Acad. Sci., Paris* **222** 373

Vorburger T V and Cosens B L 1971 in *Precision Measurement and Fundamental Constants* ed. Taylor B N and Langenberg D N (NBS Special Publication 343) (Washington DC: US Government Printing Office) pp 361ff

Wagoner R V 1970 *Phys. Rev.* D1 3201

Wallenstein R and Hänsch T W 1975 *Opt. Commun.* **14** 353

Walther F G, Phillips W D and Kleppner D 1972 *Phys. Rev. Lett.* **33B** 422

Wapstra A H and Bos K 1977 *Atomic Data and Nuclear Data Tables* **19** 177–214

Warner and Nather R E 1969 *Nature* **222** 157

Washburn E W and Bates S J 1912 *J. Am. Chem. Soc.* **34** 1341, 1515

Weber E W and Goldsmith J E M 1979 *Phys. Lett.* **70A** 95

Weichert E 1899 *Ann. Phys., Lpz.* **69** 739

Weiman C and Hänsch T W 1976 *Phys. Rev. Lett.* **36** 1170

Weirauch W 1978 in *Fundamental Physics with Reactor Neutrons and Neutrinos* (Inst. Phys. Conf. Ser. 42) ed. von Egidy T (Bristol: The Institute of Physics) pp 47–52

Wesley J C and Rich A 1971 *Phys. Rev.* A4 1341

—— 1972 *Rev. Mod. Phys.* **44** 250

Wesson P S 1973 *Q. Not. R. Astron. Soc.* **14** 9

—— 1975 *Astrophys. Space Sci.* **36** 363

—— 1978 *Cosmology and Geophysics* (Bristol: Adam Hilger)

Wheeler J H 1971 in *Gravitation*

Whiddington R and Woodroofe E G 1935 *Phil. Mag.* **20** 1109

Wilcox L R and Lamb W E Jr 1960 *Phys. Rev.* **119** 1915

Wilkinson D H 1958 *Phil. Mag.* (Ser. 8) **3** 582

Wilkinson D T and Crane H R 1963 *Phys. Rev.* **130** 852

Will C M 1981 *Theory and Experiment in Gravitational Physics* (Cambridge: Cambridge University Press)

Williams E and Park D 1971 *Phys. Rev. Lett.* **26** 1651

Williams E R, Faller J E and Hill H A 1971 *Phys. Rev. Lett.* **26** 721

Williams E R and Olsen P T 1979 *Phys. Rev. Lett.* **42** 1575

Williams J G, Dicke R H, Bender P L, Alley C O, Carter W E, Currie H G, Eckhardt D H, Faller J E, Kaula W M, Mullholland J D, Plotkin H H, Poultrey S K, Shelus P J, Silverberg E C, Sinclair W S, Slade M A and Wilkinson D T 1976 *Phys. Rev. Lett.* **36** 551

Williams J G, Sinclair W S and Yoder C F 1978 *Geophys. Res. Lett.* **5** 943

Williams R C 1938 *Phys. Rev.* **54** 558

Wilsing J 1888 *Annual Report of the Board of Regents of the Smithsonian Institution* 1888 p 635

Wineland D J, Bollinger J J and Itano W M 1983 *Phys. Rev. Lett.* **50** 628

Wineland D J and Dehmelt H 1975 *J. Appl. Phys.* **46** 919

Wing W H 1968 *PhD Thesis* University of Michigan (unpublished)

Winkler P F, Kleppner D, Myint T and Walther F G 1972 *Phys. Rev.* A **5** 83

Witteborn F C and Fairbank W M 1977 *Rev. Sci. Instrum.* **48** 1

Wolf F 1927 *Ann. Phys., Lpz.* **83** 849

Wood O R II, Patel C K N, Murnick D E, Nelson E T, Leventhal M, Kugel H W and Niu Y 1982 *Phys. Rev. Lett.* **48** 398

Woods P T and Joliffe B W 1976 *J. Phys. E: Sci. Instrum.* **9** 395

Woods P T, Shotton K C and Rowley W R C 1978 *Appl. Opt.* **17** 1048

Wyler A M 1969 *C.R. Acad. Sci., Paris* **B269** 743

—— 1971 *C.R. Acad. Sci., Paris* **B271** 186

Yagola G K, Zingerman V I and Sepetyi V N 1962 *Izmer. Tekh.* **5** 24 (Engl. Transl. 1962 *Meas. Tech.* **5** 387)

—— 1966 *Izmer. Tekh.* **7** 44 (Engl. Transl. 1968 *Meas. Tech.* **7** 1483)

Yahil A 1975 *The Interaction Between Science and Philosophy* ed. Elkann Y (London: Humanities Press) p 27

Yamazaki T, Shida K and Kanno M 1972 *IEEE Trans. Instrum. Meas.* **IM-21** 372

Yang J, Schramm D N, Steigman G and Rood R T 1979 *Astrophys. J.* **227** 697

Yennie D R 1967 in *Proceedings of the International Symposium on Electron and Photon Interactions at High Energies, SLAC* (Stanford: Calif) p 32

Zahradnicek J 1933 *Phys. Ztg.* **34** 126

Zeeman P 1918 *Proceedings of the Section of Sciences, Koniklijke, Akademie von Wetenschappen to Amsterdam* **20** 542

Zorn J C, Chamberlain G E and Hughes V W 1963 *Phys. Rev.* **129** 2566

Further reading

As has been apparent throughout this book, the topic is thoroughly interwoven with the fabric of physics. The following books, some of which have only recently become available, are particularly recommended for their coverage of the precision measurement aspect of the subject.

1   Cohen E R, Crowe K M and DuMond J W M 1957 *Fundamental Constants of Physics* (New York: Wiley Interscience)

2   Sanders J H 1961 *The Fundamental Atomic Constants* 2nd edn (Oxford: Oxford University Press)

3   Taylor B N, Parker W H and Langenberg D N (eds) 1969 *The Fundamental Constants of Physics and Quantum Electrodynamics* (New York: Academic)

4   Froome K D and Essen L 1969 *The Velocity of Light and Radio Waves* (London: Academic)

5   Ferro-Milone A, Giacomo P and Leschuitta S (eds) 1980 *Metrology and Fundamental Constants* (Amsterdam: North-Holland)

6   Cutler P H and Lucas A A (eds) 1983 *Quantum Metrology and Fundamental Physical Constants* (New York: Plenum)

The books in the AMCO series of conferences contain many papers on measurements of the fundamental physical constants. These are:

7   Hintenberger H (ed.) 1957 *Nuclear Masses and their Determination* (London: Pergamon)

8   Duckworth H E (ed.) 1960 *Proceedings of the International Conference on Nuclidic Masses* (Toronto: University of Toronto Press).

9   Johnson W H Jr (ed.) 1964 *Proceedings of the Second International Conference on Nuclidic Masses* (Vienna: Springer)

10   Barber R C (ed.) 1968 *Proceedings of the Third International Conference on Atomic Masses* (Winnipeg: University of Manitoba Press)

11   Sanders J H and Wapstra A H (eds) 1972 *Atomic Masses and Fundamental Constants 4* (New York: Plenum)

12   Sanders J H and Wapstra A H (eds) 1975 *Atomic Masses and Fundamental Constants 5* (New York: Plenum)

13   Nolan J A Jr and Benenson W (eds) 1980 *Atomic Masses and Fundamental Constants 6* (New York: Plenum)

The proceedings of the two conferences held at the National Bureau of Standards are particularly recommended as conveying the flavour of the subject:

14   Langenberg D N and Taylor B N (eds) 1971 *Precision Measurement and Fundamental Constants* (National Bureau of Standards Special Publication 343) (Washington DC: US Government Printing Office)

15   Taylor B N and Phillips W D (eds) 1984 *Precision Measurement and Fundamental Constants 2* (National Bureau of Standards Special Publication 635) (Washington DC: US Government Printing Office)

For coverage of the modern constants not considered in this book, the proceedings of the following one-day meeting provide a useful summary of the state of the art:

16   McCrea W H and Rees M J (eds) 1983 *The Constants of Physics* (London: The Royal Society)

For a discussion of uncertainties with examples from the history of fundamental constants:

17   Topping J 1962 *Errors of Observation and their Treatment* 3rd edn (London: Chapman and Hall)

18   Bridgeman P W 1931 *Dimensional Analysis* (New Haven: Yale University Press)

19   Stille U 1955 *Messen und Rechnung in der Physik* (Braunschweig: Vieweg)

20   Ellis B 1966 *Basic Concepts of Measurement* (Cambridge: Cambridge University Press)

21   Berka K 1983 *Measurement, its Concepts, Theories and Problems* (Dordrecht: Reidel)

Books concerning the SI units:

22   Rossini F D 1974 *Fundamental Measures for Science and Technology* (Cleveland OH: CRC)

23   McGlashan M L 1971 *Physico-Chemical Quantities and Units* 2nd edn (London: Royal Institute of Chemistry)

24   Vigoureux P 1971 *Units and Standards of Electromagnetism* (London: Wykeham)

25   Goldman D T and Bell R J (transl.) 1982 *SI, the International System of Units* 4th edn (London: NPL, HMSO)

Index